Laboratory techniques in biochemistry and molecular biology

7

LABORATORY TECHNIQUES IN BIOCHEMISTRY AND MOLECULAR BIOLOGY

Volume 7

Edited by

T.S. WORK – *N.I.M.R., Mill Hill, London*
E. WORK – *'East Lepe', 60 Solent View Road, Cowes, Isle of Wight*

NORTH-HOLLAND PUBLISHING COMPANY
AMSTERDAM · NEW YORK · OXFORD

Part I

W. Howard Evans

PREPARATION AND CHARACTERISATION OF MAMMALIAN PLASMA MEMBRANES

Part II

C.R. Lowe

AN INTRODUCTION TO AFFINITY CHROMATOGRAPHY

1979

NORTH-HOLLAND PUBLISHING COMPANY
AMSTERDAM · NEW YORK · OXFORD

ISBN – series: 07204 4200 1
– volume: 0 7204 4224 9

Published by:

ELSEVIER/NORTH-HOLLAND BIOMEDICAL PRESS
335 JAN VAN GALENSTRAAT, P.O. BOX 211
AMSTERDAM, THE NETHERLANDS

Sole distributors for the U.S.A. and Canada:

ELSEVIER/NORTH-HOLLAND INC.
52 VANDERBILT AVENUE
NEW YORK, N.Y. 10017

Printed in The Netherlands

Editors' preface

Progress in research depends upon development of technique. No matter how important the cerebral element may be in the planning of experiments, a tentative hypothesis cannot be converted into an accepted fact unless there is adequate consciousness of the scope and limitation of existing techniques; moreover, the results may be meaningless or even positively misleading if the technical 'know how' is inadequate.

During the past ten or fifteen years, biochemical methods have become specialized and sophisticated to such a degree that it is now difficult for the beginner, whether undergraduate, graduate or specialist in another field, to grasp all the minor but important details which divide the successful from the unsuccessful experiment. In order to cope with this problem, we have initiated a new series of Laboratory Manuals on technique. Each manual is written by an expert and is designed as a laboratory handbook to be used at the bench.

It is hoped that use of these manuals will substantially reduce or perhaps even remove that period of frustration which so often precedes the successful transplant of a specialized technique into a new environment. In furtherance of this aim, we have asked authors to place special emphasis on application rather than on theory; nevertheless, each manual carries sufficient history and theory to give perspective. The publication of *library volumes* followed by *pocket paperbacks* is an innovation in scientific publishing which should assist in bringing these manuals into the laboratory as well as into the library. In undertaking the editing of such a diverse series, we have become painfully conscious of our own ignorance but have been encouraged by our board of advisers to whom we owe many valuable suggestions and, of course, by our authors who have co-operated so willingly and have so patiently tolerated our editorial intervention.

T. S. & E. Work
Editors

Contents of parts I and II

PART I

PART II

PREPARATION AND CHARACTERISATION OF MAMMALIAN PLASMA MEMBRANES

W. Howard Evans

National Institute for Medical Research,
Mill Hill, London N.W.7, U.K.

Contents

List of abbreviations

SDS	Sodium dodecyl sulphate
EDTA	Ethylenediamine tetraacetic acid
EGTA	Ethyleneglycol bis(β-aminoethyl)ether-*N*,*N*′-tetraacetic acid
ATPase	Adenosinetriphosphatase catalysing in the presence of Mg^{2+} the reaction: $ATP + H_2O = ADP + orthophosphate$
$(Na^+$-$K^+)$-ATPase	Incremental ATPase activity in the presence of Na^+ and K^+ that is inhibited by ouabain
Tris	2-Amino-2-hydroxymethyl-1,3-propanediol

Subject index p. 261

CHAPTER 1

Introduction

This monograph describes methods and the principles underlying them, for preparing, identifying and characterising subcellular fractions containing the cell's plasma (surface) membrane and its constituent parts. The cell surface membrane is increasingly recognised as playing a crucial role in the regulation of cell metabolism, growth and communication; it interprets environmental signals and controls the nature of molecules entering and leaving the cell. The plasma membrane fraction is the in vitro model of the cell surface membrane used to correlate functional and molecular studies of surface events.

Structurally, the mammalian plasma membrane is conceptualised as a mainly fluid bilipid layer containing a variable but usually large number of metabolically dynamic proteins, enzymes and glycoproteins. Proteins bordering the cell's environment interact specifically and in small numbers with a number of biologically-active substances ('ligands') such as hormones, drugs, antibodies, toxins and infectious agents. Those proteins traversing the bilipid layer are thought to provide the vehicle enabling the cell interior to be made aware of environmental perturbations, and to constitute channels for the passage of ions and charged and hydrophilic molecules across a largely hydrophobic barrier. Further, submembraneous elements (microfilaments, microtubules) may allow direct interaction with cytoplasmic structures and events occurring deep inside the cell. The wide range of processes occurring at the cell's periphery has meant that the plasma membrane is not only being investigated to explore its biological organisation per se, but it has also featured in combination with cellular and other approaches in the experimental designs of endocrinologists, pharmacologists, immunologists, virologists, etc. The plasma membrane is also being studied extensively in attempts to relate primary metastatic lesions to cell surface modifications.

In this account, attempts are made to rationalise and generalise practical guidelines for the preparation of plasma membrane fractions. In Chapter 2, methods for cell breakage are described and assessed. Polydispersity of plasma membrane fragments produced at the crucial cell breakage step constitutes the most common pitfall limiting the recovery of plasma membranes acceptably free of contamination by cytoplasmic components. Although centrifugation is the major technique used to separate subcellular components, attention is also focussed in Chapter 3 on other complimentary methods such as free-flow electrophoresis, the partitioning of membranes between solvents, immunological approaches, and chromatographic techniques. The use of the more conventional markers to identify and characterise plasma membrane fractions, i.e. enzymic, chemical and morphological, is described in Chapter 4. To corroborate the claims made for the quality of the final product, and to monitor the separation of plasma membranes, newer approaches involving the attachment of radio-labelled markers to the plasma membrane are described. The mammalian plasma membrane, especially in cells constituting tissues and organs, is increasingly appreciated to be a highly differentiated organelle constructed of geographically distinct areas interfacing with different environments and specialised for widely disparate physiological functions. Thus, in most plasma membrane preparative procedures (except when cell membrane 'ghosts' are isolated e.g. from erythrocytes and adipocytes), the final product represents an actual subfraction of an idealised anatomical membrane. The low recovery of plasma membranes frequently reported not only may indicate the persistence of contaminating intracellular membranes, but is also likely to betray the selective purification of a restricted and unrepresentative part of the plasma membrane. Finally, in Chapter 5 the general methodological approaches described in Chapters 2–4 are further illustrated by reference to the preparation of plasma membrane fractions from a range of organs, tissues and cultured cells. However, no attempt is made to describe intricate experimental details in individual cases, since this falls well outside the brief of a single author and the space available. References are made to the original publications, which have been frequently redescribed and amplified in treatises such as Methods in Enzymology (edited by S. Fleischer and L. Packer) Vols. 31 and 32, and in the various volumes of Methods in Membrane Biology (edited by E. Korn, Plenum Press).

Inevitably, a wide range of techniques is encompassed in a description of methods for preparing and characterising mammalian plasma membranes and these are described in such detail as is applicable in the present context. More comprehensive and detailed descriptions are available in other volumes of this series, namely Density Gradient Centrifugation by Hinton and Dobrota (1976), Electrophoresis of Proteins in Polyacrylamide and Starch Gels by Gordon (1976), Immunochemical Techniques for the Identification and Estimation of Macromolecules by Clausen (1969), Techniques in Lipidology by Kates (1972) and in Affinity Chromatography by Lowe accompanying this volume.

Methods of cell breakage: assessing their suitability and efficacy

The first step in membrane preparation is breakage of the cell membrane so as to liberate the constituent organelles, membranes and soluble components. A number of methods, tailored to a wide variety of cells, tissues and organs accomplish this with varying degrees of success. In all these methods, the major objective is to effect the disruption of all cells to obtain relatively intact organelles and membrane fragments while inflicting the minimum physical (e.g., heat) and biological (e.g., enzyme inactivation) damage to the liberated components and their content. In chosing methods for the disruption of different cell types, the major subsequent requirement is the adequate separation of liberated parts. The more traditional methods that use shear homogenisation (see below), normally of soft tissues, are geared more to the preservation of the subcellular organelles, especially nuclei and mitochondria. In the isolation of intact lysozomes, the monitoring of the extent of organellar damage reached a high degree of sophistication as a wide range of quantitative biochemical parameters were developed to monitor the separations. Methods of cell rupture for preparing plasma membranes, while respecting the necessity for preserving organelles intact, must also attempt to control the extent of breakage of the cell surface membrane so as to produce a class or classes of membrane fragments of similar size and density adequately separable from other subcellular particles, organelles, etc. This requirement for limiting the polydispersity of plasma membrane fragments is especially important when differential centrifugation methods feature in the subsequent separation techniques (see § 3.1.1). Therefore, the methods of cell rupture employed are of prime importance in plasma membrane preparation, for this initial procedure represents a critical point for control of the degree of polydispersity of membrane particles and therefore, of the yield and purity of the isolated membrane fragments.

2.1. Apparatus and techniques

Disruption of cells has been effected by a variety of techniques; the most popular involve propelling cells and tissue fragments through narrow gaps or orifices in which the shearing forces generated cause cell breakage. Other methods utilise osmotic forces, rapid pressure changes or ultrasonic energy to break cells. For disrupting soft tissues the mild shearing forces generated by tissue homogenisers are generally used (Potter and Elvehjem 1936). Cultured cells are usually more difficult to disrupt and this has prompted the introduction of methods similar to those originally developed for the disruption of bacteria and micro-organisms, e.g., French press (Milner et al. 1950) Hughes press (Hughes 1951). Pressure homogenisation, originally introduced by Fraser in 1951, has also featured in the disruption of mammalian cells for isolation of plasma membrane fragments and is carried out in the apparatus described by Hunter and Commerford (1961). For homogenising large numbers of cells under controlled and reproducible conditions, cell rupturing pumps that inject cell suspensions through spring or air-loaded valves set at various pressures are currently finding favour (Wright et al. 1974; Crumpton and Snary 1974).

2.1.1. Potter-Elvehjem homogenisers

These homogenisers (Potter and Elvehjem 1936) are extensively used for disruption of tissues and organs and are available from a number of suppliers in a wide range of capacities (see Table 2.1). In its modern form, it consists of a Teflon plunger or piston attached to a motor that rotates it inside a glass cylinder. The number of revolutions per minute can be closely monitored by connecting the piston to a tachometer. A modern Potter–Elvehjem apparatus is

TABLE 2.1
Properties of tissue homogenisers.

A Potter–Elvehjem Type[a]					
Capacity (ml)	1	4	10	30	55
Grinding length (mm)	–	90	75	105	110
Chamber clearance (mm)[c]	0.1–0.15	0.07–0.13	0.1–0.15	0.15–0.18	0.15–0.23
B Dounce homogeniser[b]					
Capacity (ml)	5	14	30		
Clearance[c] (mm)					
'tight' pestle	'tight'	0.076	0.076		
'loose' pestle		0.127	0.127		

[a] Homogenisers at or approximating to these dimensions are available from Kontes, Thomas, Arnold H. Horwell, Jenkons, Braun, Tri-R instruments. Note that chamber clearance varies according to the capacity of the homogeniser. Large scale (up to 200 ml) homogenisers are available from Houston Glass Fabricating Co.

[b] Homogenisers are usually sold with 'tight' and 'loose' pestles. They are available from Blaessig, Kontes, Thomas and Braun.

[c] Clearance is the difference (mm) between the pestle diameter and the vessel's inside diameter. Radial clearance (or gap) is half this measurement.

shown in Fig. 2.1.

Two major parameters are usually varied. The first is the gap or clearance existing between the rotating pestle and the wall of the glass vessel. The clearance adopted varies according to the tissue being homogenised. Although the dimension of the gap between the plunger and container vessel is usually stated by the manufacturer (Table 2.1), this can be increased by trimming down the diameter of the Teflon pestle in a machine shop. Most commercially available Potter–Elvehjem homogenisers have a clearance of 0.10–0.15 mm. This value does not vary very much with temperature during use because of the low coefficient of expansion of Teflon. The second parameter that can vary is the control exercised over the homogenisation conditions, i.e., the number of passages up and down the vessel, the time taken and the speed of rotation. This includes also the

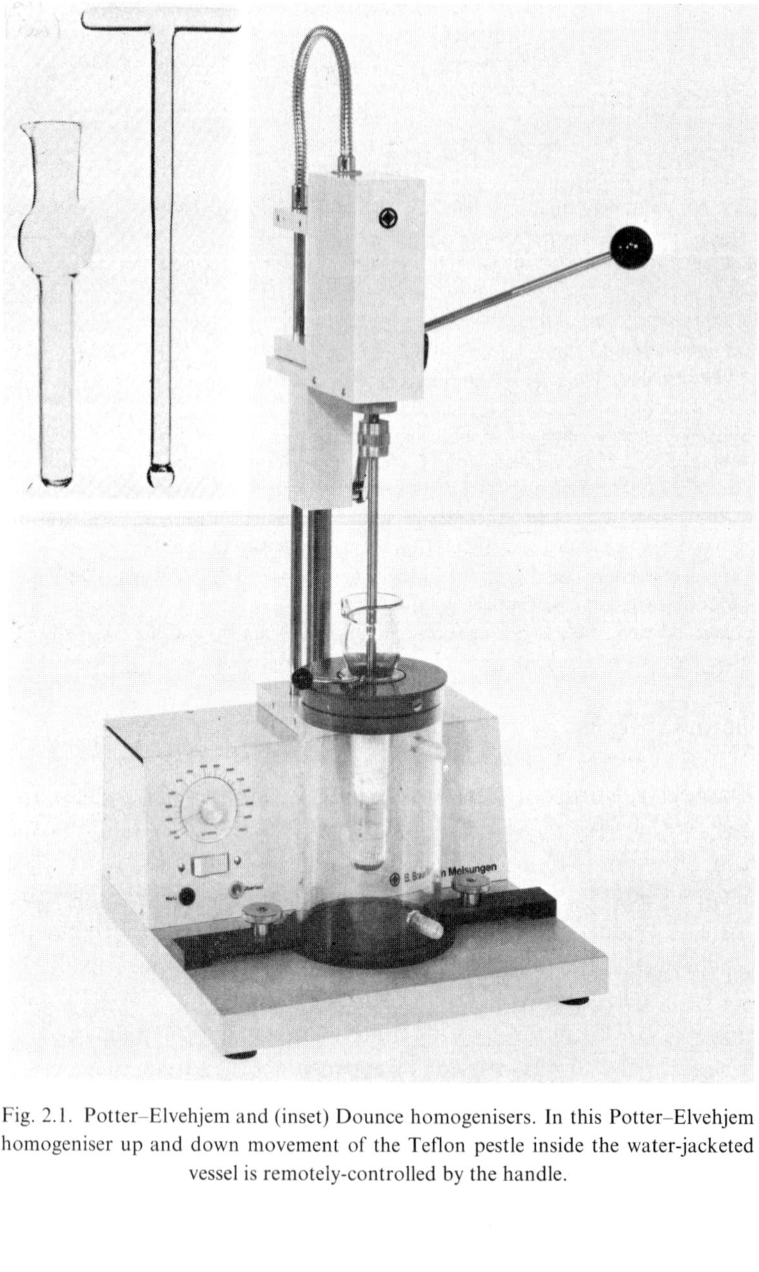

Fig. 2.1. Potter–Elvehjem and (inset) Dounce homogenisers. In this Potter–Elvehjem homogeniser up and down movement of the Teflon pestle inside the water-jacketed vessel is remotely-controlled by the handle.

tissue to homogenising medium ratio and, overall, this is probably the parameter that may vary most amongst users. Thus, a full description of the range of conditions used should include: (a) the tissue to medium ratio (§ 2.2); (b) the necessity for prior removal of connective tissue by filtration through gauze (e.g., cheese-cloth) and for chopping up the tissue into manageable portions (§ 2.3); (c) the clearance between the pestle and plunger; (d) the number of complete passes of the rotating pestle (i.e., passages down and up the vessel) and (e) the speed of rotation, as monitored by a tachometer.

The power-driven Potter–Elvehjem homogeniser can cause problems for the novice operator, since failure to keep in parallel the rapidly rotating pestle and the hand-held container can result in the sudden breakage of the glass. From a safety point of view, it is therefore recommended that a glove is used to hold the glass vessel and a separate control, e.g. a footswitch is present to free the hands to operate the homogeniser. To rotate the pestle at a constant speed, it is useful to have a high-torque (series wound) motor equipped with speed control and a tachometer. A further important consideration is to ensure that homogenisation at high rotation speeds does not result in local heating thereby damaging heat-labile tissue constituents. Homogenisers with water jackets are commercially available, but usually it is more practicable to immerse the homogeniser in ice-water for a few seconds between each passage of the plunger.

The conditions for cell disruption will vary widely between tissues and animal species. Older animals contain more collagenous tissue than younger animals, thus making complete homogenisation more difficult. Thus variation can be expected in the use of Potter–Elvehjem homogenisers to disrupt tissues from animal species of different ages and physiological or pathological states.

For homogenisation of cells of small size, e.g., platelets, a 'no-clearance' Teflon pestle (Kontes No. 12–152–A) may be used (Marcus et al. 1966). The Teflon pestle does not fit into the vessel at 20 °C, but at 4 °C it has contracted slightly. As homogenisation proceeds, local heating causes it to expand and this provides strong shearing forces.

2.1.2. Dounce homogenisers

This homogeniser (Fig. 2.1, inset), was first introduced to prepare intact nuclei from liver for metabolic experiments (Dounce et al. 1955) and has since featured prominently in many surface-membrane preparative routines. The Dounce homogeniser was first used for plasma membrane isolation by Neville (1960). Compared with the Potter–Elvehjem homogeniser, used under similar conditions (Emmelot et al. 1964) the Dounce hand homogeniser is a far more controllable instrument for tissue disruption, since the operator can obtain a 'feel' of the disruption process as the ball-tipped pestle is moved up and down the vessel. High shearing forces that are characteristic of motor-driven homogenisers are avoided, thus largely eliminating any possibility of damage due to heat generation. Furthermore, in many instances of plasma membrane preparation, the low shearing forces result in the release of large sheets of plasma membrane that can be sedimented at low centrifugal forces. Dounce homogenisers are available in various capacities and with two plungers, tight or loose fitting (Table 2.1). The clearance between the ball on the plunger and the vessel constructed of constant bore glass tubing gradually increases after prolonged use. A rapid method of assessing whether the gap is of the correct order and to compare or standardise loose-fitting homogenisers involves measuring the time taken by the plunger to descend by gravity through the water-filled constant-bore portion of the vessel. For example, the optimal gap for disruption of rat liver for plasma membrane isolation is such that the plunger will descend in 4–5 sec through a water-filled, large, type A Dounce homogeniser (Blaessig Glass Specialities).

Dounce homogenisers are frequently used to disrupt tissue and cells in hypo-osmotic media. An additional parameter involved under these conditions is the amount of swelling of cells prior to use of the homogeniser, as discussed in § 2.2.

2.1.3. Microhomogenisers

As techniques of membrane isolation are extended to include cells or tissues available only in limited amounts, e.g., biopsy material, a requirement for microhomogenisation methods is emerging. Three types of microhomogenisers, that are more efficient than freezing and thawing methods, have been briefly described (Neuhoff 1973). In the apparatus described by Eichner (1966) a loop of Nikrothal wire (Kanthal AB) of diameter 60 or 35 μm, attached to a motor (a dentist's drill in this instance) is inserted into the material suspended in a capillary tube. Homogenisation is performed under a strong lens or stereomicroscope. For larger amounts of material, 10-μl capillary tubes and a dentist's nerve-channel drill may be used. Microhomogenisation can be effected also by adding razorblade splinters to the sample and solution in a glass tube, and then shaking the tube on a whirlmixer. Blade fragments are removed from the homogenised tissue by a magnet.

Homogenisers that can handle 0.1–1 ml of previously dispersed tissue are useful for resuspending membrane pellets, and these are available from commercial suppliers, e.g., Jencons, Thomas.

2.1.4. Pressure homogenisation (gas cavitation)

This technique has been used mainly to disrupt cultured cell lines and lymphocytes. The principle of the method is based on the fact that cells, equilibrated inside a vessel ('bomb') containing a gas (usually nitrogen) at a high pressure (up to 1000 lb/in^2; 76 atm) dissolve large quantities of gas. When the gas pressure is suddenly released, gas bubbles form inside the cytoplasm, expand and rupture the membranes of the cell, converting them into populations of vesicular structures (Fraser 1951).

Since the original description of a pressure homogeniser, especially that adapted from a hydrogenation apparatus by Hunter and Commerford (1961), a number of commercial models have become available that accommodate 1–2000 ml of sample (see Table 2.2). Cell suspensions or minced tissue are usually placed in a beaker or directly into the pressure vessel (Fig. 2.2 shows the features of a cell disruption bomb). After the upper portion has been bolted on firmly, the gas (nitrogen, argon) is introduced into the vessel (all valves and fittings, etc., are provided with the apparatus). The cell suspension may be

TABLE 2.2
Examples of conditions used to disrupt cells by gas cavitation.

Tissue or cells	Conditions	Reference
Calf thymus	36 atm for 15 min	Van Blitterswijk et al. (1972)
Lymphocytes	50 atm for 15 min at 4 °C 30 atm for 15 min at 20 °C	Ferber et al. (1972)
Ehrlich ascites cells	58 atm for 20 min at 4 °C	Molnar et al. (1969)
Mouse fibroblasts	65 atm for 30 min	Miller et al. (1974)
Various categories of cultured cell lines	58 atm for 20 min	Graham (1972)
Platelets	40–120 atm for 10–30 min	Broeckman et al. (1974)
Monkey erythrocytes	40 atm for 30 min at 4 °C	Wallach and Conley (1977)
Monkey erythrocytes parasitised with *Plasmodium knowlesi*	20 atm for 15 min at 4 °C	Wallach and Conley (1977)

Conversion factor: 1 atmosphere (atm) = 14.7 lb/in^2

The following manufacture pressure homogenisers: Baskerville & Lindsay Ltd.; Artisan Industries Inc.; Parr Instrument Co.; Kontes*; Braun; Yeda Research & Development Co. Ltd.

* Kontes also make a small disruption chamber that offers special advantages for volumes of 1–15 ml.

mechanically stirred, or with the small vessels, the vessel itself may be shaken manually. After full equilibration at the required pressure (the time taken for equilibration varies according to the tissue or cell type, pressure, etc.), the pressure is released by opening the gas inlet valve and the cell homogenate is collected slowly through a delivery tube attached to the outlet valve.

Examples of the conditions used to disrupt a variety of cell types are shown in Table 2.2. Strictly, conditions have to be evolved for each type of cell. The range of parameters regulating the efficiency and reproducibility of rupture are; (a) the ratio of cells or tissue mince to medium; (b) the volume of the suspension relative to the volume of the disruption vessel; (c) the pressure applied; (d) the

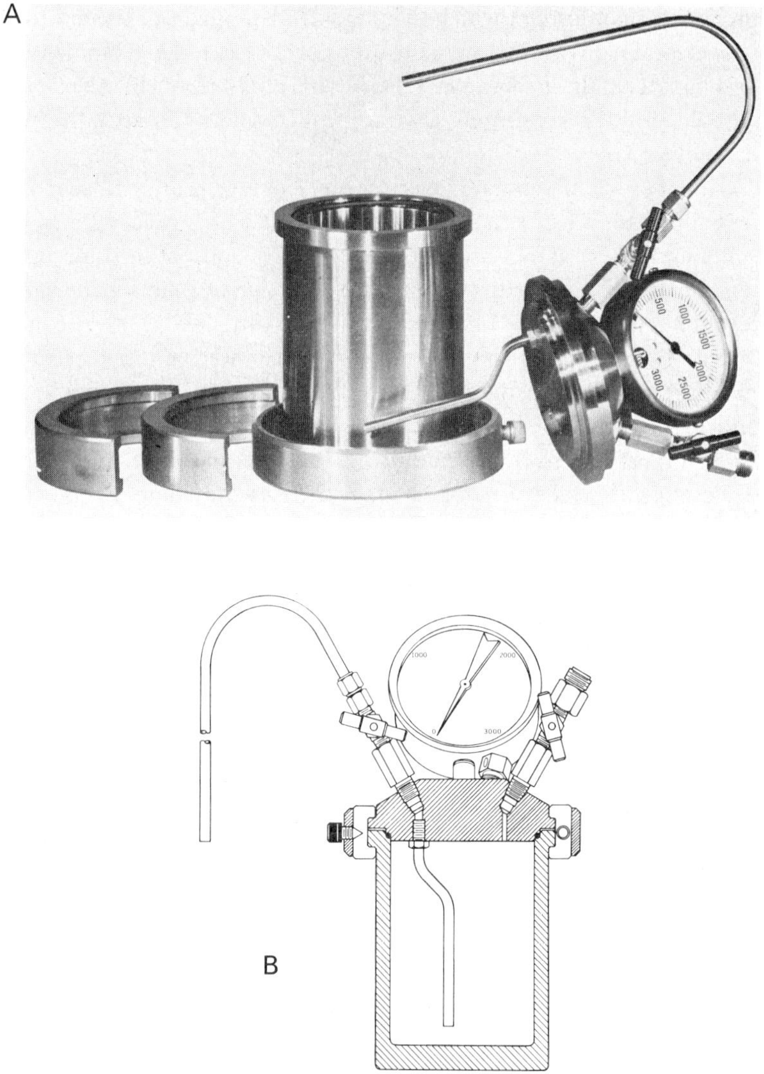

Fig. 2.2. Pressure homogeniser. (A) Disassembled; (B) Cross section. Internal volume is 920 ml. Courtesey of Parr Instrument Co. Mini-disruption bombs are also available, see Table 2.2

time and temperature at which the pressure is applied; (e) the nature of the medium, especially whether cations are present, and its tonicity. Unfortunately, these parameters are rarely described completely, and the optimum conditions have to be derived experimentally by the operator.

A number of advantages over 'shear' homogenisation methods are claimed for this method. The minimisation of physical stresses results in no heat damage, and the speed of operation and the fact that each cell is exposed only once to explosive decompression are the major advantages. Hunter and Commerford (1961) were concerned with a range of tissues and organs that were minced before introducing them into the pressure vessel, and they noted that the application of pressures of between 50–70 atm gave the most complete disruption, as judged by the number of clumped cells remaining in the homogenate. The method has found its greatest application in dealing with isolated cells, especially fibroblastic cell lines. However, operation of the cell disruption 'bomb' is not without its problems. Excessive foaming occurs as the homogenate is released from the vessel, and this may be minimised by centrifuging it. Too rapid release of the contents of the vessel can result in freezing, due to adiabatic cooling. A major and underestimated problem is to control the extent of organellar damage occurring within the bounds of achieving more than 80% cell disruption. The cell disruption process converts most of the endoplasmic reticulum and plasma membrane into small populations of vesicles while stripping of ribosomes and outer mitochondrial membranes can also occur. For example, Graham et al. (1975) found that microsomal fractions of NIL 8 hamster fibroblast homogenates prepared by pressure homogenisation had low RNA to protein ratios due to loss of ribosomes during homogenisation. Clearly, the conditions of use, especially the pressure adopted and the time of equilibration have to be equated with the quality of the final plasma membrane product for each cell type. A further problem is the subsequent separation of endoplasmic reticular and plasma membrane vesicles that possess similar densities; this was achieved by Wallach and Kamat (1966) by centrifugation on

Ficoll gradients containing divalent cations. Cells possessing a more developed cytomembrane system pose problems in separating the plasma membrane vesicles from intracellular-derived vesicles, and this proves a major disadvantage in applying the pressure homogenisation method to plasma membrane preparation from tissues.

2.1.5. Cell rupturing pumps

Motor and air-driven pumps have recently been introduced that provide a continuous method for the disruption of mammalian cells and bacteria in liquid suspension (Wright et al. 1974; Crumpton and Snary 1974). These show great potential as alternatives to the more traditional methods of cell breakage, especially from the viewpoint of plasma membrane preparation. The extent of disruption of the cells can be controlled and the method is highly reproducible from batch to batch of equivalent material. Disruption is induced by the application of controlled and variable shear forces, generated as the cell suspension passes under pressure through an orifice.

In the pump described by Wright et al. (1974) the orifice consists of a sleeve into which fits a needle valve attached to a spring. In the 'Stansted' model, used by Crumpton and Snary (1974), the orifice is constructed of a reversed cone into which fits a ball-bearing onto which a variable back-pressure is applied either by a spring (in earlier models) or an air-diaphragm valve (in later models). The shearing force is regulated by the diameter of the ball-bearing.

The 'Stansted' apparatus, which may be air or electrically driven, is shown in Fig. 2.3. It consists essentially of a high-pressure liquid pump which forces the liquid medium containing the cells to be disrupted through a back-pressure valve. The back-pressure (disrupting) valve can be adjusted finely to give a range of pressures as required by the nature of the material to be disrupted. Although the electrical pump is more suitable when large quantities of material are to be processed, the air-operated version is less expensive and simpler in construction. Air-driven machines are operated either from a compressed air supply or a nitrogen cylinder at pressures of more than 3 atm. For disruption of small quantities of cells, the

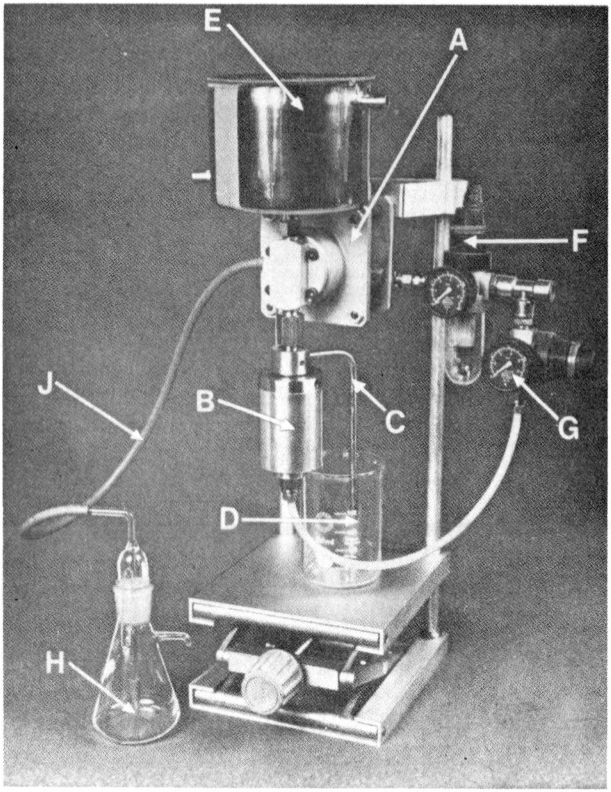

Fig. 2.3. Air-driven 'Stansted' cell disrupter. The central unit is the pump A coupled to the disrupting valve B. The outlet delivery port C is positioned for collecting the disrupted cells in beaker D. Compressed air is supplied to the pump via air regulator F, thus controlling the pumping rate, and to the disrupting valve via air regulator G which controls the disrupting pressure. Valve G is adjusted to give the required degree of disruption as judged by using various methods such as cell viability and phase-contrast microscopy. With decreasing size of the disrupting valve (no. 716 to 116) the same setting of regulator G gives increasing disruption pressure. J is the air outlet to the pump. When noxious or pathogenic material is being disrupted, the exhaust air is passed through flask H containing a suitable protective (safeguard) agent.

air-driven version has the important advantage of having a lower dead space volume.

The pump has been used to disrupt mainly single cell suspensions. Up to approximately 4 l/h of cell suspensions (cell density should not exceed 10^8/ml) can be processed. Plasma membrane fractions were prepared from human cultured lymphocytes (Crumpton and Snary 1974; Bridgen et al. 1976), mouse thymocytes (Wright et al. 1974) and mouse spleen lymphocytes (Abney et al. 1976). Additionally, platelets, primary chick fibroblasts, sperm and chondrocytes have been disrupted and in some instances plasma membrane fractions prepared (Crumpton, personal communication). The valve pressure range for breaking lymphocytes (6–8 μm cell diameter) is 1–3 atm using disrupting valve 516 (Stansted Fluid Power). In general, the smaller the cell diameter, the greater the disrupting pressure required.

Cell rupturing pumps have recently been introduced by Stansted Fluid Power Ltd. (electrical and air-driven models available) and the model described by Wright et al. (1974) is available from Research Engineers Ltd.

2.1.6. Tissue presses

A number of developments of the 'French' press (Milner et al. 1950) or 'Hughes' press (Hughes 1951) have been described. In the model described by Emanuel and Chaikoff (1957), a tissue suspension is propelled through a narrow annular space contained between a steel cylinder and a rod. The width of the annular space, at the 1 cm long piston tip can be varied between 0.02–0.04 mm according to the diameter of the selection of rods supplied with the instrument. Pressures of 130–680 atm applied by a hydraulic press, are required to propel tissue suspensions through the narrow orifices. Before use, the apparatus is cooled in a refrigerator. The Emanuel–Chaikoff press has not found wide application in the disruption of tissues and cell suspensions, probably because its purpose is fully served by Potter–Elvehjem homogenisers. The 'French' hydraulic press (available from American Instrument Co.) also continues to serve as an alternative and has been used, for example, to prepare nuclear

Subject index p. 261

and plasma membranes of chicken erythrocytes (Blanchet 1974). For disruption of small volumes, miniature cells (Cat. No. 4–3399) with a cell piston area of 0.11 in^2 are available at which pressures of 150 atm can be applied. The 'French' press has proved useful in breaking open isolated membrane vesicles to liberate content, e.g., Golgi secretory vesicles (Bergeron et al. 1973b). (See § 3.7). The X-Press Cell Disintigrator (LKB-Biotec Instruments; capacities of 5 ml and 25 ml available) has been used to disrupt rat liver, human carcinoma, rat skin and ox cornea. The shearing stress is created when ice at $-25\,^{\circ}$C is subjected to high pressure. The phase change in the crystal structure of the ice, combined with rapid expansion and contraction as the frozen material is forced through a small orifice, results in cell disruption.

A simple press for disrupting solid fibrous tissue. A simple press, originally described for disrupting pig mesenteric nodes for plasma membrane preparation (Allan and Crumpton 1970), has been developed for membrane preparation from other tissues. In this method, the chopped tissue is disrupted by pressing through a perforated plate, during which cells are sheared sufficiently to break the surface membrane. The press is constructed of stainless steel and comprises 4 pieces, i.e. a cylinder, a plunger, a perforated plate and a screw cap to position the perforated plate at one end of the cylinder (Fig. 2.4). Smaller, hand-operated versions have been used to prepare plasma membranes from mesenteric lymph nodes (Allan and Crumpton 1970) and experimental granulation tissue (Lehtinen et al. 1975). In these cases, a screwing action propels the chopped tissue across the perforated plate. A larger version that conveniently handles up to 1 kg of tissue, e.g., chopped placenta, is modified by incorporation of delrin strips on the plunger; these reduce friction and minimise jamming due to tissue accumulating in the upper portion of the cylinder. The plunger is propelled downwards by using a rack and pinion press (e.g., a mandrel press, Buck and Hickman Ltd.).

Operation of the larger press may require the removal of connective tissue accumulating at the plate between each 'refill'. However,

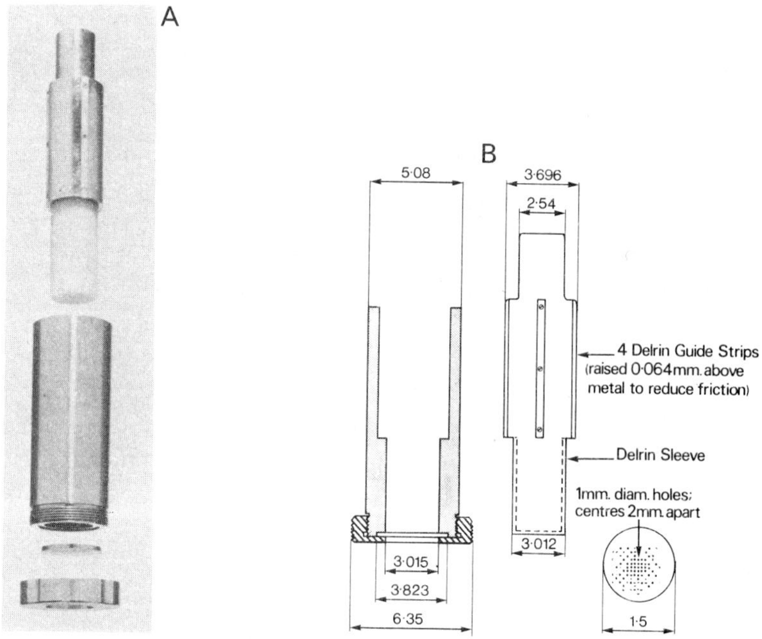

Fig. 2.4. Tissue press. (A) Exploded view; (B) Machine drawing; dimensions are in cm. The gap at the lower (narrower) end of the piston is 0.002 cm. The central portion of the plunger has guide strips to reduce friction and avoid jamming. The well at the lower end of the cylinder is two-thirds filled with chopped tissue. The plunger is pressed into the cylinder by means of a mandrel rack and pinion press. Courtesy of M.J. Crumpton.

this pad of connective tissue also plays a critical role in cellular disruption, since with 'soft' tissues, e.g., brain or liver which lack extensive connective tissue, the press is ineffective when plasma membranes are to be prepared. Hence, a balance between allowing a sufficient amount of connective tissue to effect cell disruption and easy operation needs to be maintained. For soft tissues lacking connective tissue, the plate may be precoated with connective tissue from another tissue. An important parameter determing the yield of plasma membranes is the diameter of the holes in the plate; for example, 2 mm diameter holes gave one-tenth of the yield of plasma membranes obtained by using holes of 1 mm in diameter. Lehtinen et al. (1975) disrupted up to 20 g of fresh granulation tissue by forcing 3 g amounts through holes 0.75 mm in diameter and spaced at a distance of 1.5 mm from centre to centre, and then used the milky suspension in isotonic sucrose to prepare plasma membranes.

In the general procedure, as applied to lymph nodes, placenta, spleens, tissue collected from the press is then mixed by mechanical stirring with 2 volumes of isotonic Tris-buffered saline, pH 7.3, filtered through muslin cloth, and fractionated by differential centrifugation to yield a crude microsomal fraction. Plasma membranes are then collected at a 25/37% (w/w) interface of a discontinuous sucrose gradient (§ 5.5).

Although the press provides a convenient method for preparing plasma membranes in good yield (up to 25% recovery of homogenate 5′-nucleotidase), the dispersion of tissue initially in the absence of a suspending medium may result in the occlusion of cytoplasmic soluble components inside plasma membrane vesicles. This reservation does not apply to the technique as used by Lehtinen et al. (1975). The problems associated with occluded cytoplasmic proteins, and their effect on the density of membrane vesicles, together with methods for their release are discussed in § 3.7.

2.1.7. Sonication

Although sonic or ultrasonic energy has not been used extensively to break up cells due to potential damaging effects on biological processes, ultrasonic probes have found limited use in the dispersion of membranes and subcellular aggregates. Two types of sonic generators are available: (a) diaphragms, e.g., Kinematica 'Polytron' high-frequency instruments (see § 2.1.8) that liberate low energy levels (< 16 kHz) and (b) ultrasonic machines, e.g., those available from M.S.E. & Branson that use 'probes' to liberate up to 20 kHz. The effects of sonic energy are mainly mechanical, since the radiation pressures of the sonic waves generate local shearing forces due to bubbles of gas and/or vapour that undergo cavitation. Ultrasonic energy is difficult to use in a controlled fashion on intact cells (but see below) because in addition to cell and organellar breakage, anisometric molecules, e.g., DNA, are broken, thiol groups are oxidised (Kanig and Kunkel 1957) and electrostatic and even peptide bonds may be cleaved, e.g., transplantation antigens and cyclic nucleotide phosphodiesterases (Sakai et al. 1974) are liberated from tissues. If sonication is to be used to disrupt cells or isolated fractions, the following parameters should receive careful consideration: (a) temperature control; the sonication process should be carried out in short bursts of not more than a few seconds' duration in a container surrounded by ice-water; the probe may be precooled to $-20^{\circ}C$; (b) intensity and frequency control; (c) the nature of the medium; (d) the overall length of exposure to sonic energy; (e) the concentration of the sample should be stated.

An example of the use of sonic energy is provided by Cooper and Stanworth (1976) who disrupted peritoneal mast cells and then prepared plasma membranes. Cells ($6–8 \times 10^6$), washed twice in 0.25 M sucrose and suspended in 1–2 ml of 0.25 M sucrose are pipetted into a 20-ml ice-chilled cup of an M.S.E. 100 W ultrasonicator. Using a precooled microprobe (titanium exponential probe, end diameter 3 mm) the cells were sonicated at 20 kHz (4 μm peak to peak) for 10 s. This treatment disrupted 95% of the cells without rupturing the histamine-containing granules.

2.1.8. Tissue homogenisers, e.g., 'Polytron, Ultraturrax' machines

This type of machine (see Fig. 2.5) has a wide range of uses, from cell and tissue disruption to the dispersion or emulsification of a wide range of chemicals. The rotation of blades within the open tip of the instrument results in mechanical low shear homogenisation and at higher settings some sonic energy is possibly liberated. Like the Dounce homogeniser (§ 2.1.2), this machine was used first to prepare nuclei from rat heart tissue by homogenising tissue minced in 0.32 M sucrose, 2 mM $MgCl_2$, for 2.5 min in the 'Ultraturrax' TP 18/2 tissue disintegrator (Janke and Kunkel), run at 25% of maximum voltage (Widnell et al. 1967). To prepare plasma membranes, Yunghans and Morré (1973) homogenised rat liver mince at a ratio

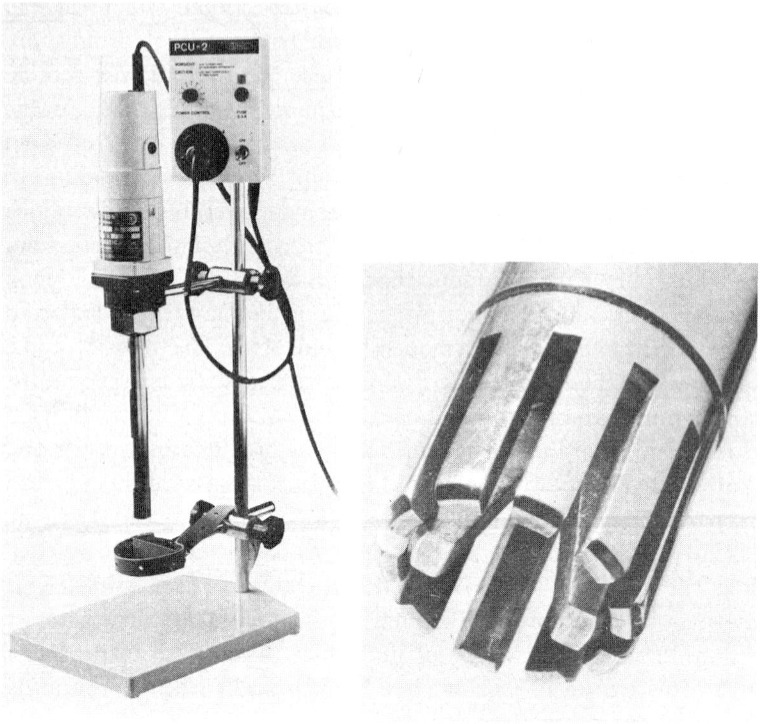

Fig. 2.5. 'Polytron' homogeniser. The probe consists of a shaft (stator) and the rotor. The circular saw serves to promote the disintegration of material (inset). Models PT 10ST (illustrated) and PT 20ST (Kinematica), processing 1–100 ml and 2–500 ml, respectively, have been used to prepare membranes. The 'Ultraturrax' is of similar construction.

of 3–4 g/15 ml $NaHCO_3$, pH 7.4, for 90 s in a 'Polytron' model 20 ST (Kinematica).

Other examples of the use of these homogenisers are as follows. Tsou et al. (1974) homogenised rat hepatoma in a 'Polytron' at setting '4' for 1 min and then prepared a plasma membrane fraction. Oda et al. (1974) used a 'Polytron' homogeniser to prepare liver plasma membranes. Fat cells were homogenised in a 'Polytron' PT-10 at a speed setting of 2.5 for 30 s to prepare a plasma membrane fraction by Chang et al. (1975). Mooseker (1976) used a Sorvall Omni-mixer (Sorvall–Dupont Instruments), setting 7, 10–15 s exposure, to release brush borders from intestinal epithelial sheets.

Homogenisation of tissue minces in 'Polytron'- or 'Ultraturrax'-type machines is used mainly with soft tissues as a substitute for the Potter–Elvehjem and Dounce-type homogenisers. It is claimed that these machines possess advantages over conventional homogenisers, especially in the reproducibility of cell breakage. Undoubtedly they offer the advantage of speed when large amounts of tissue are to be homogenised. More extensive use of tissue homogenisers of this type has been restricted by their initial high costs relative to Teflon-glass homogenisers, and the difficulty in standardising the tissue shear conditions, an important consideration when vesicularisation of the plasma membrane needs to be avoided. 'Polytron'-type machines can be calibrated by converting flashing light reflected from the rotating half-blackened shaft into an electrical signal by means of a photocell, and matching the frequency of an audiosignal generator to this signal by means of an oscilloscope (Nyquist and Morré 1971). In this way, the number of revs/min corresponding to instrument setting can be obtained.

2.2. Media

Cells and chopped tissues are usually disrupted whilst suspended in an aqueous medium. A wide range of media have been used, and the reasons determining their selection are discussed in greater detail with reference to various cell types and tissues in later sections. However, it is appropriate to make some generalisation at this point. Cell disruption is usually carried out in either hypotonic or iso-osmotic media that may contain a variety of additives. Iso-osmotic media (usually 0.25–0.32 M sucrose) are optimised for the enzymic and morphological preservation of nuclei and mitochondria. The inclusion of 1–2 mM Mg^{2+} or Ca^{2+} helps to maintain nuclear integrity (Hogeboom et al. 1948; Potter 1955). This is important, since extensive nuclear disruption may cause homogenates to become viscous due to released chromatin. On the other hand, overjudicious use of divalent cations can cause aggregation of the endoplasmic reticular vesicles that are present in microsomal fractions, e.g., addition of 8 mM Ca^{2+} to a rat liver post-mitochondrial supernatant resulted in a microsomal-like fraction being sedimented at 30 *g* for 10 min (Kamath and Rubin 1972). The use of chelating agents to remove divalent cations may lead to the inhibition of plasma membranes marker enzymes that require these ions for activity. There is also variation in cation sensitivity between various subcellular membranes, and the well-known sensitivity of some tissues, e.g., brain tissue and myelin to cations results in coacervation or aggregation of membranes and organelles, thus impeding their separation on density gradients.

2.2.1. Hypo-osmotic media

Hypotonic media feature frequently in the isolation of plasma membranes despite deleterious effects that their use may cause to intracellular organelles. A reason for the use of hypotonic media is that cells are more likely to yield on disruption plasma membrane either in the form of large ghosts or sheets, or with general reduced polydispersity that makes subsequent fractionation easier. Hence, the initial successful preparation of plasma membranes from low speed homogenates of tissues dispersed in 1 mM $NaHCO_3$ by using Dounce homogenisers (Neville 1960) has led to the adoption of similar hypotonic media to prepare plasma membranes from a wide range of tissues and tissue culture cells. However, the intercellular junctions which aid in the production of large plasma membrane fragments from the contiguous or lateral face of cells of epithelial tissues are absent in cultured cells, and chemical fixation techniques were developed by Warren et al. (1966) to stabilise or strengthen the plasma membrane to minimise fragmentation during cell breakage and subsequent manipulation (see § 2.3.4).

A large number of modifications to the simpler hypotonic media have since been described. These involve the addition of low amounts of divalent cations (0.5–2.0 mM Ca^{2+}) to the hypotonic medium (see § 2.2 for rationale), the addition of low amounts of sucrose, e.g., 0.08–0.09 M sucrose (Weaver and Boyle 1969; Hemminki and Suovaniemi 1973), and the restoration of isotonicity by addition of concentrated sucrose as soon as the cells have been satisfactorily disrupted.

The ratio of cells or tissues to medium also varies widely. In methods using iso-osmotic sucrose it is convenient to use a 10% homogenate (i.e., 10 g wet tissue weight to 100 ml medium), but lower ratios are generally employed with hypotonic media. This may reflect the necessity to minimise the consequences of organellar damage, especially nuclear damage resulting in the release of chromatin that may lead to aggregation of plasma membrane. Few direct comparisons of hypotonic and iso-osmotic homogenisation conditions have been reported. Sidhu et al. (1975) prepared plasma membrane fractions from bovine corpus luteum and reported that for the preparation of plasma membranes with hormonally-sensitive adenylate cyclase activity, the membranes prepared from hypotonic homogenates were superior to those dispersed in iso-osmotic sucrose, although this may be at the expense of isolating fully representative portions of the surface membrane (§ 5.7).

2.2.2. Medium pH

Most media used for cell breakage are weakly buffered to physiological pH, although this is hardly necessary in most cases in view of the strong buffering capacity of membranes. An important exception is is the need to buffer muscle homogenates owing to the lowering of the pH by lactic acid. The effects of the pH of the homogenising medium on membrane fragility has received little attention (see Bell et al. 1971) but it appears that slightly alkaline media (pH 7.5–8.0) help to minimise the extent of breakdown of plasma membranes to smaller fragments. In the fractionation of bovine thyroid plasma membranes, a pH of 8.6 resulted in gross simplification of the number of zones appearing in sucrose gradients, thus suggesting some aggregation of membranes (Hilderson et al. 1975). Dodge et al. (1963), Hanahan et al. (1974) and Bramley et al. (1971) have made a thorough study of the effect of pH and ionic strength on the size of membrane fragments and the extent of retention of haemoglobin in the preparation of erythrocyte ghosts. A further important parameter to control in the medium is Ca^{2+}, for in addition to its effects in maintaining integrity of nuclei, its concentration may also regulate the extent of fragmentation of the plasma membrane, thus dictating the ratio of vesicles to sheets being produced (Allan et al. 1976).

2.3. Special conditions

2.3.1. Solid tissues

Most fractionation procedures that apply to soft tissues (Claude 1946; Schneider and Hogeboom 1950) require the tissue or organ to be chopped or minced into small segments before homogenisation. Tissues or organs are transferred onto a watch glass standing on ice, or into homogenising medium. Blood is removed by washing in medium and they are then cut into small segments (1 mm thick) or a mince either manually by scissors or a blade, or mechanically by tissue choppers, e.g., Stadie-Riggs microtome, or apparatus similar to that described by McIlwain and Buddle (1953), and available from Mickle Lab. Engineering Co. and Yeda Research and Development Co. Ltd. When large amounts of material are to be homogenised, the tissue may be broken up initially in a 'Waring' blender or other machines with rotating blades, e.g., 'Ultraturrax' or 'Polytron' (see § 2.1.8). Muscle and tissues containing extensive collagen or fibrous material, e.g., thyroid glands, can present formidable problems in homogenising. Muscle tissue is usually disrupted rather drastically be mechanical means (e.g., 'Waring' blender) that can result in the alteration or destruction of organelles (Cleland and Slater 1953; Barrett and Headon 1975). A major problem with muscle tissue is the predominance of myofibrils, which on homogenisation trap other organelles, thus preventing their mutual separation by differential centrifugation. At low ionic strengths, myofibrils may also bind large amounts of RNA (Heywood et al. 1968). Myofibrils of tissues removed from animals are extensively contracted and this can hamper subsequent fractionation of the muscle cell into its components. Zak et al. (1973) have described a technique for relaxing excised muscle prior to homogenisation that involves stirring chopped muscle tissue for 10 min in ice-cold 0.1 M KCl, 5 mM $MgCl_2$, 5 mM EGTA, 5 mM sodium pyrophosphate buffer, pH 6.8. Subsequently, homogenisation can be carried out as with soft tissues.

On a more general level, the removal of collagen fibres, cell debris, etc., is normally accomplished by filtration through layers of muslin, cheese-cloth (fine mesh No. 110 or 120 or coarse mesh No. 40) or nylon bolting cloth 110 mesh (e.g., Nybolt, Swiss Silk Bolting Cloth Mfng. Co. Ltd). In the preparation of plasma membranes from smooth, skeletal and cardiac muscle, Kidwai et al. (1971a,b) have described a simple filtration device for removing cell debris, myofibrils, collagen, etc., from the homogenate under low suction pressure. This involves setting a stainless steel gauze on a rubber O-ring cushion inserted into a Büchner-funnel attached to a filtration flask. Wire meshes of different porosity are available, and mesh 40 appears to be suitable for filtration of cardiac tissue homogenates (wire gauzes are available, e.g., from Small Parts Inc; F.W. Potter and Soar Ltd; in addition to the mesh size, the gauge (s.w.g.) should also be stated, i.e., the percentage of area open for screening).

Although the removal of connective tissues, collagen fibres, basement membranes, etc., helps to simplify subsequent homogenisation and the complexity of the initial low speed pellet, their persistence can prove useful in tissue dispersion by a tissue press (§ 2.1.6).

2.3.2. Isolated cells

Cultured fibroblastic cell lines are difficult to disrupt. When, for example, tight-fitting Dounce homogenisers are used (as an alternative to nitrogen cavitation), the cells are first suspended in buffered hypo-osmotic media (e.g., 5 mM Tris-HCl, pH 7.4) for 15 min. This preliminary step may be combined with cell membrane 'stabilisation' by exposing cells (5×10^7 cells/ml) to 1 mM $ZnCl_2$ (Warren et al. 1966) as used in the disruption of neuroblastoma cells (Truding et al. 1974). Phase contrast observations indicate that the cytoplasm contracts away from the cell surface under these conditions.

Platelets are also difficult to disrupt, due partly to their small size. Techniques used to disrupt platelets include sonication (Baenziger et al. 1972), a 'no clearance' Teflon homogeniser (Marcus et al. 1966) and the 'glycerol lysis' technique (Barber and Jamieson 1970). The latter technique exploits the differential properties of platelet membranes with respect to the transport of glycerol and water by centrifuging washed platelets (1×10^{11} cells in 10 ml Tris buffer) through an isotonic glycerol gradient (0–40%) at 4000 *g* for 30 min. The pelleted glycerol-loaded platelets (intracellular concentration reaches 4.3 M) are then lysed by rapid transfer into a Tris/saline suspension medium and agitated on a 'Vibromix' stirrer. This technique has also been used to prepare plasma membranes from lymphoid cells (Table 2.4; Jett et al. 1977).

2.3.3. *Fragile cells*

Fat cells are fragile and can be disrupted by gentle liquid shear. Fat cells prepared from epididymal fat pads (Rodbell 1964) and suspended in 20 ml of 0.25 M sucrose 5 mM Tris–HCl (pH 7.4) are aspirated through a cannula (size 16) into a 30-ml plastic syringe fitted with a swinny filter holder (Millipore; Cat. No. XX 30-01200) containing a stainless steel support screen but no filter. Cells are sheared as they are propelled through the apertures, approx. 200 μm diameter, in the support screen. After repeatedly aspirating and ejecting the suspension the progress of cell breakage may be ascertained by phase-contrast microscopy (Avruch and Wallach 1971).

Leukemic and normal lymphoid cells can also be disrupted osmotically to produce a plasma membrane fraction. Washed cells suspended in Erlenmeyer flasks containing a solution of 20 mM borate buffer, pH 9.2, and 0.2 mM EDTA at a ratio of cells to solution not exceeding 1 : 200 are stirred magnetically for 10 min at room temperature. The lysed cells are quickly filtered through nylon gauze 0.8 mm mesh (Henry Simon Ltd.). Plasma membranes are then prepared from the cell homogenate by passage of the 2000 *g*/

30 min pellet resuspended in 2 mM borate buffer, 1 mM EDTA, through a column of glass beads (Superbrite 150, 3M Company; column internal diameter 2 cm; beads are first soaked in 0.1 M HCl and then extensively washed in H_2O). The plasma membranes are eluted from the column ahead of other components (McCollester 1970; Warley and Cook 1973).

2.3.4. *Chemical fixation of membranes to minimise fragmentation*

Plasma membranes sedimented at low centrifugal forces (usually in the conventional 'nuclear' fraction) are present as large membrane strips, indicative of rupture of the cell at limited areas of the surface. Increased fragmentation occurring at the initial homogenisation step, or during the various subsequent manipulations, usually leads to formation of small vesicles. Factors governing vesicle formation are relatively unexplored (see § 2.4.4), but nonetheless, pretreatment of cells with chemical fixation agents helps minimise vesicularisation of the plasma membrane. The use of fluorescein mercuric acetate (see Table 2.7), sulphydryl reagents (Warren et al. 1966) and glutaraldehyde usually lead to a partial or complete loss of enzymic activity. Pretreatment with Zn^{2+} (see Fig. 3.8) or Tris (Warren et al. 1966) is less damaging to enzymes, and both these can be removed subsequently from the isolated membranes (Zn^{2+} by using chelating agents). Warley and Cooke (1976) showed that the inclusion of 1 mM glutaraldehyde in the homogenisation medium of leukemic cells helped to stabilise the Golgi apparatus such that integrity and glycosyl transferase activities of dictyosomes were maintained. However, since the active sites of many plasma membrane marker enzymes are facing the medium (see Table 4.1), they are most susceptible to inactivation by chemically reactive reagents.

A further aspect of the pretreatment of cells is the phenomenon of 'blistering' of the plasma membrane, induced by exposure of cells to glycerol, dimethylsulphoxide, cytochalasin B and local anaesthetics. Although these reagents may act by different mechanisms, characterisation of the plasma membrane released from the 'blistered' area indicates that they are not representative of the cell surface (Vandenburgh 1977; see also § 5.6.3).

2.3.5. Use of inhibitors of proteases and ribonucleases

Subcellular fractionation should be carried out rapidly at 4 °C, to reduce the action of membrane proteases to an acceptable minimum. Metal-requiring proteases are inhibited by EDTA (2 mM dissolved in media), serine proteases by DFP (diisopropylfluorophosphate, 1 mM), and the sulphydryl proteases by *N*-ethylmaleimide (25 mM). Few reports describe the inclusion of these classes of proteolytic inhibitors during subcellular fractionations but minimal degradation of bovine renal glomerular basement membrane was reported by Freytag et al. (1976) who found that inclusion of a range of inhibitors of proteolysis had little effect on the quality of the membranes. Addition of 1 mM phenylmethylsulphonyl fluoride (PMSF) to media used in the preparation of placental plasma membranes had little effect on the integrity of the membrane polypeptides as judged by polyacrylamide gel electrophoresis (Carlson et al. 1976). Proteolysis of intestinal brush border plasma membranes was minimised by homogenising epithelial cell sheets (see § 5.4) in: 4 mM EDTA, 1 mM EGTA, 10 mM imidazol buffer, pH 7.4, containing 10 mM tosylargininemethyl ester (TAME; Sigma) and 0.1 mg/ml of soybean trypsin inhibitor. The use of Trasylol® to minimise proteolysis of pancreatic membranes is described in § 5.7.3.

Ribonuclease activity, ubiquitously present in cell and tissue homogenates, is reduced at pH 8.5 and at high ionic strengths (0.3–0.5 M KCl; 5–10 mM Mg^{2+}). Chemical inhibitors of ribonuclease activity of mammalian cell homogenates include 0.02–0.2% diethylpyrocarbonate, heparin, polyvinylsulphate, and proteinase K, 0.2 mg/ml (Merck). Addition of these inhibitors to media may cause membrane aggregation.

2.4. Assessment of efficacy of homogenising conditions

In this section, some examples will be quoted which describe the effects of varying the pretreatment and homogenisation conditions on the fractionation of membranes and the recovery of subcellular markers. Although the degree of recovery and the quality of the plasma membranes is the prime factor, an assessment of the suitability of the homogenisation conditions can frequently be obtained in the initial stages of subcellular fractionation. Comparative details of the efficacy of various homogenisation conditions are not often presented in the literature but the following examples illustrate the types of preliminary experiments to be used when preparing plasma membranes and especially in trouble-shooting for causes of low recovery of plasma membrane marker components.

2.4.1. Brain tissue

Relatively small changes in the homogenisation conditions can modify the distribution and yield of subcellular markers. Two examples illustrate this in brain homogenates. Whittaker and Dowe (1965) compared Potter–Elvehjem and Dounce homogenisers and the Emanuel–Chaikoff press for producing a fraction containing intact nerve-ending particles (synaptosomes), estimated by their acetylcholine content (Table 2.3). The Potter–Elvehjem homogeniser operating at the higher speed gave the best recovery of nerve-ending particles, and this has since become the preferred method for dis-

TABLE 2.3

Comparison of the effect of four homogenisation conditions on the recovery of acetyl choline in a rodent cerebral synaptosomal fraction.

Type of homogeniser	Clearance (mm)	Speed (revs/min)		Recovery of acetylcholine in synaptosomal fraction (%)	
Potter–Elvehjem[a]	0.25	840	1400	70	83
Dounce	0.12	–		46	
Emanuel–Chaikoff press[b]	0.027	–		82	

[a] Twelve up and down strokes of the pestle.

[b] Piston press hydraulic homogeniser (Emanuel and Chaikoff 1957).

ruption of cerebral tissues to produce a nerve-ending fraction and synaptosomal plasma membranes (§ 5.2). Although good yields were obtained by using the Emanuel–Chaikoff press, this method has found little application.

Hajos et al. (1974) and Coackley (1974) compared the use of Potter–Elvehjem and Dounce homogenisers in the preparation of glomerular complexes from rat cerebellum. Here, the requirement was to devise the gentlest procedure that would yield undisrupted complexes free of other components. A theoretical treatment that compared the homogenisation conditions employed reinforced the view that Dounce homogenisation yielded the better endproduct, presumably due to the approx. 100-fold lower shear forces that were applied for a shorter interval. The use of the Potter–Elvehjem homogeniser resulted in the further breakdown of the glomeruli to yield nerve-ending-like particles.

2.4.2. Lymphocytes

Lymphocytes from a range of sources have been used to prepare a plasma membrane fraction by a variety of methods. It is useful to compare the yield and quality of the plasma membranes by procedures that vary mainly according to the conditions used to disrupt the cells. Table 2.4 summarises the properties of plasma membranes prepared by using the methods of cell disruption already described. The gas-cavitation and Potter–Elvehjem methods result in a higher incidence of nuclear damage. Since small lymphocytes and thymocytes are essentially large nuclei surrounded by a thin layer of cytoplasm encased in a plasma membrane, nuclear damage is an excellent criterion to monitor in the disruption of these cells. Table 2.4 indicates the potential of cell rupturing pumps for use with lymphoblastoid cell lines, and similar results may be obtained by using lymphocytes teased from minced spleens and lymph nodes.

2.4.3. Liver tissue

Liver plasma membranes originating mainly from the lateral side and bile canalicular complex of the hepatocyte are isolated from a

TABLE 2.4

Comparison of the effects of various disruption methods on the properties of lymphocyte plasma membranes.

Cell type	Disruption method[c]	Medium	Nuclear[a] breakage (%)	Plasma membrane: Yield (% protein recovered)	Plasma membrane: Yield (% 5′-nucleotidase recovered)	Plasma membrane: Relative specific activity[b]	Reference
Pig lymphocyte	Tissue-press	10 mM Tris 0.15 M NaCl	4	1.3	15	12	Allan and Crumpton (1970)
	N_2-cavitation (50 atm, 15 min)	2 mM HEPES[h] 0.13 M NaCl 0.5 mM $MgCl_2$	30	1.7	45	25	Ferber et al. (1972)
Human tonsils	Dounce	10 mM Tris 0.2 mM DTT[d]	–	0.7	8	13	Lopes et al. (1973)
	Potter–Elvehjem	10 mM Tris 0.15 M NaCl	30	0.9	12	13	Demus (1973)
Calf thymus	N_2-cavitation (36 atm, 15 min)	Hank's[e] medium	4	0.7	5	8	Van Blitterswijk et al. (1973)
Human lymphoblastoid cell line	'Stansted' Cell rupturing pump	Eagle's[e] medium	9	0.6	41	49	Snary and Crumpton (1974)
Human lymphoma cell line	Glycerol uptake followed by hypotonic lysis	90% glycerol in Eagle's[e] medium; transfer to 10 mM Tris, 1 mM $MgCl_2$ 1 mM $CaCl_2$	0	3.4	17[f]	250[g]	Jett et al. (1977)

[a] Nuclear damage was assessed by calculating the amount of homogenate that remained in the supernatant after sedimenting nuclei at low centrifugal forces.
[b] Enzymic activity in the plasma membrane in relation to homogenate activity.
[c] These methods are discussed in the relevant sections.
[d] Dithiothreitol.
[e] For composition see, e.g., Gibco-Biocult catalogues.
[f] Alkaline phosphodiesterase (EC 3.1.4.1) (see § 4.2.1).
[g] Relative to whole cells.
[h] *N*-2-hydroxyethylpiperazin-*N*′-2-ethanesulphonic acid.

low-speed pellet of the filtered tissue homogenate (Neville 1960; Emmelot et al. 1964). Using the enzyme 5′-nucleotidase as a marker for this part of the hepatocyte plasma membrane, it is clearly advantageous to sediment into the pellet as much as possible of the homogenate activity, before fractionating further. The effect of various homogenisation conditions on the amount of homogenate 5′-nucleotidase activity recovered in the low speed pellet is shown in Table 2.5. The highest recoveries of enzyme in the low-speed pellet were obtained by using loose-fitting Dounce or Potter–Elvehjem homogenisers with a minimal number of strokes compatible with complete cell rupture. Application of higher shear forces or nitrogen pressure homogenisation fragment the plasma membrane into small vesicles that do not sediment at these low-speeds; the plasma membrane vesicles remaining in the supernatant then have to be separated from the endoplasmic reticulum vesicles.

TABLE 2.5

Effect of homogenisation conditions on the recovery of 5′-nucleotidase in a low-speed pellet of rat liver homogenates.

Type of homogeniser	Clearance (mm)	Conditions: Pestle speed (revs/min)	Conditions: Number of strokes	Enzyme in pellet[c] (%)
Potter–Elvehjem	0.33	900	3	57
Potter–Elvehjem	0.10	2900	2	48
Potter–Elvehjem	0.10	2900	15	34
Dounce	0.15	–	10	55
Emanuel–Chaikoff press[a]	0.038	–	–	33
	0.017	–	–	8
Gas-cavitation[b]	–	–	–	6

[a] Liver was minced before passing through the stated aperture.

[b] 57 atm for 20 min.

[c] Homogenate was centrifuged at approx. 600 *g* for 10 min.

Unpublished data of R.H. Hinton and W.H. Evans.

The general approach discussed here applies to most soft tissues, since the lateral plasma membrane is more resistant to fragmentation and vesicularisation by high shear forces or pressure than other areas, especially microvilli (§ 2.4.4). Thus, the sedimentation of the lateral plasma membranes at low-speed constitutes a convenient way of rapid purification. The effectiveness of various homogenisation conditions applied to a variety of tissues can be relatively simply assessed by methods similar to those applied to rat liver.

2.4.4. Brush borders of columnar epithelial cells

Brush border complexes of the apical plasma membrane of intestinal cells and the proximal tubules of kidney are usually released by gentle homogenisation, e.g., in a loose fitting homogeniser (Thuneberg and Rostgaard 1968; Wilfong and Neville 1970). More intensive homogenisation, e.g., in the 'Ultraturrax' machine, causes vesicularisation of the brush border complex (Fig. 2.6). A range of conditions that retain the intactness of the brush border complex (approx.

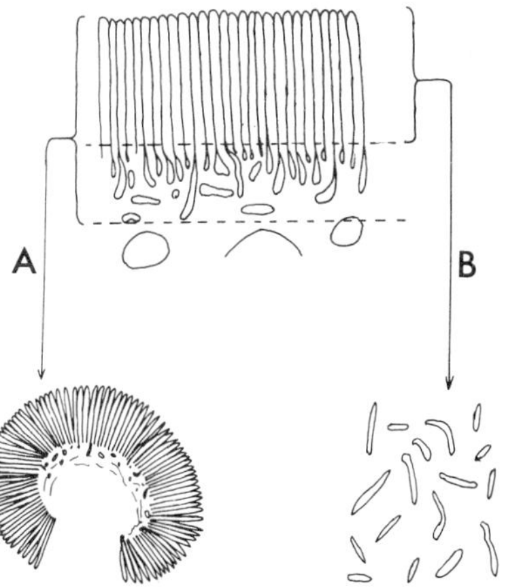

Fig. 2.6. Diagramatic representation of fragmentation of kidney or intestinal epithelial brush borders. (A) Brush border complexes released by gentle homogenisation in loose-fitting Dounce or tissue press (see Table 2.6). (B) Microvilli produced by homogenisation in the 'Ultraturrax' machine. (Figure provided by Dr. A.J. Kenny).

10–30 μm in diameter and easily observed by phase contrast microscopy) thus ensuring its sedimentation at low centrifugal forces were investigated by George and Kenny (1973) and are summarised in Table 2.6. The tissue press arrangement (as described in § 2.1.6) yielded homogenates in which the brush borders had remained relatively intact.

The processes underlying the fragmentation of brush border complexes into vesicles have been investigated (Booth and Kenny 1976). The rate of vesicularisation of the brush borders increased rapidly with increasing temperature and was independent of the composition of the medium provided that hypo-osmotic swelling did not occur. However, cross-linking agents, e.g., glutaraldehyde and sulphydryl reagents, limited the process of vesicle formation. The molecular mechanisms underlying vesicularisation to microvilli of brush borders is not fully understood but is probably related to the interaction of actin with the cytoplasmic face of the membrane.

TABLE 2.6

Effect of homogenisation conditions on fragmentation of rat kidney proximal tubule cells.

Method of homogenisation	Brush border sedimented at 300 g/5 min[a] (%)
Tissue press, metal plate, 33 holes/cm^2, 2.7 mm diameter	82
Tissue press, wire gauze, 12 mesh/cm^2, apertures 0.6 mm	71
Tissue press, wire gauze, 20 mesh/cm^2 and 56 μm mesh nylon	22
Kidneys chopped with 12 razor blades spaced 0.3 mm apart	44
Dounce homogeniser; pestle clearance 0.2–0.5 mm, 5 strokes	62
Potter–Elvehjem; pestle clearance 0.2 mm, 1000 revs/min, 1 passage	18

[a] Calculated on the basis of aminopeptidase activity (a brush border plasma membrane enzyme) sedimented.

Table modified from George and Kenny (1973).

2.4.5. *Murine fibroblasts – effects of pretreatment*

In this example (Table 2.7), the effects of varying the pretreatment of cells on the breakage and the subsequent fractionation and recovery of plasma membranes are illustrated. Hubbard and Cohn (1975) compared the recovery of plasma membranes of mouse L-cells (a fibroblastic cell line) prepared by three widely used methods. In each case the cells were disrupted in a tight-fitting Dounce homogeniser,

TABLE 2.7
Analysis of plasma membranes of mouse L-cells prepared by three different methods.

Isolation procedure	Protein recovered (%)	Acid-insoluble ^{125}I recovered (%)	Relative specific activity
FMA I	1.6	16	11
FMA II	1.7	23	14
Tris–HCl	0.7	8	11

Mouse cells were iodinated by the lactoperoxidase procedure (Hubbard and Cohn 1972) (see § 4.6) and plasma membranes prepared by 3 methods. The properties of the plasma membranes were assessed by comparing the recovery relative to cell homogenates of protein, ^{125}I attached to plasma membranes and the activities of chemical (cholesterol) or enzymic (alkaline phosphodiesterase) markers. In procedure FMA I, cells were exposed to fluorescein mercuric acetate (2 mM dissolved in 20 mM Tris–HCl, pH 8.0), and the cells, after swelling in distilled water, were broken by homogenisation in a tight-fitting Dounce homogeniser. Broken cells were adjusted to 30% (w/v) sucrose (by addition of an equal volume of 60% (w/w) sucrose) and then centrifuged onto a 45% (w/w) sucrose cushion to remove ghosts from cell debris and soluble protein. The plasma membranes were isolated finally from the 55 : 60 : 65% (w/w) sucrose interface after centrifugation (Klenk and Chopin 1969).

In procedure FMA II, the Dounce-homogenised cells, treated as in I, were centrifuged at low speed (1500 *g*/30 s) to remove unbroken cells and the subsequent low speed pellet was washed in 0.08 M NaCl before centrifuging into a discontinuous sucrose gradient. The plasma membranes were collected at a density of approx. 55% (w/w) sucrose.

In the Tris–HCl procedure, cells were loosely suspended in 10 mM Tris–HCl pH 8. After 5 min the cells were broken by homogenisation in a stainless steel Dounce homogeniser (clearance 0.06 mm) and the plasma membrane 'ghosts' were collected from a discontinuous sucrose gradient at a density of approx. 40% (w/w) sucrose.

but in procedures FMA I and II, the cells were first 'strengthened' by treatment with fluorescein mercuric acetate, and in the last procedure cells were pre-exposed to an alkaline Tris–HCl buffer. The subcellular fractionation conditions employed to prepare plasma membranes varied slightly between procedures FMA I and II (see table for explanation). The highest recoveries judged on the basis of the recovery of plasma membrane protein and iodine, and the relative specific activity of a plasma membrane marker, were obtained by using the fluorescein mercuric acetate pretreatment method. However, a major disadvantage of this technique is the inactivation of cell surface enzymes, and cholesterol was used as a plasma membrane marker, as opposed to alkaline phosphodiesterase used in the Tris–HCl pretreated cells. Clearly, for preparation of plasma membranes with enzyme and other functional activities, the Tris–HCl pretreatment method, despite its lower recovery of plasma membrane, is the preferred method. It was also noted that plasma membrane ghosts prepared from Tris–HCl pretreated cells showed a tendency to break up into vesicles during the isolation procedure; this contributes to the lower recovery of membranes and can result in variations in recovery being obtained by different investigators. The results in Table 2.7 led Hubbard and Cohn (1975) to calculate that about 3.2% of the total cellular protein of the L-cell is in the plasma membrane. The Tris–HCl pretreatment of cells, also exploited by Atkinson and Summers (1971), is discussed in greater detail in § 5.6.

General methods for the preparation of plasma membranes

The prime objective is to prepare representative and/or topographically identified fragments of the plasma membrane free from both intracellular organelles and membrane systems, and the soluble proteins present in the cytoplasm and the cell's environment. The method of cell breakage constitutes by far the most critical process in achieving this objective. To quote De Duve (1967) "the best homogenate is the one which lends itself most successfully to fractionation. Or, to put it colloquially, the proof of the homogenate is in the fractionating".

Although each tissue and cell type may pose specific problems when attempting to recover plasma membranes of acceptable purity, some general principles for separation apply. In the following sections, the practical applications and limitations of procedures for the separation of plasma membranes are described.

3.1. Separation of membranes by centrifugation

In almost all cases described, preparation of plasma membranes has entailed subjecting the cell homogenate to at least one centrifugation step. The preparation of plasma membranes from a wide range of tissues and cell populations by centrifugation methods has been achieved despite the similar sedimentation and density properties of plasma membranes compared with other subcellular components (Fig. 3.1). However, in resolving the functional heterogeneity of the plasma membrane in which different regions of the cell surface may have similar physical and morphological properties, but radically

Subject index p. 261

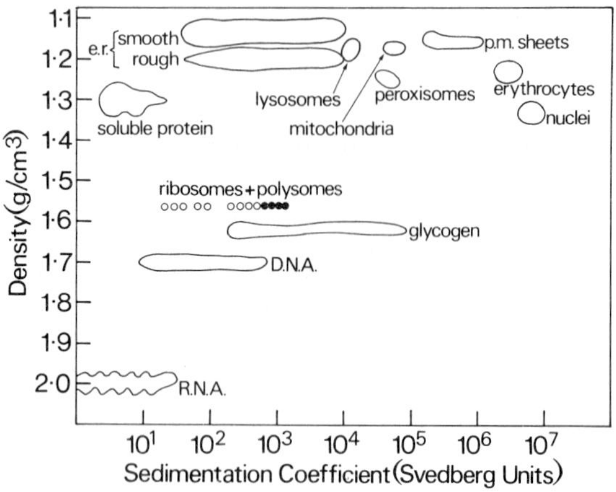

Fig. 3.1. Approximate sedimentation coefficient and banding densities of mammalian subcellular components. Plasma membrane (p.m.) vesicles are also found in the smooth endoplasmic reticulum (e.r.) fraction. Redrawn from Anderson et al. (1966).

different functions, centrifugation methods have been supplemented by other techniques such as electrophoresis, affinity separation, etc., that are described in later sections.

Centrifugation methods exploit differences in weight (differential centrifugation, differential pelleting) or density (density gradient or isopycnic centrifugation). In many cases, both weight and density parameters are exploited simultaneously to effect separation (rate-zonal centrifugation).

The theoretical basis of centrifugal separations will not be discussed (see Price 1973; Hinton and Dobrota 1976). However, it is necessary to define the Svedberg unit (S) used in Fig. 3.1,

$$s = \frac{2r^2(\rho - \rho')}{9\eta} = \frac{v}{\omega^2 x}$$

when

r = particle radius (assumes a spherical particle)

ρ = particle density (g/cm^3)

ρ' = density of suspending medium (g/cm^3)

v = sedimentation rate (cm/s^{-1})
η = viscosity of suspending medium (poises)
ω = angular velocity of rotor
x = distance of particle from axis of rotor (cm)
s = sedimentation coefficient 1×10^{-13} s = Svedberg unit (S)

3.1.1. Differential centrifugation

This is usually the first step in fractionating the homogenate and is usually carried out in angled rotors. Fig. 3.2 summarises, with examples, the possibilities open to plasma membrane fragments in homogenates and serves to emphasise, once more, how critical the homogenisation procedure and the subsequent stages are when pellets are resuspended for washing and centrifuging at higher speeds, in determining the size and shape of plasma membrane fragments. The production of polydisperse plasma membrane fragments that may sediment into all fractions (Fig. 3.2, example 5) is a central problem, and will account for failure to prepare plasma membranes in acceptable yield.

The pellets produced by centrifugal forces that yield the three major particulate fractions of mammalian tissues and cells are multilaminate and complex. Pellets are usually washed at least once by recentrifugation of the resuspended pellet under the same gravitational force to improve the homogeneity of the fraction, since before centrifugation particles near the bottom of a homogenate have a 'head' start over particles near the top of the tube. Multilaminate pellets are not seen clearly when using opaque centrifugation tubes (e.g., polypropylene) and hence it is advisable to use transparent tubes and pots, e.g., glass, polycarbonate or cellulose nitrate. Transparent tubes also enable closer control to be exercised over the removal of the supernatant covering loose pellets. In many instances, the top layer of a multilaminate pellet can be selectively removed by gently swirling the medium in the tube and then removing it by aspiration.

In plasma membrane preparation, it may be worth deviating from the standard centrifugal forces used to pellet the classical nuclear,

1. Production of membrane ghosts
   Adipocyte 'ghosts'

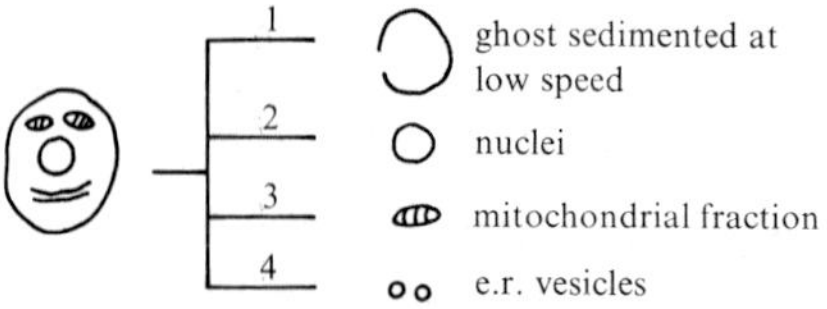

2. Production of large plasma membrane particles
   HeLa plasma membranes

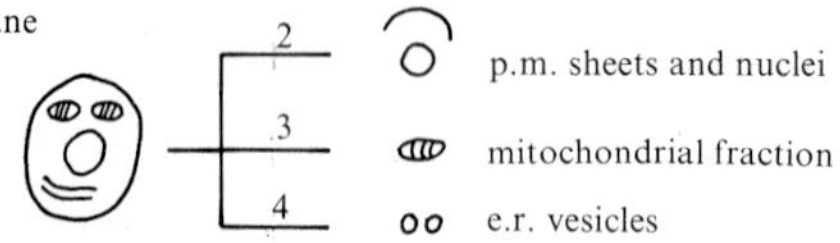

3. Production of large plasma membrane sheets and small vesicles; 'biphasic' distribution
   Tissues e.g. liver, corpus luteum, thyroid, kidney

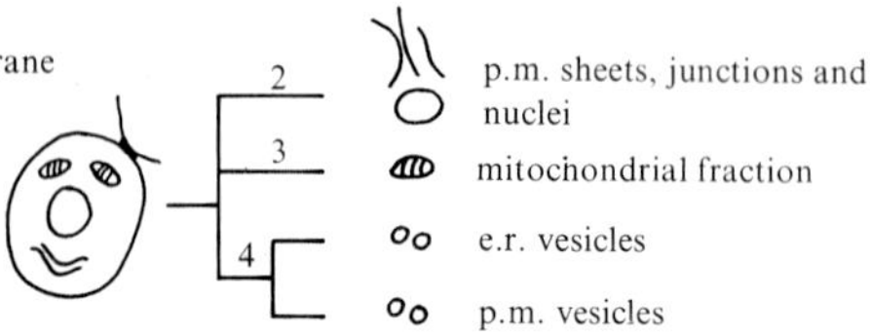

4. Production of plasma membrane vesicles
   Cultured cells disrupted by gas cavitation

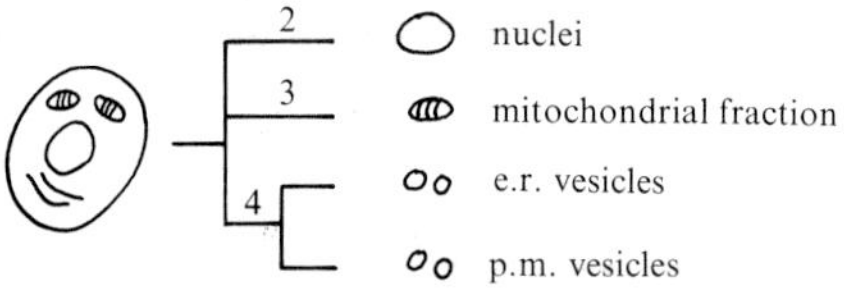

5. Production of polydisperse p.m. particles
   Unsuccessful 'multiphasic' preparation.
   Very low yield if p.m. collected from 4.

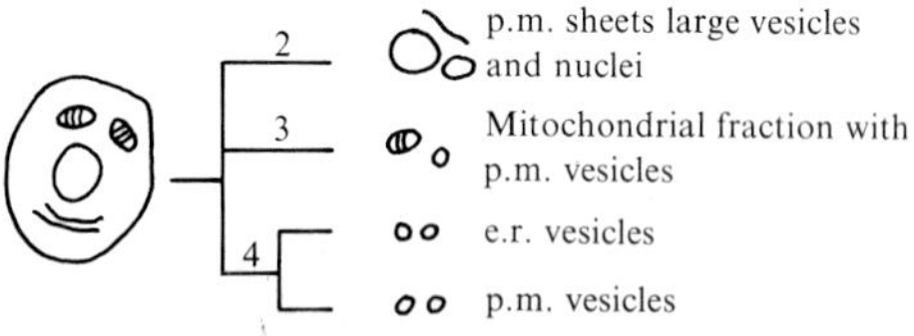

KEY

1. 100 $g \times 5'$ very low speed; unbroken cells also sediment.
2. 1000 $g \times 5 - 15'$ nuclei sediment.
3. 5 – 10,000 $g \times 20'$; mitochondria, lysozomes etc. sediment.
4. $>10{,}000\ g \times 20'$; e.r. Golgi and p.m. vesicles sediment.

 nuclei

 mitochondria

 plasma membrane (p.m.) sheets or vesicles

 endoplasmic reticulum (e.r.) vesicles

mitochondrial and microsomal fractions, and to prepare additional 'cuts'. An example of this when fractionating corpus luteal homogenates is described in § 4.1.3.

3.1.2. Centrifugation in density gradients

Most plasma membrane preparative routines also entail a density gradient step in which membranes may or may not be centrifuged to their isopycnic point. In general, 'washed' pellets, prepared by differential centrifugation, usually the 'low speed' or nuclear and 'high speed' or microsomal fractions are suspended in a separating medium and subjected to centrifugation in gradients of varying densities of sucrose, or less frequently Ficoll, Dextran, Metrizamide, etc. (§ 3.1.5). Practical aspects of the construction of density gradients will be described first, before discussing the densities of subcellular particles and some theoretical parameters determining particle density. Finally, techniques for specifically modifying the density of various subcellular components will be considered.

3.1.3. Continuous and discontinuous density gradients

When the location and properties of plasma membranes in a fraction to be separated by density gradient centrifugation are poorly understood, it is important to carry out initial investigations in continuous gradients. Fig. 3.3 indicates that discontinuous gradients used in initial stages may produce bands that can be arbitrary and misleading. By selecting the correct sucrose concentrations on the basis of the information first obtained by centrifugation of the sample into a continuous gradient (Case 1), the three fractions A, B and C may be collected subsequently from discontinuous sucrose interfaces (Case 2). Injudicious choice of sucrose concentrations, or the use of a multi-step discontinuous gradient may resolve the fraction into 6 components; such additional fractionation is artifactual, for the discrete subfractions produced are merely 'cuts' of symmetrical peaks and their

◁ Fig. 3.2. Effects of particle size on subcellular distribution of plasma membrane fragments. See text for explanation.

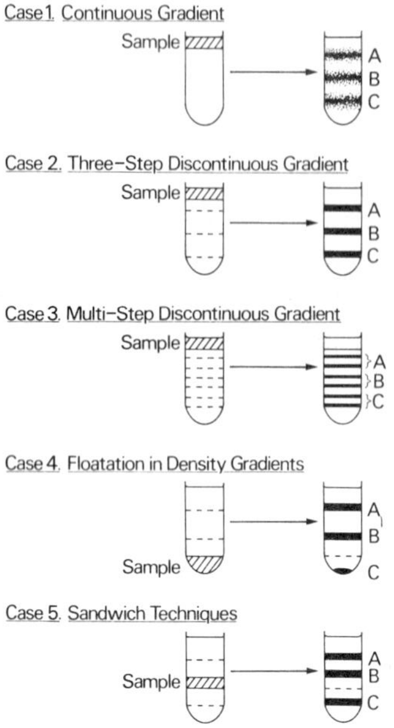

Fig. 3.3. Density-gradient fractionation of membranes. Continuous and discontinuous gradients, creation of artifacts. See text for explanation

separate collection and analysis is unnecessary (Case 3). Floatation of membrane fractions involves the initial suspending of the components in a high sucrose concentration (using Potter–Elvehjem or 'Ultraturrax'-type homogenisers) and the floatation of fractions A and B (Case 4). Collection of fractions as pellets at the bottom of density gradient tubes should ideally be avoided, since most contaminating material (e.g., bacteria) will also sediment to the bottom of the gradient. Sandwich-type gradients are advantageous, since the component parts move in opposite directions (Case 5). This is certainly a better procedure to adopt when large amounts of protein are loaded onto gradients, since it minimises the possibility of material becoming entrapped in band B.

Separation of membrane fractions is usually carried out in continuous sucrose gradients (e.g. 10–50%, w/v, sucrose), and the sucrose discontinuities are then selected at concentrations that exceed those of the leading edge of the band to be collected but are below the trailing edge of heavier components. The compact bands present at the interfaces may be rapidly collected by a needle attached to a syringe, removing the uppermost bands first. Overloading of gradients should be avoided, evidence of which is normally provided by material becoming attached to the walls of the tubes; the use of sandwich gradients in which components move in opposite directions (Fig. 3.3, Case 5) helps to minimise cross-contamination of fractions.

Separations in sucrose density gradients are usually complete after centrifugation for approx. 3 hr at 70,000–100,000 *g* when most subcellular particles have approximated close to their isopycnic point. However, the true isopycnic point or median density of a fraction should be ascertained on continuous gradients after centrifugation overnight at similar gravitational forces.

3.1.4. Construction and unloading of gradients

Discontinuous sucrose gradients are easily made by layering from a pipette zones of solutions of decreasing densities. Intermixing at the interfaces during loading is minimised by holding the centrifuge tube at a 45° angle and allowing the lighter sucrose solution to run slowly from the pipette for the minimum distance along the inner side of the tube. Practice in making gradients may be obtained by using sucrose solutions containing dyes of different colours. Linear continuous gradients (in which the concentration of solute is proportional to volume) are more laborious and time-consuming to construct. Many types of commercial apparatus are available for the construction of linear as well as concave or convex gradients; the advent of zonal rotors has made available many of the larger-volume, programmable varieties that can generate linear or exponential gradients as ancillaries to the technique (e.g., available from IEC, ISCO, LKB, M.S.E., Sorvall) (see Hinton and Dobrota 1976). Most laboratories, however, generate linear sucrose gradients suitable for rotor swing-out tubes (10–60 ml) from two-cylinder generators that can be constructed from cylindrical columns (Fig. 4.4). Owing to the high viscosity of sucrose solutions, vigorous stirring in the mixing chamber is essential. This can be achieved by attaching a pipette to a compressed air source, or N_2 cylinder and positioning the tip at the bottom of the mixing chamber. Mechanical or magnetic stirrers may also be used. If a multi-channel peristaltic pump is available (e.g., Pharmacia Peristaltic Pump P3, Gilson Minipuls II) then identical linear gradients for a number of tubes (usually 3) can be constructed simultaneously from one two-cylinder generator.

For unloading and quantitating the distribution of components in a gradient, fractionators are available commercially (e.g., ISCO) that use a syringe to pierce the centrifuge tube near the bottom (cellulose nitrate tubes are desirable; thick-walled polycarbonate tubes are difficult to pierce) and pump heavy sucrose into the tube to displace the gradient slowly upwards into a narrow channel. Fractions are then collected on a fraction 'cutter' and absorbance (usually at E_{280} or E_{250}) or turbidity is recorded, either automatically en route to the fraction cutter or manually. A simpler alternative for unloading gradients involves collecting the drops from a small hole pierced at or near the bottom of the tube. If the centrifuge tube is to be re-used or is thick-walled, then a thin hollow cylinder held vertically in a clamp is lowered into the gradient and heavy sucrose is slowly pumped into the bottom of the centrifuge tube, so displacing the gradient through an adaptor fitted to the top of the tube and channeling the gradient into collecting tubes. A further simple method that may not yield such critical separations as the methods described above but is satisfactory for well-separated bands, involves carefully inserting a needle (e.g. 18G internal diameter 0.76 mm, Smith and Nephew-Southalls Ltd.) filled with a concentrated solution of sucrose and connected to a syringe by a two-way valve (Horwell Ltd.; Becton–Dickinson) or a peristaltic pump to withdraw the fractions from the bottom of the gradient. As already indicated, bands located at discontinuous interfaces are easily and rapidly collected by using a syringe.

3.1.5. Selection of gradient materials

An ideal gradient material should be freely soluble in water, physiologically and chemically inactive, very dense, nonviscous, transparent in visible and ultraviolet light, cheap and possess negligible osmotic pressure. The physical properties of some gradient materials are summarised in Table 3.1. Sucrose is usually the material of choice, since it approximates most closely to the requirements. Tables relating sucrose concentration to viscosity and density are given in Appendix I. Commercial sucrose is frequently acceptable for centrifugation, being substantially less expensive than reagent or analytical grade materials (see also § 4.1.4). Ficoll (Pharmacia Fine Chemicals), a synthetic co-polymer of sucrose and epichlorohydrin, has found limited use in plasma membrane separations (§ 3.1.7). Its major disadvantages are expense and high viscosity; also its molecular weight may vary from batch to batch. Properties of Ficoll solutions are included in Appendix I. Caesium salts (chloride or sulphate) have found only limited use in membrane preparations. Silica sols (Lindquist et al. 1974; marketed under the trade name Ludox HS, [du Pont

TABLE 3.1

Some physical properties of gradient materials used in density gradient centrifugations.

Substance	Concentration (%)	Density[a]	Viscosity[b]	Refractive index at 20 °C
Sucrose	65 w/w	1.3260 (4 °C)	56.5 (20 °C)	1.4532
Ficoll	46.5 w/w	1.1629 (4 °C)	1020[c]	1.3764
^2H_2O	100	1.105 (20 °C)	–	1.3384
CsCl	60 w/w	1.790 (20 °C)	–	1.4074
Silica sols	40.1 w/w	1.295 (25 °C)	27 (25 °C)	–
Metrizamide	50 w/v	1.27 (22 °C)	200 (22 °C)	–

[a] Density expressed as g/ml at stated temperature.
[b] Viscosity expressed as centipoise (cP).
[c] Viscosity of a 30% w/v solution.
For full details of the properties of ranges of sucrose and Ficoll concentrations, see Appendix I.

Nemours] and Nyacol [colloidal silica grade 2040, Nyanza Inc.]), glycerol, sodium diatrizoate (Winthrop Labs.; Tamir et al. 1974, 1976), Urografin (methyl glutamine salt of *N,N′*-diacetyl-3,5-diamino-2,4,6-triiodobenzoic acid – a dense radiologically contrasting agent available from Schering Chemicals Ltd.), are other materials advocated for separation of cells and large subcellular particles e.g. nuclei, synaptosomes. For further information on materials used in density gradients see Hinton and Dobrota (1976). Recently, Metrizamide (2-[3-acetamido-5-*N*-methyl-acetamido-2,4,6-tri-iodobenzamido]-2-deoxy D-glucose; mol.wt. 789, Nyegaard & Co.) has been advocated as a density gradient medium with useful advantages claimed over sucrose and Ficoll, but although its main use has been confined to nucleic acids, proteins, ribonucleoproteins and cells, the separation of nuclei, mitochondria and lysosomes has also been reported (Munthe-Kaas and Seglen, 1974; Rickwood and Birnie 1975). Metrizamide is an off-white powder which dissolves readily in water or dilute solutions of salts; its solutions can be stored indefinitely at −20 °C and are stable at room temperature within the pH range 3–8 but exposure to light or high temperatures results in the release of iodine. Compared to Ficoll, Metrizamide is denser and less viscous at all concentrations. Other advantages claimed are its non-ionic nature and low degree of hydration in solution. Disadvantages of Metrizamide are its large molar adsorption in the ultraviolet region and a tendency to react weakly with proteins.

3.1.6. Separation parameters of the major subcellular organelles and membrane systems

The distribution of protein between the subcellular organelles and membranes, the gravitational force required to pellet them and their equilibrium (isopycnic) densities in sucrose gradients are shown for liver cells, fibroblasts and brain cells in Table 3.2 and the median density peaks of the marker enzymes and RNA are summarised in Table 3.3. These results are obtained by analytical fractionation of either total homogenates (rat embryo fibroblasts) or post-nuclear or microsomal fractions in zonal or swing-out rotors, and have yielded

TABLE 3.2

Distribution of protein, pelleting forces and equilibrium densities of subcellular components of various cells.

Fraction	Protein (%)	Pelleting force[a] (g · min)	Equilibrium density range in sucrose-H_2O gradients (g/cm³)
Erythrocytes	–	1×10^4	≈1.20
Liver cells		1×10^3	≈1.20
Nuclei	15	$5–10 \times 10^3$	1.32
Mitochondria	25	1×10^5	1.18–1.21
Lysosomes	2	4×10^5	1.20–1.22
Rough e.r. membrane	12	3×10^6	1.15–1.25
Smooth e.r. membrane	8	1×10^7	1.10–1.20
Plasma membrane			
sheets	0.5	1×10^4	1.16–1.18
vesicles	1.5	1×10^7	1.12–1.15
Golgi membranes	2	1×10^7	1.12–1.14
Peroxisomes	2.5	4×10^5	1.22–1.24
Soluble proteins	30	$>10^8$	≈1.30
Fibroblasts			
Nuclear fraction	15	1×10^4	1.30
Mitochondrial/lysosomal fraction	34	5×10^5	1.16
Microsomal fraction	17	1.2×10^7	1.15–1.17
Supernatant protein	32	$>10^8$	1.30
Plasma membranes	2	$1 \times 10^4–10^7$[a]	1.13–1.15
Brain cells			
Neurons	10–15[b]	1×10^4	1.17–1.22
Glial cells	35–40[b]	1×10^4	1.16–1.19
Nuclei			
cortex	5	1×10^4	1.30
cerebellum	11		
Mitochondria (whole brain)	15	6×10^4	1.14–1.19
Synaptosomes	2.3	4.2×10^5	1.14–1.18

Subject index p. 261

TABLE 3.2 *(continued)*

Fraction	Protein (%)	Pelleting force[a] (g · min)	Equilibrium density range in sucrose-H_2O gradients (g/cm^3)
Synaptosomal plasma membranes	0.5–1	–	1.12–1.16
Glial membranes	–	–	1.12–1.17
Synaptic vesicles	0.1	–	1.08
Lysosomes	–	–	1.15–1.19
Myelin			
adult rat	9–15		1.10
bovine white matter	39	–	1.10

[a] With plasma membranes, the pelleting force depends critically on the particle size produced by the homogenisation conditions.

[b] The relative protein content of neurons and glial cells is very much a matter of dispute; for example, it will vary according to the age of the animal and the region of the brain used.

Data compiled from Leighton et al. (1968); Evans (1970); Amar–Costesec et al. (1974) (liver); Tulkens et al. (1974) (fibroblasts); Abood (1969); Whittaker (1970); Norton and Podulso (1972,1973); Cotman (1972); Gurd et al. (1974) (brain). The Table gives an indication of the approximate range of sedimentation forces and equilibrium densities within which the bulk of the components are distributed.

useful information about the distribution on gradients (mainly sucrose) of the major enzymic and chemical markers in these cells. Strategically, such analytical fractionation would appear a logical step before proceeding to preparative fractionation of the various organelles and membranes, but the preparation of 'plasma membrane fractions' from a wide range of tissues and cell lines has been accomplished initially in the absence of this data. However, in improving the quality and representativeness of the plasma membrane fraction, such information is invaluable, since a more thorough analysis of likely contaminants can be made. The extension of the analytical fractionation approach to small amounts of normal and diseased tissues (Peters 1976) should provide further information, useful in

TABLE 3.3
Median densities in sucrose–H_2O gradients of some marker enzymes and chemicals in subcellular fractions of rat liver, fibroblasts and brain.

Subcellular location	Marker	Density g/cm^3		
		Liver[a]	Fibroblasts[b]	Brain[c]
Protein	–	1.164	1.15	1.15
Plasma membranes	5′-Nucleotidase	1.137	1.144	1.148
	Cholesterol	1.141	1.144	–
	Alkaline phosphodiesterase I	1.144	–	–
	Alkaline phosphatase	1.144	–	–
	Acetyl-cholinesterase	–	–	1.148
Mitochondria				
inner membrane	Cytochrome oxidase	1.180	1.161	1.18
outer membrane	Monoamine oxidase	1.133	–	1.17
Endoplasmic reticulum	NADH-cytochrome *c* reductase	1.149	1.166	–
	glucose-6-phosphatase	1.168	–	–
	RNA	1.201	1.156	–
Lysosomes	Acid phosphatase	1.157	1.189	–
	Cathepsin	–	1.198	–
	N-Acetyl-*β*-glucosaminidase	1.187	1.206	1.199
Golgi apparatus	Galactosyl-transferase	1.127	–	–
Peroxisomes	Catalase	1.220	1.184	1.150

[a] A liver microsomal fraction was determined. Thus plasma membrane markers exhibiting a bimodal nucleo-microsomal distribution are represented by the lower median density peak only. The low median density of monoamine oxidase probably corresponds to that of released outer mitochondrial membranes; Amar–Costesec et al. (1974).

[b] Cultured rat embryo fibroblast homogenates; Tulkens et al. (1974).

[c] A post-nuclear supernatant of rat brain homogenates; Bretz et al. (1974).

the preparation of plasma membranes from non-hepatic tissues.

Analytical fractionation has been directed primarily at the resolution of lysosomal enzymes and indeed, this approach was instrumental historically in their discovery (de Duve 1965, 1975). The technique has led to an emphasis of the postulate of biochemical homogeneity of subcellular markers, and to an attempt to explain or resolve the heterogeneity of lysosomal enzymes in the context of latency. The bimodal nucleo-microsomal distribution of the plasma membrane markers of tissues has resulted in the analytical approach being less than comprehensive due to attention being confined to postnuclear or 'microsomal' fractions. For example, much confusion surrounded the preparation of liver plasma membranes at a sucrose density of 1.16–1.18 by the technique of Neville (1960) and the demonstration by analytical fractionation of liver 'microsomal' fraction (Table 3.3) that the 5′-nucleotidase peak was at 1.14. Information on the heterogeneity of plasma membrane markers is, in general, still less complete than that available for subcellular components yielding a single symmetrical subcellular distribution peak.

Tables 3.2 and 3.3 (see also Fig. 3.1) emphasise the difficulty that will be encountered in completely separating by density gradient fractionation the various subcellular organelles and membranes. The median densities of 'light' plasma membranes, Golgi membranes, outer mitochondrial membranes and smooth endoplasmic reticular vesicles are similar and these components in sucrose gradients are bound to overlap. This has led to attempts to modify selectively the density of one group of subcellular particles so as to improve separation (§ 3.1.9). Heterogeneity, revealed by analytical fractionation of subcellar components of 'homogeneous' populations of cultured cells, predicts that a far more complex situation will be obtained in tissues that contain many cell types. For example, in brain tissue, attempts have been made to resolve different populations of the synaptic vesicles (see § 5.2) containing either cholinergic, γ-aminobutyric acid or adrenergic transmitters on the basis of their modal equilibrium/density (Bretz et al. 1974). Neuronal and glial subcellular organelles will also be expected to show heterogeneity.

3.1.7. Theoretical parameters determining particle density

The density of a subcellular particle is determined by the composite density of its component parts. Thus, the density of a membrane strip is determined mainly by that of the protein and lipid components plus the contribution from hydration. Closed membrane vesicles present a more difficult case, for the density of the internal solvent compartment (and any solutes or other contents, such as soluble cytoplasmic proteins or very low density lipoproteins) may modify the overall density. Since density gradient centrifugation is carried out in hypertonic media, the permeability of a vesicle to the gradient material is an additional factor. Sucrose molecules permeate across vesicles of the microsomal fraction, whereas Ficoll (estimated molecular weight 400,000) does so less rapidly (Nilsson et al. 1973). Vesicles will contract on prolonged centrifugation in sucrose gradient. The lower osmotic pressure of Ficoll has found use when it is necessary to minimise changes in the isotonic conditions inside closed vesicles, and perhaps one of the best examples of this is in the separation of brain synaptosomes and mitochondria. Sucrose-Ficoll gradients have been introduced to explore the possibility of increasing small density differences between membrane particles. However, this strategy has not been altogether successful, for, as shown in Table 3.4, the inclusion of two different Ficoll concentrations in sucrose gradients did not markedly modify the relative median isopycnic densities of membrane marker enzymes. Dextran gradients have proved useful in separating sealed from permeable plasma membrane vesicles, since rapid entry of Dextran into vesicles can increase their density. For example, Walsh and Crumpton (1977) prepared lymphocyte vesicles impermeable to Dextran of 10,000 molecular weight, by centrifugation at 100,000 *g* for 2 hr a vesicular fraction into a Dextran T10 (Pharmacia) discontinuous gradient. Sealed vesicles were collected at a 1.005–1.03 g/cm^3 interface, whereas permeable vesicles had a density of 1.05–1.06 g/cm^3.

Wallach (1967) and Wallach and Kamat (1966) exploited the observation that changes in the pH or divalent ion concentration may selectively alter the isopycnic density in Ficoll gradients of plasma

TABLE 3.4
Effect of Ficoll on isopycnic density of membrane-bound marker enzymes.

Enzyme	Subcellular location	Median density (g/cm^3)		
		Sucrose-H_2O[a]	Ficoll-Sucrose (0.5 M)	Ficoll-Sucrose (0.25 M)
5′-Nucleotidase	Plasma membrane	1.153	1.104	1.083
NADH-cytochrome *c* reductase	Endoplasmic reticulum	1.162	1.116	1.096
Glucose-6-phosphatase	Endoplasmic reticulum	1.182	1.121	1.106
Mono-amine oxidase	Outer mitochondrial membrane	1.146	1.110	1.091

Data from Beaufay et al. (1974).

[a] These values are approx 0.01 density unit higher than those shown in Table 3.3, owing to prolongation of centrifuge time from 3 to 10 hr. This difference suggests that complete equilibrium was not achieved at the end of 3 hr, or that some change in the physico-chemical properties of the microsomes had resulted from longer exposure to high sucrose concentrations.

membrane vesicles produced by gas cavitation of cultured cells. However, this technique has not been extended successfully to tissue cells, possibly due to aggregation occurring when negatively charged groups are neutralised. The inclusion of Pb^{2+} salts in media used to fractionate liver microsomal components can also lead to aggregation as well as difficulties in the estimation of enzymes inhibited by Pb^{2+}.

3.1.8. Use of zonal rotors to prepare plasma membrane fractions

Zonal rotors, developed by Anderson and colleagues (Anderson 1966) have featured in the analytical fractionation and preparation of membranes. There are a number of extensive accounts of the theory, design and practical use of zonal centrifugation (e.g., Anderson 1966; Reid 1971; Price 1973; Hinton and Dobrota 1976) and this section is confined to their application to plasma membrane preparation. Zonal rotors are hollow closed cylinders divided into a number of sector-shaped compartments by vertical septa which radiate from the axial core to the rotor wall (Fig. 3.4). Zonal rotors are operated when filled with density gradients in which membrane separation occurs. A number of types of zonal rotors are available

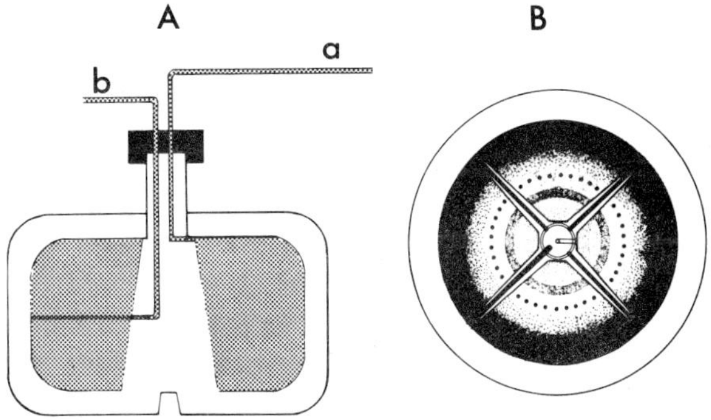

Fig. 3.4. (A) Cross section and (B) Plan view of a typical (e.g., B-XIV) zonal rotor. Channel (a) exits at the centre of the rotor, whereas channel (b) continues through the vanes to exit at the wall of the rotor. Separations (as shown in B) of sample introduced into the centre of the rotor may be observed in 'A'-type rotors. (See Table 3.5 and Appendix 2.)

commercially (Appendix II) and, from a membrane preparation viewpoint, it is useful to distinguish between low speed zonal rotors used for rate-zonal separations and high speed zonal rotors used primarily for isopycnic separations. Overall, the most popular 'high speed' zonal rotor is the B XIV; the very high cost of titanium rotors has to be weighed against the increased centrifugal force available and their resistance to caesium salts.

Some advantages of zonal rotors Compared to swing-out rotors, zonal rotors give increased resolutions for two reasons. First, they are loaded (as a sharp initial zone sample) and unloaded while revolving, so that membrane zones are stabilised by the centrifugal field at times when disturbances might occur. Second, particles sediment in sector-shaped compartments so that the wall effects encountered in tubes are eliminated. When it is necessary to fractionate large amounts of tissue homogenate so as to obtain increased amounts of plasma membranes, the large capacity of zonal rotors is a major practical advantage, e.g. in the preparation of membrane specialisations (§ 5.9). Zonal rotors can thus eliminate the need for multiple tubes and successive runs. Since the chamber of the rotor is filled with a single uniform density gradient, the sedimentation of particles in each sector is thus identical. Zonal rotors may be filled with discontinuous and/or continuous gradients constructed by using gradient makers available as ancillary equipment. Gradient density profiles may be concave, convex or isokinetic (i.e., particles sediment with constant velocity (Steensgaard 1970). However, from a membrane preparation viewpoint, complex gradients are mainly of theoretical interest, and for isopycnic separations linear or discontinuous gradients are usually satisfactory.

Some disadvantages of zonal rotors The major disadvantage of the zonal rotor results from human rather than technical reasons. Their use may appear tedious and technically complex with a requirement for ancillary equipment. Zonal rotors are also more expensive than conventional rotors. Their use occupies the centrifuge longer owing to the time taken to load and unload the rotor, and multi-operator use is frequently discouraged. Troubles with leakage in the rotating seals, prevalent with earlier models, have now been largely corrected. The speed of preliminary analytical experimental work in zonal rotors is handicapped in comparison to swing-out rotors since only one gradient design can be explored at a time. Exploration of optimal gradient designs in six-place swing-out rotors (e.g., Beckman SW 27, total volume 225 ml; see Appendix II) prior to scaling-up for preparative runs in a zonal rotor has frequently not proceeded further. Owing to the large capacity of most zonal rotors (Appendix II), there is a need for a preparative zonal rotor of smaller capacity, e.g., 100–300 ml.

Examples of plasma membrane preparation by centrifugation in zonal rotors On the whole, zonal rotors have not been fully exploited for plasma membrane preparation. In contrast, analytical fractionation of post-nuclear supernatants of tissue homogenates has been extensively carried out (see § 3.1.6). However, much of this work is carried out in the Beaufay rotor (volume 48 ml), which is not manufactured commercially; the B XIV rotor, although of larger capacity, can be used as a substitute. For preparative work, the low speed zonal rotors that separate particles of sedimentation coefficient range 10^4–10^8, have been used to prepare fractions enriched in plasma membranes from liver (see below) and kidney (Taylor et al. 1971; Price et al. 1972). High speed 'B' type zonal rotors have been used to prepare plasma membranes from baby hamster kidney (BHK) cell homogenates (Graham 1972; Graham et al. 1975; see also § 4.6.1) and homogenates from kidney (Kempson et al. 1974), brain (Cotman et al. 1968), beef liver (Gibbons et al. 1976), and bovine adrenal glands (Schlegel and Schwyzer 1977). When rodent livers are dispersed in hypertonic media by using a loose-fitting Dounce homogeniser, the pellet obtained by low-speed centrifugation of the filtered homogenate is multilaminate, consisting of at least three layers. The method of Neville (1960) resolves the pellet into its component parts by differential and density gradient centrifugation and the adaptation of this procedure to the zonal rotor presents an elegant example of its usefulness (El-Aaser et al. 1966; Anderson et al. 1968).

The separation in a sucrose gradient of the plasma membranes (identified by a peak of 5′-nucleotidase activity) from a less rapidly sedimenting mitochondrial band and a more rapidly sedimenting band containing nuclei and cell debris in 3 types of zonal rotors is shown in Fig. 3.5. In the transparent A XII zonal rotor, the resuspended low speed pellet can be observed to fractionate into 4 bands, and the plasma membranes (identified enzymically and by phase-contrast microscopy) are adequately separated from the less rapidly sedimenting mitochondria, whereas the nuclei have packed against the outer edge of the rotor. Similar but unobserved separations are obtained by using the Sorvall–DuPont SZ-14 reorientating rotor, and in the B XV type rotor (Pfleger et al. 1968; Weaver and Boyle 1969). The separation of plasma membranes as large fragments occurs rapidly (45 min) at relatively low centrifugal forces, and thus the high speed potential of the B-type rotors is not fully exploited in such rate-zonal separations. A large volume sucrose cushion is required when using B XV rotors, and for reproducible separations it is helpful to measure the total *g* force applied by using an integral attachment. However, an important advantage of the B-type zonal rotors, illustrated in the example in Fig. 3.5C, is the facility with which the rate-zonal step can be immediately followed by an isopycnic banding step, which produces a much narrower membrane band than the rate-zonal separation.

The rate-zonal method described here is applicable generally to large strips of plasma membranes (primarily from tissues in which intercellular junctions minimise vesicularisation of the surface membrane) that sediment at low centrifugal forces and is well worth considering for bulk preparation of plasma membranes from tissues.

3.1.9. Plasma membrane density perturbation

The use of digitonin to increase selectively the density of liver plasma membranes was described by Thines–Sempoux et al. (1969) and Amar–

Costesec et al. (1974). The rationale underlying this approach assumes that cholesterol (which binds digitonin) is primarily a plasma membrane constitutent and is absent from intracellular membranes and organelles. Addition of digitonin to rat liver microsomal fractions increased the median density of plasma membrane markers without modifying that of endoplasmic reticular markers.

An example of the preparative use of digitonin in plasma membrane isolation is provided in the purification of intestinal epithelium basolateral plasma membranes by Lewis et al. (1975). In homogenates of intestinal epithelial cells, baso-lateral plasma membranes and mitochondria exhibit similar sedimentation characteristics and overlapping densities. These two components were separated after exposure of the 'mitochondrial' fraction to 3 mg digitonin/mg fraction protein (Fig. 3.6). The practical considerations are as follows. Digitonin is poorly soluble in aqueous media and is first deposited as a thin film in glass tubes by the evaporation of ethanol solutions. The appropriate amount of plasma membrane containing fraction in 0.25 M sucrose, pH 7.4, is placed in the tube and dissolution of digitonin occurs during incubation at 4 °C for 15 min with mild agitation. The membranes are then loaded onto sucrose gradients. The ratio of membranes to digitonin described for intestinal cell epithelial fractions may not apply generally, and preliminary experiments are necessary to determine the optimal ratios to obtain separations as shown in Fig. 3.6.

Some important practical consequences of treatment of membrane fractions with digitonin are described by Lewis et al. (1975). (Na^+-K^+)-ATPase activity is inhibited by digitonin, and after treatment of the intestinal mitochondrial fraction as described, only 20% of the original activity remained. Succinic dehydrogenase activity was completely inhibited. On the contrary, dicyclohexylcarbodiimide-insensitive Mg^{2+} activated ATPase was activated approximately 250% by digitonin at 3 mg/mg fraction protein. The effects of digitonin on mitochondrial are complex and probably relate to their permeability and intactness. The following enzymes were unaffected by the treatment: monoamine oxidase, sucrase, trehalase, esterase,

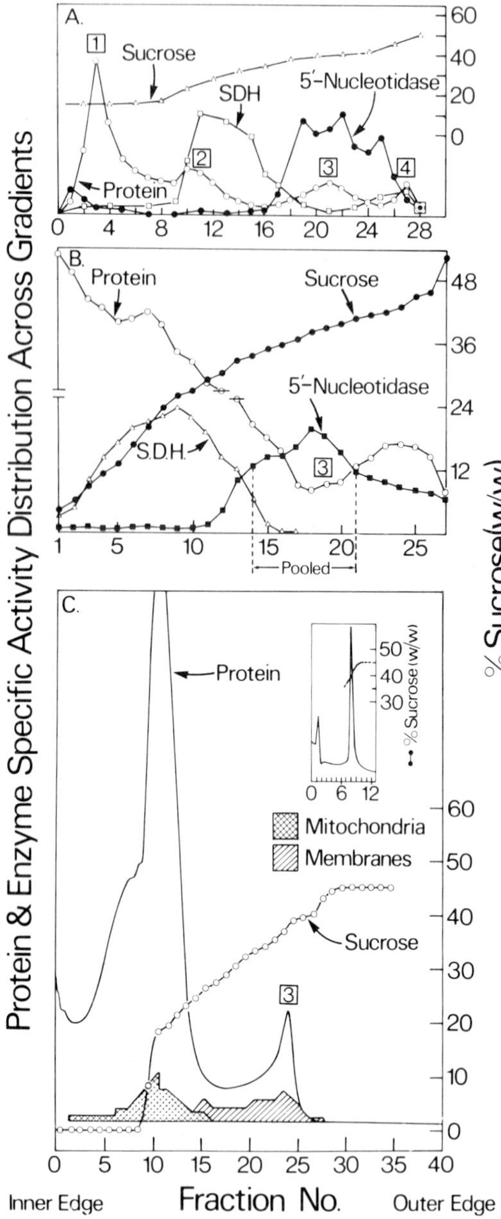
A.
Sucrose
SDH
5'-Nucleotidase
Protein
1
2
3
4
B.
Protein
Sucrose
5'-Nucleotidase
S.D.H.
3
Pooled
C.
Protein
% Sucrose (w/w)
Mitochondria
Membranes
Sucrose
3
Protein & Enzyme Specific Activity Distribution Across Gradients
% Sucrose(w/w)
Inner Edge
Fraction No.
Outer Edge

glucose 6-phosphatase and arylsulphatase. The effects of digitonin described above are probably by no means universal and relate to fraction purity; clearly, its use as a plasma membrane perturbant should be subject to an independent examination of its effects on marker enzyme activities.

Tulkens et al. (1974) showed that inclusion of 0.03% digitonin in the sucrose-EDTA solutions, used to resuspend a microsomal fraction from rat embryo fibroblasts, increased the median density on sucrose gradients of the following plasma membrane marker components: 5′-nucleotidase, acid pyrophosphatase, cholesterol and inosine diphosphatase from 1.13 to 1.19–1.20 g/cm^3. The amount of digitonin added by Tulkens et al. (1974) was such that the molar ratio of digitonin to cholesterol was close to one. Huggins and Carraway (1976) showed that suspension of mammary gland microsomes in 0.03% digitonin selectively increased the density on sucrose gradients of plasma membranes and helped to separate them from Golgi membranes.

◁ Fig. 3.5. Comparison of the separation of liver plasma membranes in 3 types of zonal rotors. (A) A-XII rotor. A low speed liver homogenate fraction resuspended in 75 ml 6% sucrose is introduced into a gradient containing 300 ml 6% (w/v) sucrose, 100 ml 24% (w/v) sucrose, a 200 ml linear 24 to 35% (w/v) sucrose gradient and a 600 ml linear 36 to 54% (w/v) sucrose gradient resting on a 60% (w/v) sucrose cushion. The centrifuge is then accelerated from 500 to 3,800 revs/min. After 45 min, four peaks of protein are observed and the plasma membrane band (peak 3) is found between 37 to 44% (w/v) sucrose. Modified from Evans (1970). (B) SZ-14 rotor. The gradient composition was similar to that in (A). The fractions are loaded and unloaded at 2500 revs/min and separation occured in 20–30 min after centrifugation at 4000 revs/min. (Sheeler and Evans, unpublished.) (C) B-XV rotor. The low speed pellet resuspended in 200 ml 8% sucrose is introduced into the rotor and centrifuged at 5000 revs/min into a 500 ml 19 to 35% sucrose (w/w) gradient resting on a 800 ml 45% (w/w) sucrose cushion until $\omega^2 t = 2 \times 10^8$ (determined by using a Beckman integrator). After removing the centripetal 750 ml the rotor is then accelerated to 20,000 revs/min until $\omega^2 t = 3.5 \times 10^9$, when plasma membranes band isopycnically at 38% (w/w) sucrose (shown in inset) (Anderson et al. 1968).

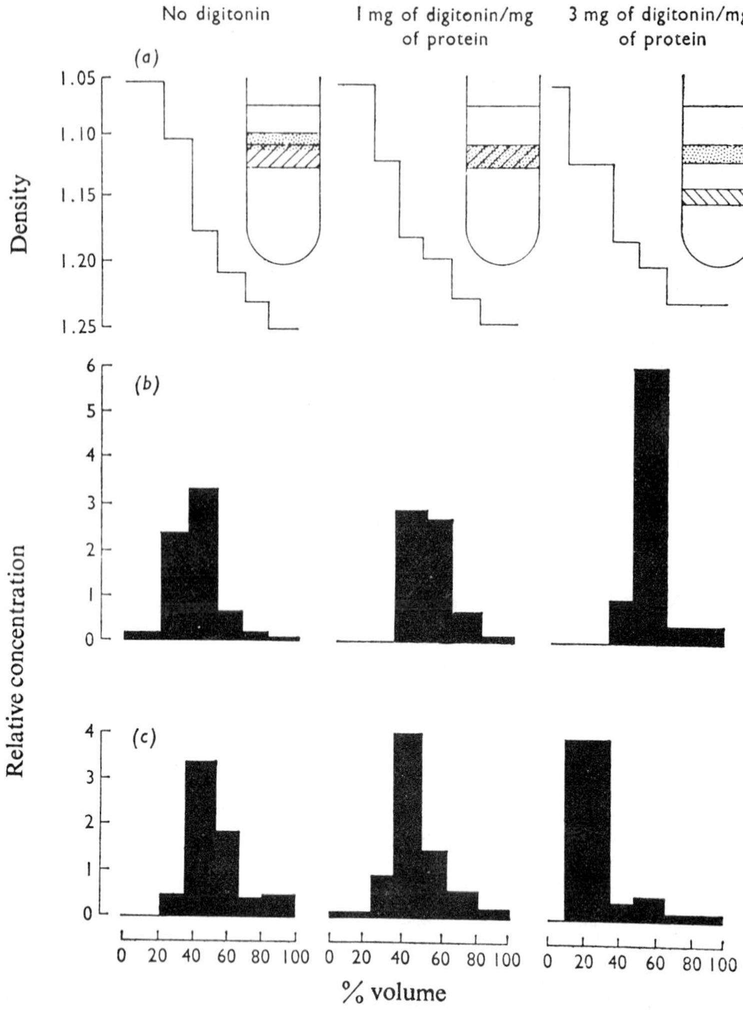

Fig. 3.6. Effects of digitonin on the separation of intestinal mitochondrial and baso-lateral plasma membrane components. Centrifugation of a 'mitochondrial' fraction was for 16 hr at 30,000 *g* on 18–56% (w/w) linear sucrose gradients. In the absence of digitonin or after treatment of the 'mitochondrial' fraction with 1 mg digitonin/mg protein, both components showed overlapping densities. However, at 3 mg digitonin/mg protein baso-lateral and mitochondrial components were separated. The separation was confirmed by measurement of (a) appearance of bands and density profile of the gradient, (b) dicyclohexylcarbodiimide-sensitive adenosine triphosphatase (mitochondrial marker; Beechey et al. 1967) and (c) (Na^+-K^+)-ATPase activities (marker for baso-lateral plasma membranes) across the gradient. Reproduced from Lewis et al. (1975) by permission of the Biochemical Journal.

3.1.10. Perturbation of the density of intracellular components

In the preparation of synaptosomal plasma membranes from cerebral cortex, a major problem is the contamination of the membranes by mitochondria (see Chapter 5, § 5.2). To minimise this contamination Davis and Bloom (1970) suggested the use of formatazan deposits to increase the buoyant density of the mitochondria. Cotman and Taylor (1972) incubated an osmotically shocked 'mitochondrial' fraction of rat cerebral cortex in a minimum volume of sodium phosphate buffer (40 mM, pH 7.4), 50 μM $CaCl_2$, 60 mM succinate and 1 mM *p*-iodonitroneotetrazolium violet (I.N.T.) at 3 °C for 20 min. This was then washed twice by pelleting in iso-osmotic sucrose to remove trapped salts that may interfere with the subsequent density gradient steps. The procedure allowed synaptic plasma membranes to be collected on a cushion of 1.0–1.1 M sucrose and the number of mitochondria persisting at this density was much reduced.

A number of lysosomal density perturbants (lysosomotropic agents) have been described in which cells (mainly hepatocytes) are first 'overloaded' with either inorganic compounds (iron salts, silica) or organic macromolecules (synthetic polymers; Triton WR 1339) before homogenisation. The use of Triton WR 1339 has allowed lysosomes ('tritosomes'), normally present in the mitochondrial fraction at a median density of 1.22, to be separated as a sharp peak corresponding to a density of 1.11 g/cm^3 (Wattiaux et al. 1963). Rats injected intra-muscularly with an iron-sorbitol-citric acid complex yielded a 'heavy' liver lysosomal fraction (Arborgh et al. 1974).

Golgi cisternal and secretory elements of liver have been prepared and separated by using density perturbation techniques. Ehrenreich et al. (1973) prefed rats with 50% ethanol in H_2O; this results in the accummuation of very low density lipoproteins in the Golgi apparatus. The microsomal fraction, prepared in the normal way, is suspended in 1.15 M sucrose and loaded at the bottom of a discontinuous sucrose gradient. Centrifugation for approx. 13×10^6 *g* × min yields by floatation Golgi fractions of sucrose density > 1.041 and < 1.173 g/cm^3.

In general, conditions for selectively modifying membrane or vesicle density have to be carefully selected. The effects of the perturbant on membrane markers that may be used to assess the density change require careful monitoring as shown by the effects of digitonin on intestinal membranes.

3.1.11. Combined affinity-density perturbation approaches

Although experimentally in their infancy, attempts to exploit ligand-receptor interactions to modify subsequently the density of membranes are mentioned mainly for their potential. The rationale, outlined by Wallach et al. (1972) assumes that receptors are unevenly distributed on the plasma membrane and that cell disruption will produce populations of right-side-out plasma membrane vesicles that will contain different amounts of ligand-binding sites. The differential combination of the ligand (covalently attached to a heavy atom or molecule) with receptors on the vesicles should then allow fractionation techniques to separate them according to their density (Fig. 3.7). Preliminary results of the technique, applied to lymphocyte plasma membrane receptors that combine with Concanavalin A (approx. 10^7/lymphocyte) were reported (Wallach et al. 1972). The density perturbant selected was coliphage owing to its stability, shape and small size, and it was coupled to ^{125}I-labelled Concanavalin A by using glutaraldehyde. Lymphocyte plasma membrane vesicles treated with this affinity density perturbant had their median

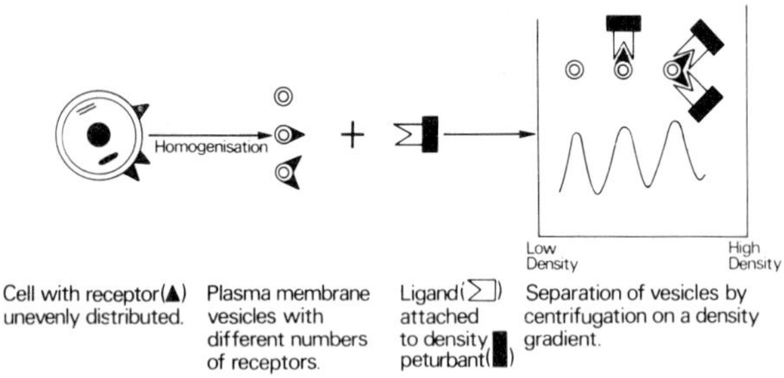

Fig. 3.7. Diagrammatic representation of the separation of plasma membrane vesicles by affinity density perturbation. Redrawn from Wallach et al. (1972).

density in CsCl gradients increased from 1.18 to 1.30–1.34 g/cm^3; the equilibrium density of free coliphage was 1.5 g/cm^3. A 90% dissociation of the complex was obtained by addition of excess α,α-trehalose (3 × 10^4 mol sugar/mol conjugated Con A). The distribution of plasma membrane vesicles became multi-shouldered, and hence the affinity density separation was not fully successful.

A further example in which the ligand was immunoglobulin and the density perturbant was latex beads has been described in which erythrocyte vesicles were separated (Lim et al. 1975). Latex spheres (consisting of 53% methylmethacrylate, 30% 2-hydroxyethylmethylacrylate, 10% methacrylic acid and 7% ethylenedimethacrylate) are synthesised by using an aqueous emulsion copolymerisation technique (Molday et al. 1975). They are then tagged with the fluorochrome dansyl ε-lysine using a cyanogen bromide reaction and are subsequently coated with goat anti-rabbit IgG antibodies by a two-step glutaraldehyde reaction. Erythrocyte ghosts coated with a rabbit anti-human erythrocyte antiserum react with the IgG-latex complexes and the isopycnic density is increased. For example, the density in sucrose gradients of erythrocyte membrane vesicles is 1.169 g/cm^3; that of erythrocytes sensitised with rabbit anti-erythrocyte antibodies and labelled with immunolatex conjugates is 1.20 (or 1.212 g/cm^3 when treated with higher concentrations of latex-conjugates). The latex spheres had a density of 1.232 g/cm^3; this can be changed by varying their composition.

In view of the wide range of ligands available, e.g., antibodies, lectins, peptide hormones, substrate analogues for plasma membrane ectoenzymes, and the potential for varying the size and density of perturbants, this basic approach remains ripe for practical exploitation.

3.2. Separation of membranes in liquid-polymer phase systems

In addition to methods that exploit weight and density differences between subcellular components, counter-current distribution in two-phase polymer systems has recently been used successfully in the purification of plasma membranes. The theory and practical application of the partitioning properties of membranes and macromolecules between two or more polymer phase systems has been thoroughly reviewed by Albertsson (1971). The factors governing the partitioning of macromolecules, cell organelles and membranes between two-phase polymers are complex and include size, density, hydrophobicity and surface charge (Walter et al. 1976). A wide range of liquid-polymer two-phase systems has been investigated by Albertsson (1971) for the separation of a variety of biological substances and particles; two non-ionic polymers (polyethyleneglycol and dextran), dispersed in water, have been successfully used in the purification of mammalian plasma membranes whose distribution appears to be dependent on the following parameters: surface properties of the membranes, i.e., the size and nature of the charged groups on the surface, the concentration of the polymers and the ionic composition of the phases. Presumably, the presence on plasma membranes of higher concentrations of negatively charged groups is one of the parameters that favours their separation at the interface, relatively free from nuclei, mitochondria, intracellular membranes, etc. under the stated conditions. The procedure for preparing stock solutions of the two-phase system is as follows: 200 g of 20% (w/w) Dextran 500 (Pharmacia) in distilled water, 103 g of 30% (w/w) polyethyleneglycol (mol. wt. 6000–6100) (Carbowax 6000, Union Carbide) in distilled water, 333 ml of 0.22 M sodium phosphate buffer (pH 6.5) and 179 ml distilled water are mixed by inversion in a separating funnel and allowed to stand for 48 hr at 4 °C. The polymers separate into two phases (the top one being polyethyleneglycol, and the lower one dextran) which are collected separately. To store the phases, they may be filtered (e.g., through Millipore DAWP (0.65 μm) filters) and kept as stock solutions at 4 °C or frozen. The fraction to be resolved is suspended in the upper phase that is then mixed with the lower phase. After low speed centrifugation the plasma membranes are collected from the interface.

Examples of the use of two-phase polymers to prepare plasma membranes

The use of this method to prepare a plasma membrane fraction from mouse L-cells grown in suspension culture is shown in Fig. 3.8; the plasma membrane fraction was satisfactory on the basis of enzymical and morphological evidence (Brunnette and Till 1971). However, care must be exercised in applying this method to cultured cells of different origin and growth states, and it is important to ensure that the material obtained at the interface is enzymically and morphologically characterised in each case. For example, Gruenstein et al. (1975) prepared plasma membrane fractions by the two-phase

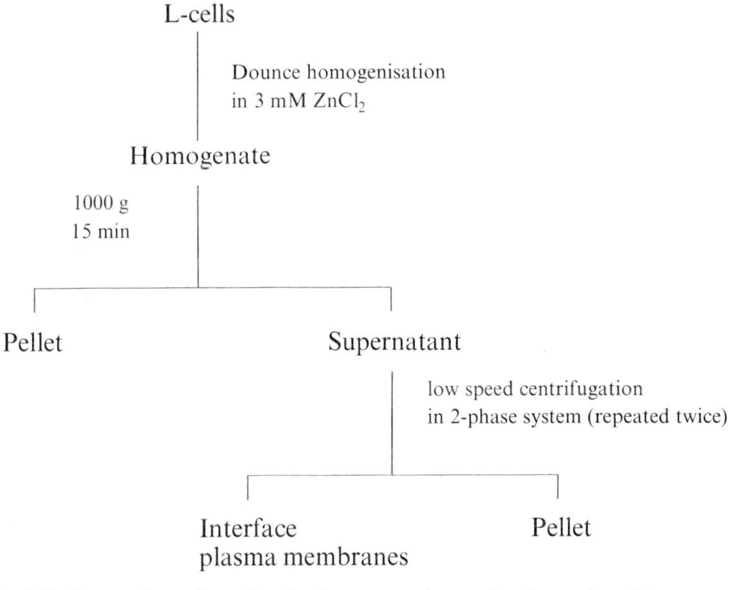

Fig. 3.8. Preparation of an L-cell plasma membrane fraction using the aqueous two-phase polymer system. Cells (approx. 8×10^8) were stabilised by exposure to 40 ml 3 mM $ZnCl_2$ and homogenised in a tight-fitting Dounce homogeniser (50–200 strokes). See text for construction of two-phase polyethyleneglycol/dextran system and the method of suspending fractions in the polymer phases. Plasma membranes collected at interface were washed several times by low speed centrifugation in H_2O. Approx. 7% of the homogenate protein is recovered in the plasma membrane fraction, and the procedure takes less than 2 hr.

technique from HeLa and 3T3 cells, trypsinised and released from substratum by means of a rubber policeman; this gave interfacial plasma membrane fractions that differed substantially in their purity. The membranes prepared from 3T3 cells were shown to be appreciably contaminated by mitochondrial and vesicular remnants, whereas the interfacial fraction from HeLa cells was a plasma membrane fraction of acceptable purity.

Lesco et al. (1973) and Brivio–Haugland et al. (1976) compared the properties of liver plasma membranes prepared by the two-phase polymer system (in a procedure almost identical to that of Fig. 3.8) and by conventional methods using density gradients (§ 5.1). They showed that the protein yield of plasma membranes was similar in both methods, and that the 5′-nucleotidase and (Na^+-K^+)-ATPase were similarly increased, relative to the homgenate. Phase contrast microscopy showed that liver plasma membranes prepared by the two-phase polymer system had fragmented less than those prepared by density gradient centrifugation. Briefly, the low speed pellet prepared from rat liver homogenates (§ 5.1) or L-cells dispersed by Dounce homogenisation (Fig. 3.8) are gently suspended in 30 ml of the polymer top phase and 10-ml aliquots pipetted onto 3 × 50 ml lower phase cushions in polycarbonate centrifuge tubes. The tubes are gently shaken by hand inversion and then centrifuged in swinging bucket rotors at 2000 *g* for 15 min. The plasma membranes collect as a thin white sheet at the interface. Repeating this procedure 3 times yields liver plasma membranes with similar properties to those prepared by the longer and more elaborate density gradient procedure of Ray (1970). The plasma membranes are recovered from the interface with a syringe and the excess polymers are removed by repeated washing by low speed centrifugation in iso-osmotic sucrose solutions.

A two-phase Freon 113-Triton X-100 system was used to prepare brain synaptic junctional complexes (§ 5.9.3).

General comments The two-phase polymer system is a useful adjunct to separation procedures exploiting only particle weight and density.

The results shown in the examples, as well as those obtained with plasma membranes prepared from kidney brush borders (Glossmann and Gips 1974) indicate that the technique has an important advantage in being more rapid than conventional methods and it dispenses with the requirement for high speed density gradient centrifugation in swinging bucket rotors. The requirement to limit the amount of starting material so as to obtain optimum results, emphasised by Lesko et al. (1973), is not a serious drawback, since the method can be scaled up.

In liver, the two-phase polymer method yields low amounts of plasma membranes originating from limited areas of the hepatocyte surface, a drawback shared with all methods that prepare membranes from a low speed 'nuclear' pellet. This limitation was emphasised by Chang et al. (1975) who showed that rat liver sinusoidal plasma membranes, labelled by perfusion of liver with ^{125}I-wheat germ agglutinin (§ 4.6.2) were not recovered in high yield by the two-phase polymer method.

The two-phase aqueous polymer method also shows potential for methodological development and its use, for example for the separation of plasma membranes from the components of heterogeneous microsomal fractions, is relatively unexplored. A potentially powerful extension of this method appears in its use for the affinity partitioning of plasma membrane receptors (Flanagan et al. 1975). This technique is an extension of affinity chromatography, in which a specific ligand is coupled to a solid matrix to selectively adsorb a macromolecule. In affinity partitioning a ligand is covalently attached to one of the polymer systems. For example, Flanagan et al. (1975) attached to the polyethyleneoxide of one phase a quaternary ligand specific to the acetylcholine receptor and showed that this was applicable on a small scale to the purification of membranes enriched in acetylcholine receptors. These experiments suggest that this approach, incorporating two-phase polymer systems with specific ligands, may be generally applicable to the purification of surface membranes containing bio-specific receptors and intact cells (Flanagan and Barondes, 1975).

3.3. Plasma membrane isolation using immunoabsorbents

Luzio et al. (1976) have described an immunological technique for isolating rat fat cell plasma membranes. Cells are first coated with rabbit anti-rat erythrocyte antibodies, homogenised and the plasma membranes then allowed to interact specifically with an immunoabsorbant prepared by covalently coupling donkey anti-rabbit globulin antibodies to cellulose (Fig. 3.9). Using this technique, 25–40% of plasma membranes (adjudged on the basis of 5′-nucleotidase and adenylate cyclase activities) were recovered from cell homogenates within 30 min. The practical details are as follows (Fig. 3.10): fat cells from the epididymal pads of 4 rats are incubated in 20 ml of Krebs-Ringer bicarbonate buffer, pH 7.4, containing

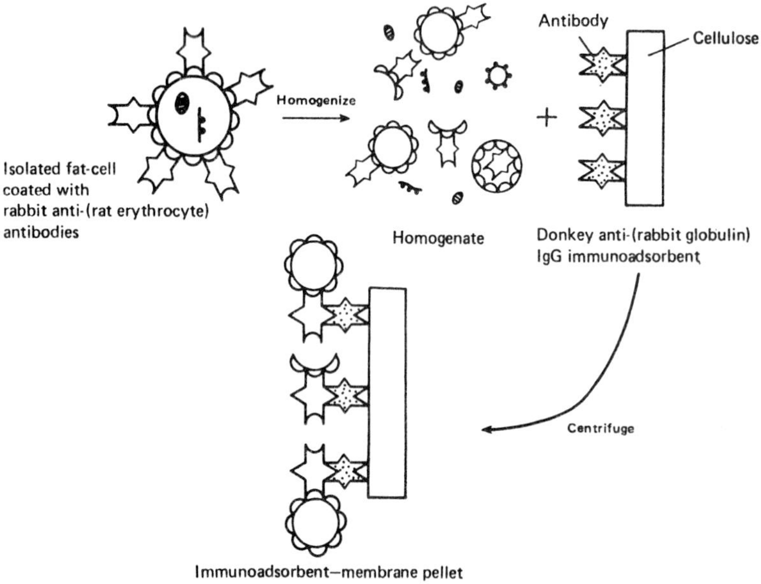

Fig. 3.9. Diagrammatic representation of the use of donkey (anti-rabbit) globulin IgG immunoabsorbent in the preparation of plasma membrane fragments from fat cells coated with rabbit (anti-rat) erythrocyte antibodies. From Luzio et al. (1976). Reproduced by permission of the Biochemical Journal.

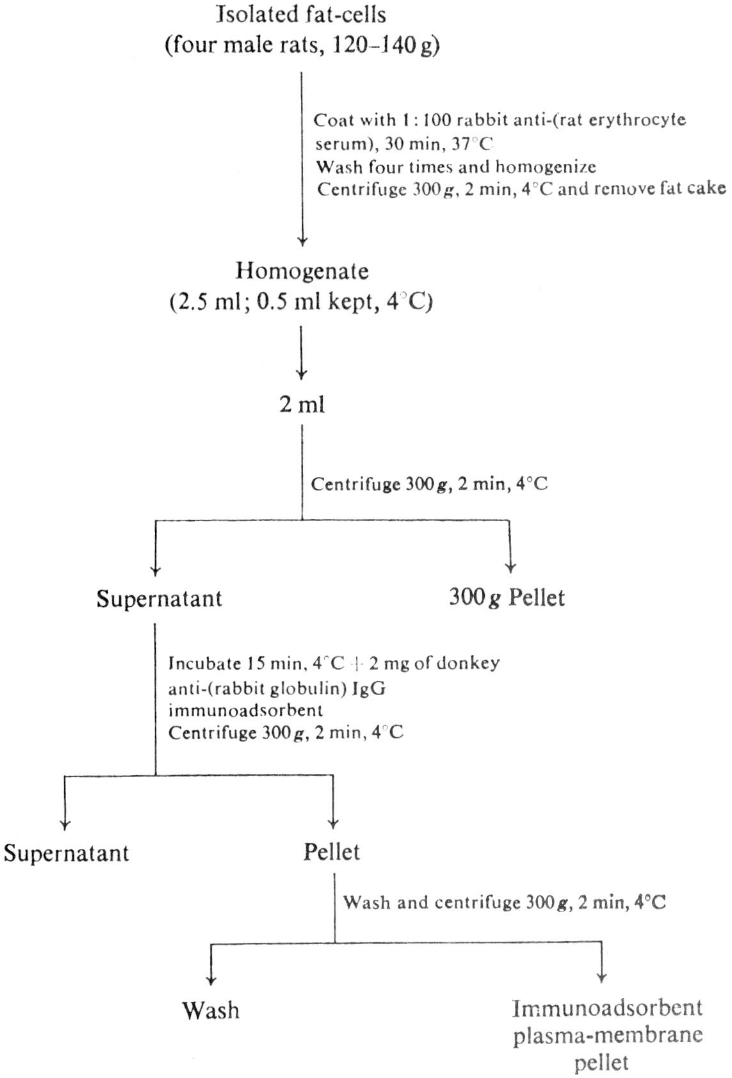

Fig. 3.10. Fractionation scheme for the immunological preparation of rat fat cell plasma membrane. From Luzio et al. (1976).

1.3 mM Ca^{2+}, 40 mg/ml bovine serum albumin and 0.2 ml of rabbit anti-rat erythrocyte serum. After extensive washing by low speed centrifugation in buffers containing 1 mg albumin/ml, the cells are homogenised at 1400 revs/min in a Potter–Elvehjem homogeniser (clearance 100 μm), or by repeated aspiration through a needle (§ 2.3.3). The 300 *g*/2 min supernatant of the homogenate (after removal of the overlying fat cake) is incubated with 2 mg immuno-absorbent, prepared by coupling of IgG to a diazonium derivative of powdered cellulose (Whatman CC41; Miles and Hales 1968) or to activated Sepharose (immunoabsorbents should be washed with homogenising buffer before being added to cell homogenates). Two centrifugations at 300 *g*/2 min produces a pellet of plasma membrane and immunoabsorbent. DNA may absorb non-specifically to the pellet, and this is minimised by removal of nuclei by low speed centrifugation before incubation of the homogenate with the immuno-absorbent (Fig. 3.10).

This rapid and simple method has also been applied in preparing hepatic and pancreatic plasma membranes and shows further potential when antisera against specific plasma membrane proteins are available. In developing immunological methods for plasma membrane preparation, it is necessary to fully characterise the antisera, and only vesicles of a right-side-out configuration are recovered (Fig. 3.9). Problems in removing membranes from immunoabsorbents can be a major drawback.

3.4. Free-flow electrophoretic separation of membranes

This method exploits differences in electrophoretic mobility of sub-cellular particles suspended in a buffer moving in an electric field. These differences are mainly due to differences in chemical composition, especially of glycoprotein composition, between plasma membranes and intracellular membranes and organelles.

The application of free-flow electrophoresis (as opposed to methods that utilise support media) in the separation of subcellular particles owes much to the development of the apparatus described by Hannig (1972), which was available from Desaga, Germany and Brinkmann

Instruments Inc., U.S.A. The most recent model is available from Bender and Hobein, Germany (Fig. 3.12). The principle of free-flow electrophoresis is illustrated in Fig. 3.11; the sample to be fractionated flows down a rectangular separation chamber (10 × 50 cm, 0.7 mm thick). The vertical velocity of the sample is equal to that of the flowing buffer. The particles in the sample orientate themselves horizontally with a velocity that depends on the current applied across the chamber and the electrostatic properties of the particles. The sample, suspended in buffer, is fed continuously into the top of the separation chamber and bands containing particles of similar mobility separate in the electric field and are collected from the bottom of the chamber by a manifold of collecting tubes. A pulse-free peristaltic pump ensures constant injection of sample into the separating chamber, and the separated fractions are then collected at the lower end of the chamber via a multi-channel peristaltic pump and tubes leading into a fraction collector. The separation is highly dependent on the maintenance of effective uniform cooling so as to

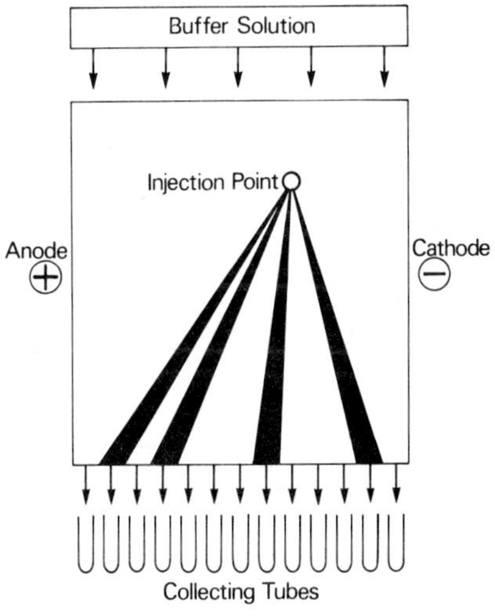

Fig. 3.11. Diagrammatic representation of the electrophoretic separation with 4 bands obtained in a free-flow apparatus. Plasma membranes are found in the band nearest the anode.

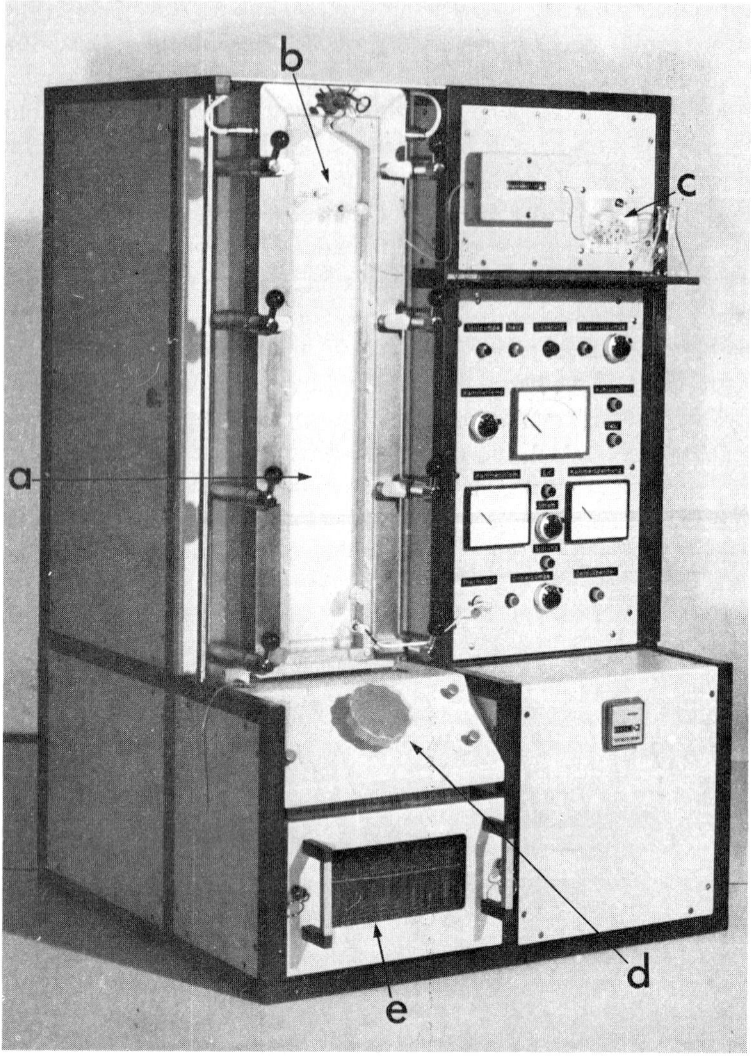

Fig. 3.12. Free-flow electrophoresis apparatus (Model FF5, Bender and Hobein). (a) separation chamber; (b) opening for sample introduction; (c) peristaltic pump; (d) multi-channel peristaltic pump; (e) fraction collector. Photograph by courtesy of Dr. K. Hannig.

minimise convection currents. The specific conductivity of the separation buffers used (see Table 3.5) usually varies between 300–1500 $\mu\Omega^{-1}/cm^{-1}$; the use of high conductivity buffers is not recommended due to the ensuing lower electrophoretic mobility of the components to be separated. In the separation of subcellular membranes and organelles, the correct osmolarity is obtained by the addition of sucrose. Complexing ions, e.g., EDTA, are to be avoided generally since they reduce the electrophoretic mobility.

A number of general practical comments that may aid in developing the optimal separation of various subcellular particles have been made (Stahn et al. 1970; Hannig 1972). The density of the membrane suspension injected into the apparatus should not be lower than that of the separation buffer, and it is usual to wash or dialyse the sample in the buffer used. Proteins absorbed to membranes should first be removed by washing to uncover the charges of the membranes. The presence of high concentrations of salts should be avoided in order to preserve low ionic strengths. It is also necessary to use buffer systems that are not dissociated by electrolysis, e.g. triethanolaminoacetic acid. Aggregation and precipitation of membranes can occur during the separation, and the disruption of nuclei resulting in the liberation of nucleic acids and chromatin, may exacerbate this. Thus, special care is necessary in the preparation of cell or tissue homogenates, and in the subsequent centrifugation steps. Earlier examples of the freeflow electrophoresis apparatus were notoriously unreliable and dangerous due to difficulties in preventing leakage of buffer from the separating chamber, and especially in controlling the temperature in view of the high voltages applied. Satisfactory results were obtained with the FF4 apparatus (see Table 3.5) and the FF5 apparatus incorporates further technical improvements.

Subfractionation of plasma membranes by electrophoresis Table 3.5 summarises the conditions used to prepare various subcellular fractions from mammalian tissues. The preparation of kidney plasma membranes is discussed in greater detail to emphasise some practical

TABLE 3.5

Examples of the separation of cell membranes by free-flow electrophoresis.

Fraction	Separation buffer used	Electrophoresis conditions	Other comments	References
Lysosomes	10 mM triethanolamine 10 mM acetic acid 1 mM EDTA, 0.33 M sucrose adjusted to pH 7.4 with NaOH	Desaga 85 V/cm 4 °C buffer flow rate 20 ml/hr Fractions of 2.2 ml/hr collected. 10 mg homogenate protein processed/hr	Crude lysosomal fraction first prepared by centrifugation	Stahn et al. (1970)
Synaptosomes	Michaelis-veronal buffer, pH 7.15	Brinkmann FF-3 1200 V, 55 mA	Synaptosomes compared in purity with those prepared by centrifugation	Ryan et al. (1971)

Erythrocyte membrane ghosts	10 mM triethanolamine 10 mM acetic acid 0.1 mM $MgSO_4$ adjusted to pH 7.4 with NaOH	Desaga FF4 100 ± 10 V/cm 100 mA 5°C Fractions of 2.3 ml/hr collected. Sample injection 3.8 ml/hr		Heidrich and Leutner (1974)
Kidney basal and brush border plasma membranes or Kidney pinocytotic vesicles	8.5 mM triethanolamine 8.5 mM acetic acid 0.28 M sucrose adjusted to pH 7.4 with NaOH	Desaga FF4 90 ± 10 V/cm 85 mA 6°C Fractions of 2 ml/hr collected	See § 3.4.1	Heidrich et al. (1972) Bode et al. (1976)
Gastric mucosal membranes	8 mM Tris 8 mM acetic acid 0.25 M sucrose adjusted to pH 7.4 with NaOH	FF5 120 ± 10 V/cm 147 mA 7.4°C Fractions of 4 ml/hr collected	A microsomal fraction was subfractionated to yield 3 bands containing different marker enzyme activities	Saccomani et al. (1977)

considerations and the biological implications of the separations obtained.

Starting from a crude brush-border preparation from kidney homogenates (§ 5.3), Heidrich et al. (1972) prepared by free-flow electrophoresis two functionally distinct areas of the kidney proximal tubule cell plasma membrane. The crude brush border fraction was prepared in 0.25 M sucrose buffered with 100 mM triethanolamine-HCl, pH 7.4, to correspond closely to the ionic conditions used in the separation buffer in the free-flow electrophoresis apparatus. The electrode buffer was 10 times as concentrated as the separation buffer and the conductivity of the buffer in the electrophoresis chamber was 5.0–10^{-2} $\mu\Omega^{-1}$. The problem of aggregation of membranes during electrophoresis was minimised by centrifuging the sample at 500 *g* for 10 min before injection into the apparatus; the effect of this procedure on yield and its variability were not reported. Using the conditions indicated in Table 3.5, two slightly overlapping peaks of (Na^+-K^+)-ATPase and alkaline phosphatase that were characteristic of plasma membranes from the baso-lateral and luminal (brush border) areas of the cell surface, respectively, were obtained (Fig. 3.13). Purer fractions were obtained by re-electrophoresis of the pooled peaks. This examination of the heterogeneity of the membranes present in the kidney brush border fraction has been extended further to allow the identification of fractions in the electrophoretic profile that contain pinocytotic vesicles (Bode et al. 1976).

General comments Free-flow electrophoresis promises to be a useful method for resolving heterogeneous membrane fractions prepared by centrifugal methods. The procedure has so far been restricted in its use to a few laboratories and mainly to kidney membranes, and therefore it is difficult to evaluate the general reproducibility of the method. The high costs of the instrument may be a factor restricting its general application. Although the method has been used routinely in accomplishing the separation of different areas of the plasma membrane described above that are difficult to obtain by other methods (§ 3.1.9 for separation of similar regions by alternative

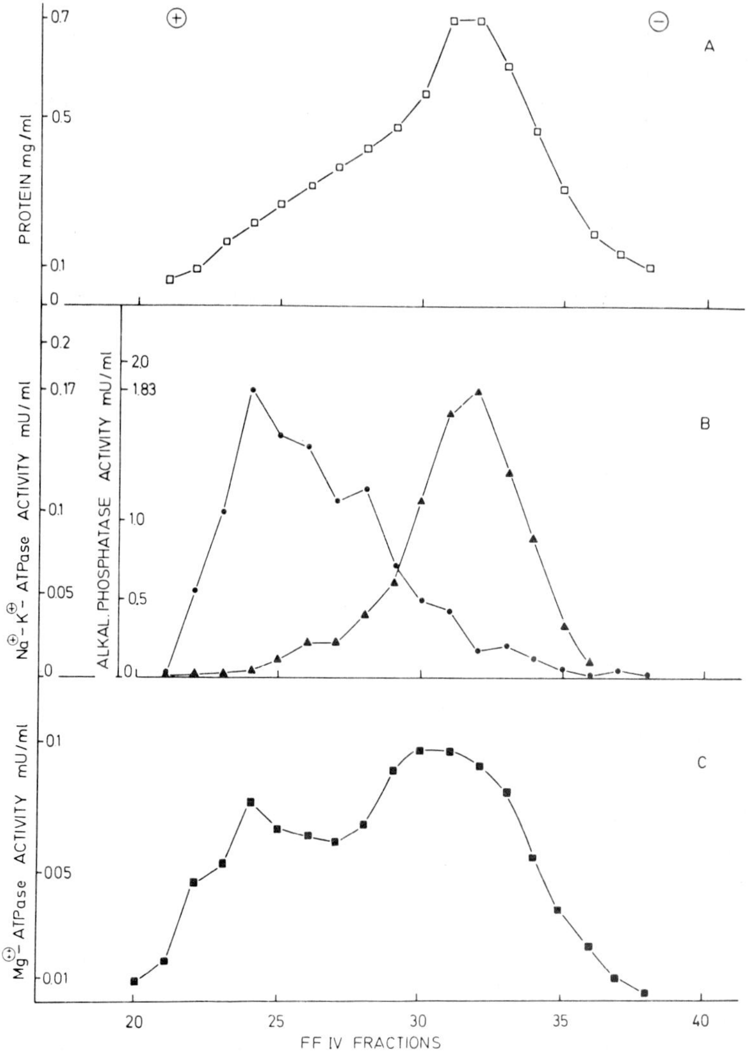

Fig. 3.13. Separation of a kidney crude brush border fraction in a free-flow electrophoresis apparatus (model FF4). Heidrich et al. (1972). (A) Protein distribution; (B) distribution of (Na^+-K^+)-ATPase activity (●----●), and alkaline phosphatase activity (▲----▲); (C) distribution of Mg^{2+}-ATPase activity. All the material was deflected towards the anode (reproduced by permission of the Journal of Cell Biology).

methods), a number of practical limitations are worth noting. Absorption of soluble proteins to membranes during the preliminary centrifugation of tissue homogenates must be minimised to optimise the electrophoretic separation. Gravitational settling of membranes during electrophoresis can be another problem. An intrinsic major limitation of the method is the low volume of sample that can be processed but there are possibilities of continuous operation and recycling. There is also variation in operating conditions resulting in the membrane separation spectrum varying by 3–4 tubes from run to run, thus requiring enzymic monitoring on each run.

3.5. Membrane separation by chromatography on columns of glass beads

A method that shows potential for the separation of membrane fragments of disparate size or shape involves chromatography on columns of glass beads of controlled pore size. Successful separation of classes of guinea pig cerebral cortex synaptic vesicles was achieved by Nagy et al. (1976) using this method. The experimental details pertaining to controlled pore glass chromatography are as follows. The chromatographic columns (1.2 cm internal diameter × 100 cm) are packed wet. The absorbing sites on the glass surfaces are blocked by treating the glass beads with a 1% solution of a high molecular weight non-ionic water-soluble polyethyleneglycol, of molecular weight 20,000 (Carbowax 20M, Union Carbide). The glass beads (CPG-10-3000, 324 nm mean pore size, 120–200 mesh) are available from Electro-Nucleonics Inc. Chromatography is carried out at 4 °C at 1 atm pressure at a flow rate of 3 ml/min.

Glass bead columns prepared as described were used to fractionate 4 ml samples (up to 1–2 mg protein) of cerebral hemisphere synaptic vesicles and soluble proteins previously released from synaptosomes. Sucrose, 0.3 M, is used to elute the fractions. Although some charge interactions between the vesicles and the polyethyleneglycol-coated glass beads was observed, the present method shows potential, for example, in the separation of populations of large and small plasma membrane vesicles; the availability of glass beads with varying mean pore sizes may help to increase the utility of this approach.

3.6. Preparation of right-side-out (RSO) and inside-out (IO) plasma membrane vesicles

It is generally assumed that plasma membrane vesicles are isolated predominantly in the RSO configuration. Since plasma membrane glycoproteins are mainly, if not exclusively, located on the outer surface of cells (Nicolson and Singer 1974) then the RSO vesicles will have the terminal sugar moieties (mainly sialic acid and galactose) on the outside. The motivation of studies that aim to prepare RSO and IO plasma membrane vesicles is to identify those proteins that are exposed on both sides of the lipid bilayer and to locate the position in the membrane of enzymes and lipids. Such vesicles are also useful in studies on the transport (e.g., of nucleosides) across plasma membranes (Chung and Hochstadt 1976) and raise questions such as the permeability of the vesicles to compounds of various molecular weights.

The preparation of RSO and IO vesicles from erythrocytes has provided a prototype system in which to use labelling techniques (§ 4.6) to identify proteins that traverse the lipid bilayer (Boxer et al. 1974; Steck 1974). However, this technique (see below) has not been as successful with plasma membranes of eukaryotic cells, e.g., fat cells (Bennett and Cuatrecasas 1973). Vesicles of the IO configuration are prepared from e.g., mouse L-cells by latex ingestion (see below), and this approach shows promise. Methods for separating IO and RSO vesicles proceed on the assumption that both populations are produced on cell breakage and are present in plasma membrane vesicular fractions and that RSO components are retarded on passage through ligand Sepharose support columns. Examples of this approach are provided by Walsh et al. (1976) who retarded pig lymphocyte plasma membrane RSO vesicles on Concanavalin Sepharose and antilymphocytic serum–Sepharose supports, and

Zackowski and Paraf (1974) who investigated murine plasmacytoma plasma membrane vesicles. However, the validity of this approach depends on the plasma membrane fraction used, being largely free of vesicles derived from intracellular membranes, since these may also behave as IO vesicles.

Erythrocyte inside-out and right-side-out vesicles The technique, described by Steck and Kant (1974), is as follows. Washed erythrocyte ghosts are suspended in 0.5 mM phosphate, pH 8. To the portion to be made into RSO vesicles is added sufficient $MgSO_4$ to bring the suspension to 0.1 mM in $MgSO_4$; samples to be made into IO vesicles receive no cations. After centrifugation at 20,000 *g* for 1 hr and removal of supernatants, the pellets are left overnight at 4 °C, a procedure that promotes vesicle resealing. Pellets are resuspended in 0.5 mM phosphate pH 8.0 (containing additionally 0.1 mM $MgSO_4$ for the RSO vesicle fraction) by repeated passage through a 27 gauge hypodermic needle attached to a plastic syringe. The suspensions are layered on a 2–3-fold excess volume of 1.03 g/cm^3 Dextran 110, dissolved in 0.5 mM phosphate pH 8.0 and centrifuged at 100,000 *g*, for 4 hr (e.g., Beckman S.W.27 rotor). IO and RSO vesicles in their separate tubes are removed from the top of the Dextran cushions and are then washed in 10 mM Tris–HCl, pH 7.4.

The efficiency of preparation of these vesicles should be checked routinely by determination of the ectoenzyme acetylcholinesterase activity. Activity of RSO vesicles should be at least 10 times higher than the IO vesicles, and this difference should be eliminated when the fractions are pretreated in 0.1% Triton X-100.

It is also possible to separate IO and RSO erythrocyte vesicles by free-flow electrophoresis (§ 3.4).

Inside-out membrane vesicles from mouse L-cells Hunt and Brown (1975) took advantage of the polystyrene latex bead method, initially described by Wetzel and Korne (1969) and modified by Heine and Schnaitman (1971) and Charalampous et al. (1973), to prepare IO membrane vesicles from monolayer cultures of L-cells. Briefly, L-cells were suspended at a density of 10^5 cells/ml in medium 199 (Flow Labs) containing 20% foetal calf serum, antibiotics and 10^4 polystyrene beads (0.714 μm diameter, Dow Chemical Co.) per cell. The cells were incubated with stirring at 37°C for 1 hr; the rate of bead ingestion was checked by dark-field microscopy. Under the above conditions, cells ingested approx. 500 beads each, and more than 90% of the cells in suspension were found to have taken up beads. Cells were then cooled to 4°C and excess beads were removed in the supernatant following successive 500 *g*/10 min centrifugation steps in phosphate-buffered saline. Cell lysis was carried out by suspending between 10^7–10^8 cells in phosphate-buffered saline containing 0.25 M sucrose, 1 mM EDTA and equilibrating them at 51 atm of N_2 for 20 min in a nitrogen bomb (§ 2.1.4) and rapidly returning them to atmospheric pressure. Virtually all cells were lysed under these conditions; the lysate was diluted with an equal volume (15 ml) of 60% sucrose dissolved in phosphate-buffered saline, 1 mM EDTA, and then layered in 10-ml portions at the bottom of discontinuous sucrose gradients prepared in 30-ml tubes. The gradients, containing layers of 30, 25 and 10% sucrose in phosphate-buffered saline, 1 mM EDTA, were centrifuged at 80,000 *g* for 90 min and the vesicles were found at the 10–20% sucrose interface. Approx. 10^9 IO vesicles were obtained from 10^7 cells.

3.7. Removal of non-membraneous protein from isolated fractions

The process of cell or tissue homogenisation displaces organelles and membrane systems from their natural environment into an artificial environment. This can result in varying degrees of organellar disruption and the relocation or occlusion of soluble or cytoplasmic proteins, inside plasma membrane vesicles generated during homogenisation. Occlusion of soluble proteins, or their non-specific attachment to membranes is especially favoured by the use of high tissue to medium ratios. In secretory tissues, e.g. liver, pancreas or parotid, the persistence inside membrane vesicles of compounds destined for secretion is a further problem, and can be used to assess, for example, the contamination of plasma membranes by intracellularly-derived membrane vesicles. Elimination of non-membrane proteins in plasma membrane fractions is important for establishing accurate values for enzyme specific activities and protein: phospholipid ratios.

Emmelot et al. (1964) found that washing a liver plasma membrane fraction in isotonic saline solubilised up to 30% of the associated protein, thereby increasing the specific activities of membrane enzymes. To minimise the persistence of non-membranous proteins in plasma membrane fractions prior to enzymic determinations, and especially polyacrylamide gel analysis, the following procedure is followed. Plasma membrane pellets are resuspended by using a small Dounce homogeniser fitted with a 'tight' plunger (§ 2.1.2) in an excess of glass-distilled H_2O; after a few minutes, the membrane dispersion is made isotonic by addition of the appropriate amount of concentrated NaCl and membranes are sedimented by centrifugation. The protein content of the supernatant is then determined.

The presence of secretory products entrapped inside membrane vesicles, especially Golgi apparatus-derived vesicles that frequently contaminate plasma membrane fractions, constitutes a problem of a different magnitude. Table 3.6 indicates the relative effectiveness of some procedures adopted to release into supernatants the content of liver Golgi vesicles. The use of the French press (§ 2.1.6) at the correct pH is most effective and is related to the breaking open of the vesicles as shown by electron microscopy (Bergeron et al. 1973b). With pancreatic tissue, Meldolesi et al. (1971) found that media of pH 7.5–8.0 were most effective in the release of protein from zymogen granules, but the resilience of the proteins inside parotid gland secretory granules is illustrated by the failure to remove them by treatment with high (0.5 M) or low (0.05 M) NaCl, or sonication in the presence of sodium bromide (0.25–0.50 M) (Castle et al. 1975).

To assess the extent of contamination of subcellular fractions by secretory proteins, Castle et al. (1975) collected secretory products from parotid glands following the administration of isotopic precursors, e.g. [^{3}H]leucine, and included such labelled protein in the

TABLE 3.6
Efficacy of procedures for removal of intravesicular components.

Procedure	Label released into supernatant (%)
Freezing and thawing; 10 times	10
Alkaline hypotonic shock; pH 8.0	38
Sonication (20 s) in alkaline hypotonic media	53
Alkaline hypotonic shock followed by passage through French press	
(a) pH 7.0	51
(b) pH 8.0	74
(c) pH 8.4	84

Golgi apparatus components of rat liver were radioactively labelled and the isolated membrane fraction treated as described. The distribution of radioactivity between pellet and supernatant formed on centrifugation was determined.
Table modified from Bergeron et al. (1973a).

dispersing medium during the subsequent homogenisation of unlabelled tissue. In this way, the persistence of non-membranous components in subcellular fractions, or as bands on polyacrylamide gels can be assessed.

3.8. Conditions for storage of plasma membranes

These conditions are dictated by the membrane properties to be investigated. For morphological investigations, membrane fractions should be dehydrated, fixed and stained as soon as the preparation is completed to help to preserve unit membrane structure. However, for negatively stained preparations, frozen membranes may be used. Ideally, all estimates of enzymic activities should be carried out on freshly prepared fractions resuspended as stock suspensions at 1–2 mg membrane protein/ml of 0.25 M sucrose. If membranes are to be stored overnight, this is best done at 4°C as a compact pellet covered with 0.25 M sucrose in a closed tube; the pellet is resuspended next day. When analysis cannot be carried out within 1–2 days of membrane preparation, membranes may be stored as a suspension in 0.25 M sucrose (buffered with 5 mM Tris–HCl, pH 7.4) at −70 °C for months, although −20 °C may suffice for shorter periods. It is advisable to partition membrane suspensions for storage into small (e.g., 0.1 ml) aliquots that can be rapidly thawed when required. Rapid freezing is obtained by inserting tubes into solid CO_2/methanol mixtures and then transferring them to −70 °C cabinets. In general, membrane-bound enzymes that are not lipid-dependent, e.g., 5′-nucleotidase and leucylnaphthylamidase, maintain their activity after numerous freezing and thawing cycles, whereas there may be a decrease in some lipid-dependent enzyme activities such as glucose 6-phosphatase $(Na^+$-$K^+)$-ATPase or histocompatibility antigenic activity. Variation in the case of a single enzyme between tissues and species is to be expected. Enzyme substrates are sometimes incorporated in storage media, e.g., Lelievre (1973) stored murine plasmacytoma plasma membranes by freezing them in buffered 50% glycerol, 1 mM ATP, 1 mM AMP, to protect 5′-nucleotidase. Alkaline phosphatase of kidney plasma membranes doubled in activity after storage at 4 °C for 8 hr (Glossman and Gips 1974).

No special advantages accrue by freeze-drying plasma membranes, except for chemical or gravimetric estimations. However, before freeze-drying it is important to ensure that membranes are freed from sucrose by repeated washing by centrifugation or by dialysis against H_2O, a process that may impair their integrity.

The following are indicators that membranes are of poor quality: a) decaying smell; b) aggregation and clumping (likely to occur on storage in H_2O or hypotonic media); c) poor resolution on polyacrylamide gel electrophoresis; bands are lacking their usual sharpness; d) presence of higher amounts (5–10% of total phospholipid) of lysophospholipids shown by thin-layer chromatography of chloroform/methanol extracts of membranes. Methods for limiting proteolysis of membranes are discussed in § 2.3.5.

Identification of subcellular fractions, especially those containing plasma membranes

4.1. General features

In monitoring the isolation and establishing the purity of mammalian plasma membranes, the greatest reliance has so far been placed on enzymic markers, predominantly hydrolases attacking phosphate esters, peptides and saccharides. Earliest clues of the usefulness in plasma membrane isolation of the prototype marker enzyme 5′-nucleotidase emerged from observations combining histochemical studies on liver sections (e.g., Essner et al. 1958) with the measurement of enzyme activity in plasma membrane fractions containing canalicular and junctional structures present in low-speed (nuclear) pellets of liver homogenates. Subsequently, a number of other plasma membrane enzymic and supporting markers have emerged to help determine the relative purity and yield of plasma membranes from a wide range of mammalian cell types. The use of marker enzymes in plasma membrane isolation studies is usually complemented by morphological monitoring and more recently by the attachment to the cell surface of intact cells of covalently bound or biologically specific markers that are radioactively labelled. The following features and practical considerations apply generally to plasma membrane markers.

Exclusiveness of plasma membrane markers It is often assumed that plasma membrane marker components reside solely in these membranes. However, the absence of biosynthetic activities in the plasma membrane argues that it is unlikely to be the case for endogenous

Subject index p. 261

markers, since they must arise by transfer from their intracellular sites of synthesis. Hence, sensitive techniques can identify plasma membrane markers on intracellular membranes, e.g., 5′-nucleotidase on endoplasmic reticular membranes (Widnell 1972) and Golgi membranes (Farquhar et al. 1974); insulin receptor binding activity on Golgi membranes (Bergeron et al. 1973b), adenylate cyclase activity stimulated by F^- and a GTP analogue in mitochondrial fractions (Zinder et al. 1976).

The use of applied markers (covalently-bound radioactivity or strongly-bound radioactively labelled ligands; see § 4.6.1 and § 4.6.3) is also subject to qualifications depending mainly on how they are used. Although they are initially restricted to the plasma membrane after brief periods of application, viable intact cells may interiorise these markers. Indeed, such studies can yield clues of the relationships between the plasma membrane and lysosomes (Werb and Cohn 1972; Steinman et al. 1976).

Representiveness of plasma membrane markers It is an incorrect assumption frequently made that plasma membrane markers are evenly distributed on the surfaces of cells permanently structured to form tissues and this may also apply to free-living cells. Clearly, in assessing the recovery of chosen plasma membrane markers, their regional disposition on the cell's surface membrane should be borne in mind, and this is emphasised in sections dealing with the isolation of plasma membrane markers from tissues (Chapter 5). Applied markers can provide a more reliable estimate of the yield of plasma membranes, but again it is only an assumption that the sites at which they bind are uniformly distributed around the cell. Considerations of the exclusiveness and representativeness of plasma membrane markers are the practical consequences of the fact that the surface membrane of most animal cells is a complex differentiated organelle. The biochemical and morphological factors underlying this differentiation are retained in the isolated plasma membrane fragments and can thus be utilised to identify the individual parts of this complex organelle.

Plasma membranes and the hetereogeneity of cells Solid organs and tissues consist of many cell types. Liver, despite being considered as a relatively homogeneous cellular tissue, contains approximately equal numbers of parenchymal and reticulo-endothelial cells; kidney and intestinal tissue consist of a variety of epithelial cells of various shapes and at different developmental stages; an extreme cellular and functional complexity is found for example in the central nervous system and the lung (Johannesen et al. 1977). Vascular and connective tissue add to this complexity to extents that vary according to the tissue, age of the animal, etc. Such cellular complexity poses the principal question of identifying the cellular origin of plasma membrane fragments derived from a tissue or organ homogenate. Cellular complexity may be resolved by first dissociating the organ or tissue into its constituent cell types and then separating the major categories of cells before fractionating. Such approaches are in general in their infancy, but despite the problems associated with cell surface damage caused by enzymic treatments used for tissue dissociation and problems with the yield and viability of separated cells, some progress is being made in this direction.

4.1.1. Latency of plasma membrane markers

Plasma membranes are isolated as populations of membrane strips and vesicular profiles. In plasma membranes identified as closed vesicles, the membrane may form a physical barrier limiting access of substrates to the inside of the vesicle. Hence, membrane-bound marker enzymes located on the inside of the vesicles may appear absent or at least show diminished activity. The classical case of the phenomenon of latency in subcellular fractions obscuring the identification of categories of cell components was that leading to the discovery of lysozomes (De Duve 1975). The lysosomal marker enzymes, e.g., acid phosphatase, β-glycerylphosphatase, are located at the inner boundary of the lysosomal membrane and to ensure access of substrates, it is necessary to disrupt the isolated fractions by treatment with Waring blenders, hypotonic media, freezing and thawing cycles and detergents.

Clearly, knowledge of the orientation or disposition of enzymes on membranes is an important prequisite in selecting and correctly using plasma membrane marker enzymes, i.e., it is useful to know whether the active site of the enzyme faces the environment (ecto-enzyme) or the cytoplasm. Fortunately, the most widely used plasma membrane marker enzymes are ectoenzymes with their active sites facing the environment or medium of the intact cell. Table 4.1 summarises the known disposition of enzymes likely to be useful in the characterisation of plasma membranes. Adenylate cyclase activity is probably the most difficult to determine accurately, especially the hormonally-activated component (see § 4.2.1).

Latency is generally overcome by carrying out the enzymic determinations in the presence of 0.1% Triton X-100. For example, lysosomal enzymes and Golgi glycosyl 1-transferases are now routinely determined in the presence of Triton X-100. However, some enzymes that are closely associated with lipids may lose activity when pretreated with detergents, or when enzymic activity is measured in detergents, e.g. (Na^{+}-K^{+})-ATPase or adenylate cyclase, especially the hormonally activated component. In using detergents to reveal latency, consideration should also be given to observations that certain ranges of detergent concentrations activate enzymes. The effects of 3 detergents on the 5′-nucleotidase activity of liver plasma membranes are shown in Fig. 4.1. Thus, activation by detergents per se does not give unambiguous evidence about the localisation on a plasma membrane vesicle of the substrate binding site. Prior to the use of detergents in membrane studies, subcellular fractions were examined for enzymic latency by passage through narrow orifices (e.g., French press) that tear open membrane vesicles, or by mechanical disruption in an 'Ultraturrax' machine (Jones and Ernster 1962); the second procedure is more convenient with small samples.

4.1.2. Yield of plasma membranes

The amount of plasma membrane protein or marker enzyme recovered from homogenates will vary according to the tissue or cell types under investigation. Morphometric methods that estimate

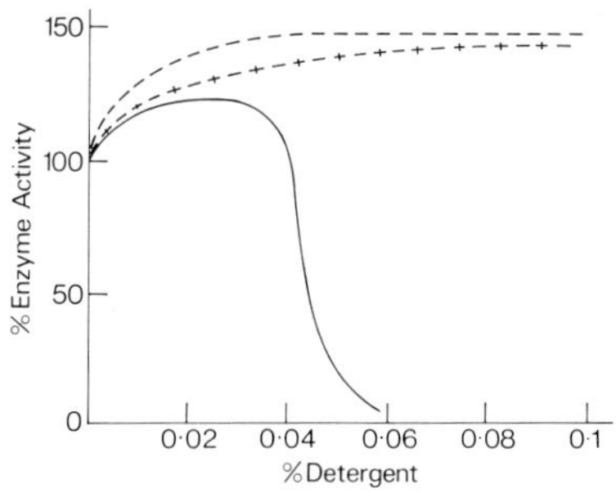

Fig. 4.1. Effects of detergents on 5′-nucleotidase activity at 37 °C of rat liver plasma membranes. Triton X-100[a] (+ – + – +), Alipal CO-433[b] (– – – –), SDS (————). At detergent concentrations above 0.1%, appreciable membrane solubilisation occurs. [a] Iso-octylphenoxypolyethoxyethanol (Rohm and Haas); [b] Sulphated alkylphenoxypoly(ethyleneoxy)ethanol (G.A.F. Corp.).

the relative amounts of the various organellar and membrane components in a cell section (§ 4.5.6) have been most frequently applied to hepatocyte intracellular components. The results of Wiebel et al. (1969) were used by Lauter et al. (1972) to calculate that 1.1% of total liver protein should be recovered as plasma membranes, i.e., 2 mg of membrane protein/g wet weight of rat liver tissue. For similar calculations of the expected yield of neuronal as opposed to synaptosomal plasma membranes (see § 5.2.1). Protein recoveries in the plasma membrane fraction are subject to qualification regarding fraction purity and the presence of attached and occluded non-membranous protein (see § 3.7).

Most plasma membrane yields are expressed as the amount of one or more enzyme activities recovered from the homogenate; usually 5–50% of the homogenate activity. As already indicated (see also § 5.2.1) the calculations are frequently subject to the qualification that the enzyme marker selected may be unevenly distributed on the cell surface and that specific areas of the plasma membrane have been purified. For example, in hepatocytes, where the plasma membrane accounts for approximately 1% of the protein, an enzyme marker exclusively located in the plasma membrane should theoretically

TABLE 4.1

Orientation on the membrane of some commonly used marker enzymes in subcellular fractions.

	Enzyme	Cell type	Location[a]	Reference
(A) Plasma membranes	5′-Nucleotidase	Hepatocyte	Ectoenzyme	Trams and Lauter (1974)
				Gurd and Evans (1974)
				Essner et al. (1958)
		HeLa, KB	Ectoenzyme	Trams and Lauter (1974)
		Fat cell	Ectoenzyme	Newby et al. (1975)
		Lymphocyte	Ectoenzyme	Misra et al. (1974)
		Neuroblastoma	Ectoenzyme	Trams and Lauter (1974)
		Neuroglioma	Ectoenzyme	Stefanovic et al. (1975)
		Polymorphonuclear leukocyte	Ectoenzyme	DePierre and Karnovsky (1974)
		Macrophage	Ectoenzyme	Edelson and Cohn (1976)
	Nucleotide pyrophosphatase	Hepatocyte	Ectoenzyme	Evans (1974)
				Bischoff et al. (1976)
		Lymphocyte	Ectoenzyme	Abney et al. (1976)
		Thymocytes	Ectoenzyme	Monneron (1974)
	Aminopeptidase	Hepatocyte	Ectoenzyme	Trams and Lauter (1974)
	Alkaline phosphatase	HeLa cells	Ectoenzyme	Lin et al. (1976)
		Polymorphonuclear leukocyte	Ectoenzyme	DePierre and Karnovsky (1974)

	Sialidase	Hamster embryo fibroblasts	Ectoenzyme	Schengrund et al. (1976)
	Adenylate cyclase	Hepatocytes	Cytoplasmic surface	Trams and Lauter (1974)
		Mouse embryo trophoblasts	Cytoplasmic surface	Vorbrodt et al. (1977)
(B) Endoplasmic reticulum	NADPH-cytochrome *c* reductase	Hepatocyte	Cytoplasmic surface	DePierre and Dallner (1975)
	Glucose 6-phosphatase	Hepatocyte	Luminal surface	DePierre and Dallner (1975)
	β-Glucuronidase	Hepatocyte	Lysosomal luminal surface	DePierre and Dallner (1975)
(C) Golgi apparatus	5′-Nucleotidase	Hepatocyte	Luminal surface of secretory vesicles	Farquhar et al. (1974)
	Adenylate cyclase	Hepatocyte	Cytoplasmic surface	Cheng and Farquhar (1976)

[a] Ectoenzymes have their active site facing the environment of the cell; in plasma membrane vesicles of the right-side-out confugation (§ 3.6) they face the medium.

be purified 100-fold; in most cases the extent of purification is much less than this, usually 10–80-fold. However, the bile canalicular region that accounts for approximately 10% of the hepatocyte surface membrane and contains high activities of plasma membrane marker enzymes may be purified nearly 100-fold on the basis of certain plasma membrane marker enzymes (Wisher and Evans 1975). Hence, extremely high specific activities of plasma membranes marker components and low protein recovery may provide clues that a small specific area of the plasma membrane was purified.

4.1.3. Expression of results

The distribution of subcellular components in the major fractions, nuclear, mitochondrial, microsomal and supernatant, is usually expressed in histogram form as used by De Duve (1959, 1975, for reviews). In such schemes, the specific activities and protein recovery of enzymes in the specified fractions dictate the height and width, respectively, of the fraction histograms and together give an immediate indication of the distribution and recovery of subcellular components. The expression of plasma membrane marker distribution and recovery in histogram form did not find favour, at least initially. This is probably for two reasons; firstly, De Duve postulated that each marker enzyme should be restricted to a single intracellular site and since many plasma membrane marker enzymes are sedimented in both nuclear and microsomal fractions, such a bimodal distribution was thought to complicate the analytical approaches to plasma membrane isolation. Secondly, the low degree of recovery of plasma membrane marker enzymes and/or protein amounted to an embarrassment when analysed in histogram form and was thought to detract from the major difficulties overcome in at least isolating some plasma membrane fragments free from contamination by other subcellular components. However, this approach is still a valuable one especially when attempting to isolate and characterise a plasma membrane fraction from non-hepatic tissues as shown in the following example. Powell et al. (1976) analysed subcellular fractions of bovine corpus luteum and compared the distribution of a variety of marker enzymes to determine whether a receptor for prostaglandin $F_{2\alpha}$ was located on the plasma membrane. Fig. 4.2 shows a histogram of the fractions prepared by differential centrifugation, and it is clear that the 5′-nucleotidase and prostaglandin receptor distribution correspond closely. This example also indicates how it is possible to veer away from the traditional 4 fractions and it constitutes a rapid method for preparing plasma membranes from the corpus luteum (§ 5.7).

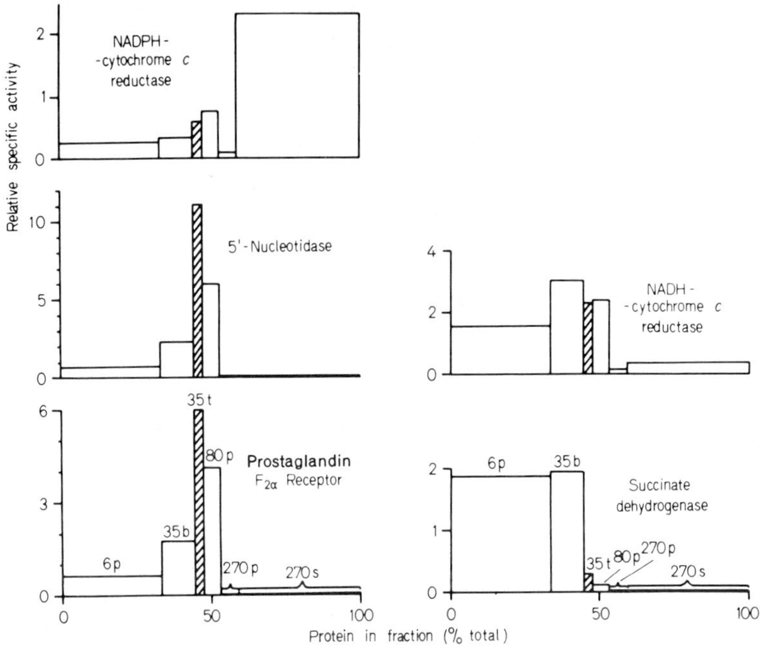

Fig. 4.2. Distribution histogram of components of bovine corpus luteum homogenates showing co-purification of 5′-nucleotidase and prostaglandin $F_{2\alpha}$ receptor by differential centrifugation. The relative specific activity plotted on the ordinate is the activity per mg in the fraction/the activity per mg protein in the homogenate. The shaded bars represent the fraction most enriched in plasma membranes; 6p, 6000 *g* × 10 min pellet; 35b, bottom layer of the 35,000 *g* × 30 min pellet; 35t, top layer of the 35,000 *g* pellet; 80p, 80,000 *g* × 90 min pellet; 270p, 270,000 *g* × 3 hr pellet; 270s, 270,000 *g* supernatant. Fractions 35t and 80p may be further purified by density gradient centrifugation. Powell et al. (1976). Reproduced by permission of Federation of European Biochemical Societies.

4.1.4. Other practical considerations in determining membrane marker components

The amount of membranes (generally measured as membrane protein) to be used in the assay of marker components is usually determined by trial and error and will depend mainly on the sensitivity of the assay and the purity of the membrane fraction. In most cases it is between 2–100 μg protein, and it is convenient to ensure that this amount of protein is contained in a small and easily handled volume (e.g., membranes may be stored at 1–5 mg membrane protein/ml in 0.25 M sucrose). Values for cell or tissue homogenate marker activities are susceptible to handling errors and special care should be taken to minimise these, since relative specific activities are critically dependent upon these values being approximately correct. Pipettes with wide sampling orifices should be used, especially in dealing with tissue homogenates dispersed by using loose Dounce homogenisers. Strictly, a straight line correspondence between the amount of membrane protein and the marker activity should be established as well as (in the case of enzyme markers) the linearity of the reaction with time. Most enzyme assays proceed linearly for 5–30 min, but for longer times, substrate depletion may occur; this is a strong possibility when using radioactive substrates to which carrier has not been added.

Most subcellular fractionation methods contain a final purification step carried out in sucrose gradients. Reagent grade sucrose contains various impurities, including a ribonuclease (Roth 1967). As a substitute for expensive highly purified sucrose, reagent grade sucrose may be treated with activated charcoal to remove ultraviolet absorbing materials, including ribonuclease. The determination of protein and enzyme activities in aliquots collected directly from sucrose gradients can yield erroneous values, especially at sucrose concentrations above 20%, w/v. For example, Hinton et al. (1969) showed that a number of commonly used subcellular marker enzymes were reduced in activity in a linear relationship to sucrose concentration. No interference by sucrose was detected in the estimation of inorganic phosphate by the Lowry and Lopez (1946) method.

Generally, reduction in enzyme activities caused by sucrose are reversed on dilution and therefore, membrane solutions in hypertonic sucrose should first be pelleted by high speed centrifugation and then resuspended at 1–5 mg membrane protein/ml in iso-osmotic sucrose.

The correct cofactors should be present in reaction mixtures. 5′-Nucleotidase activity requires divalent cations; thus EDTA included in sucrose solutions used in membrane preparation will inhibit the enzyme (Johnsen et al. 1974). Problems of this nature may be overcome by adding an excess of the appropriate divalent cation; for example EGTA inhibition of the activity of bovine adrenal cortex adenylate cyclase was reversed by addition of Ca^{2+} (Glossmann and Gips 1976). A further factor in determining marker enzyme activities is the necessity to measure activity at or near the pH optimum. Since membranes are exposed over a period of hours to media of varying tonicity and ionic strength, it is important to establish how resistant chosen marker enzymes are to experimental manipulation. The problem of determining the level of attached or occluded protein in isolated plasma membranes is discussed in § 3.7.

Finally, plasma membrane enzyme specific activities and composition may vary with the sex and age of the animal (Solyom and Trams 1973; O'Bryan and Lowenstein 1974).

4.2. Enzymic markers and their estimation

Methods for determining a number of commonly used plasma membrane enzyme markers in addition to some enzymes used to determine the extent of contamination of plasma membrane fractions by intracellular components are now described.

4.2.1. Plasma membrane enzymic markers

4.2.1.1. 5′-Nucleotidase (EC 3.1.3.5) This enzyme, hydrolysing 5′-ribonucleotides to ribonucleosides and orthophosphate, is by far the most commonly determined plasma membrane marker. Analysis of the pH activity curves of 5′-nucleotidase of human, rat and mouse liver plasma membranes (Song and Bodansky 1967; Evans and Gurd 1973), smooth muscle of small intestine (Burger and Lowenstein 1970) and heart (Edwards and McGuire 1970) indicate a pH optimum of 7.4, with a second optimum at pH 10 in the presence of Mg^{2+}. The enzyme shows similar kinetic properties when isolated from a number of tissues and organs (Riemer and Widnell 1975). Cell sap of liver, and possibly of other tissues and cells, also contain a soluble 5′-nucleotidase with a pH optimum of 6.5 (Itoh et al. 1968; Fritzson and Smith 1971), that differs from the membrane-bound enzyme in having different kinetics and a preference for IMP over AMP (Van den Berghe et al. 1977). Activity of the soluble enzyme is inhibited by ATP and ADP. The enzyme purified from rodent liver membranes is a glycoprotein of apparent mol. wt. 140,000 that may consist of two subunits (Evans and Gurd 1972). A definitive physiological role for the ectoenzyme remains to be established. In some cases, the enzyme activity is present at low or variable amounts on mammalian plasma membranes (e.g., human polymorphonuclear leucocytes, in contrast to those of guinea pigs and rabbits which contain little demonstrable activity; Strauss and Burrows 1975) and lymphocyte subpopulations vary in enzyme activity (Uusitalo and Karnovsky 1977). Enzyme activity is also extremely low in L-cells (Hubbard and Cohn 1975).

The enzyme is usually measured by following the liberation of phosphate from AMP or other nucleotide 5′-monophosphates. The incubation medium contains in 2 ml (final concentration), 100 mM KCl, 10 mM $MgCl_2$, 50 mM Tris–HCl buffer, pH 7.4, 10 mM potassium sodium tartrate, 5 mM AMP and 5–50 μg membrane protein. After 15 min at 37 °C the reaction is stopped by addition of 1 ml 25% (w/v) trichloroacetic acid, and phosphate released is determined by suitable methods, e.g. King (1932), Martin and Doty (1949) (§ 4.3.7). Tartrate is included to inhibit AMP breakdown by non-specific phosphatase activity.

A more rapid method measures liberated adenosine spectrophotometrically in a deamination reaction catalysed by excess adenosine deaminase (Ipata 1967; 1968). In a quartz cell (volume

1.8 ml) are included (final concentration) 10 mM $MgCl_2$, 100 mM KCl, 10 mM potassium sodium tartrate, 20 μl of an adenosine deaminase suspension (adenosine aminohydrolase, EC 3.5.4.4, from calf intestine, e.g., Sigma, Type 1) diluted 1 : 100 before use, and membrane suspension. Hydrolysis of 1 μmol AMP causes a decrease in absorbance of 0.494 units at 265 nm. This assay is most conveniently carried out in a recording spectrophotometer with the enzyme reaction mixture maintained at 37 °C.

A highly sensitive method measures the liberation of [^{3}H]adenosine from [^{3}H]AMP (Avruch and Wallach 1971). In 500 μl final concentration are contained 200 μM AMP, 50 mM Tris–HCl buffer, pH 7.4, and tracer [^{3}H]AMP (20,000 cpm). The reaction is commenced by addition of 10 μl membrane suspension and terminated with 100 μl 0.15 M $ZnSO_4$. Excess AMP remaining unhydrolysed is removed by precipitation with 100 μl 0.15 M $Ba(OH)_2$ which removes 98% of added AMP but only 11% of adenosine. The supernatant is then added to 5 ml of a scintillant and radioactivity is determined.

4.2.1.2. Alkaline phosphodiesterase and nucleotide pyrophosphatase
These enzyme activities are probably due to the same plasma membrane ectoenzyme as demonstrated in liver (Evans et al. 1973; Bischoff et al. 1975) and lymphocytes (Abney et al. 1976). The enzyme is a glycoprotein of apparant mol. wt. 130,000 and is also present in hepatic endoplasmic reticulum (Bischoff et al. 1976). Alkaline phosphodiesterase I activity (oligonucleate 5′-nucleotidohydrolase, EC 3.1.4.1) is determined by measuring the hydrolysis of thymidine *p*-nitrophenylphosphate (TPN) in a cuvette containing (final concentration) 100 mM Tris–HCl buffer, pH 8.9, and 0.5 mM TPN. The formation of *p*-nitrophenol is followed at 37 °C by the increase in absorption at 400 nm. The hydrolysis of 0.1 μmol substrate produces an increase in absorbance of 1.20 units (Razzel 1963).

Nucleotide pyrophosphatase acts on a range of substrates. Using UDP-[^{3}H]galactose as substrate (Sela et al. 1972; Mookerjea and Yung 1975), the incubation mixture contains in 100 μl (final con-

Subject index p. 261

centration) 10–80 μg protein, 30 mM cacodylate buffer, pH 8.6, UDP-[^{3}H]galactose 20 μM (100,000 cpm), 5 mM $MnCl_2$ and 5 μl of a 10% solution of Triton X-100. The reaction is stopped by addition of 100 μl ethanol and 50 μl acetic acid, and after centrifugation, portions of the supernatant are applied to Whatman 3 MM paper. Reaction products are separated by descending chromatography for 18 hr in ethanol/M sodium acetate buffer, pH 3.8 (7.5:3, v/v). The distribution of ^{3}H radioactivity is determined by cutting the paper into 0.5 cm strips which are then suspended in a toluene-based scintillant. A preliminary indication of the expected position of the reaction products by reference to the standards UDPgalactose, galactose and galactose 1-phosphate may be obtained by rapidly scanning the paper strip in, e.g., Packard Radiochromatogram Scanner, Model 7201. Galactose 1-phosphate may be hydrolysed further to galactose by a membrane-bound phosphatase, and this has to be taken into account.

An alternative non-radioactive method for measuring hydrolysis of sugar nucleotides by nucleotide pyrophosphatase is described by Bachorik and Dietrich (1972). The liberated hexose 1-phosphate is acid hydrolysed after treatment of the incubation system with charcoal (Ogowa et al. 1966). To measure the hydrolysis of ATP to AMP and pyrophosphate (designated ATP pyrophosphohydrolase; EC 3.6.1.8; see Fig. 4.3), the reaction is carried out in duplicate

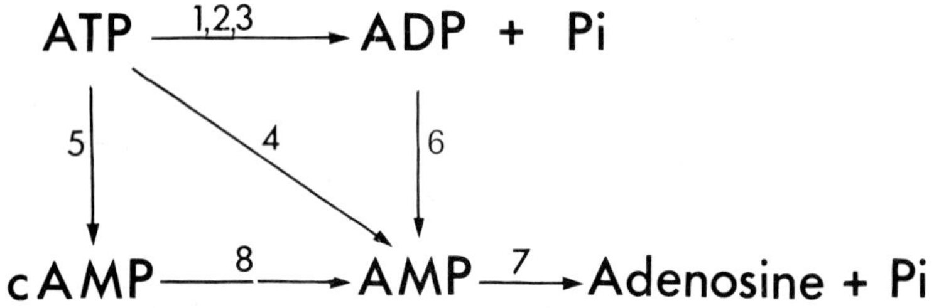

1. Mg^{2+}-ATPase
2. (Na^+-K^+)-ATPase
3. Ca^{2+}-ATPase
4. Nucleotide pyrophosphatase
5. Adenylate cyclase
6. Nucleotide diphosphatase
7. 5′-Nucleotidase
8. Cyclic nucleotide phosphodiesterase

Fig. 4.3. Enzymes acting on nucleotides in mammalian plasma membrane fractions.

in paired tubes, one of which contains an inorganic pyrophosphatase and the differential phosphate release is determined (Lieberman et al. 1967).

NADH-nucleotide pyrophosphatase (dinucleotide nucleotidohydrolase, EC 3.6.1.9) is determined by measuring the hydrolysis of NADH to reduced nicotinamide mononucleotide and 5′-adenylic acid. Lauter et al. (1972) incubated membrane samples for 30 min at 37 °C in 0.2 ml containing (final concentration) 75 mM aminopropanol buffer (pH 9.0), 0.5 mM NADH, 10 mM KCl and 0.05 mM Co^{2+}. The reaction is terminated by addition of 0.3 ml 0.5 M $HClO_4$ and the liberated inorganic phosphate (simultaneously and stoichiometrically released from AMP by the action of indigenous 5′-nucleotidase) is measured in the supernatant.

4.2.1.3. Nucleotide triphosphatases (EC 3.6.1.3) These enzymes are difficult to measure accurately in plasma membranes due to the many enzymes acting on ATP and its hydrolysis products (see Fig. 4.3). The (Na^+-K^+)-ATPase, inhibited by ouabain, is a more specific plasma membrane enzyme and highest activities of this enzyme are obtained in excitable tissues. Activity is determined in a 1 ml reaction mixture containing (final concentration) 30 mM Tris–HCl buffer, pH 7.4, 30 mM KCl, 3 mM $MgCl_2$, 3 mM ATP (Tris salt), 100 mM NaCl (ATP solutions should be freshly prepared). The basal Mg^{2+}-activated component is measured in the same reaction mixture but NaCl is replaced by equimolar choline chloride. The reaction is stopped by addition of 2 ml of 6% (w/v) trichloroacetic acid and liberated orthophosphate is determined. The increased enzyme activity measured in the presence of Na^+ and K^+ is eliminated by inclusion in the medium of 10^{-3} to 10^{-4} M ouabain (Strophanthin g) or 10^{-3} M Scillaren (Sandoz; Boyer and Reno 1975). These concentrations are well in excess of those (approx. 10^{-7} M) shown to give half maximal inhibition of sodium transport and of (Na^+-K^+)-activated ATPase of red blood cell ghosts and neuronal membranes (Skou et al. 1971).

The Ca^{2+}-ATPase activity of plasma membranes (e.g., renal baso-lateral plasma membranes; Kinne–Saffran and Kinne 1974) is determined in a similar way in the presence of 5 mM $CaCl_2$ and 2 mM ouabain. Additional phosphate release over that obtained in the absence of divalent cations is believed to be due to a Ca^{2+}-activated ATPase.

4.2.1.4. Leucyl-naphthylamidase (EC 3.4.11.1) The enzyme is an aminopeptidase catalysing the general reaction L-amino-acyl-naphthylamide→L-amino acid + naphthylamine. Species and tissue differences restrict the value of this enzyme as a plasma membrane marker. For example, the enzyme activity is increased approximately 10-fold in liver plasma membranes (higher in the bile-canalicular plasma membranes), but human platelet plasma membranes show no increase over the homogenate value (Barber and Jamieson 1970). Kidney and intestinal brush borders are rich in this class of hydrolytic enzymes. The enzymes are glycoproteins (Wacker 1974) of apparent molecular weight 140,000 (Wacker et al. 1976) and are located on the outer surface of a range of mammalian cells (Aoyagi et al. 1976). The metabolic significance of aminopeptidases remains to be established. Activity is measured in a 1 ml volume containing (final concentration) 0.1 M phosphate buffer, pH 7.0, 0.6 mM L-leucyl-β-napthylamide and 5–10 μg membrane protein (Goldberg and Rutenberg 1958). After 30 min at 37 °C, the reaction is stopped with 0.5 ml 40% (w/v) trichloroacetic acid. After bench centrifugation, 1 ml of the supernatant is transferred to another test tube and liberated β-naphthylamine is diazotised for 5 min by adding 1 ml of 0.1% sodium nitrite. Excess sodium nitrite is decomposed by adding 1 ml of ammonium sulphamate (0.5%). After a further 5 min, 2 ml of *N*-(1-naphthyl)-ethylenediamine dihydrochloride (0.5 mg/ml, dissolved in 95% ethanol) is added and the liberated dye is measured at 560 nm. The release of 1 μg leucine produces an increase in absorption of 0.037 units. A spectrophotometric method has also been described by Mellors (1969).

4.2.1.5. Nucleotide cyclases Adenylate cyclase (ATP pyrophosphate-lyase (cyclising), EC 4.6.1.1) is one of the few enzymes that currently fits into accepted dogma of plasma membrane function and peptide hormone action. Although adenylate cyclase is frequently regarded as a useful plasma membrane marker enzyme, its use is subject of a number of pitfalls. For example, there is increasing evidence that the enzyme is not confined solely to the plasma membrane, but is present in other subcellular fractions. Zinder et al. (1976) described the preparation of a secretory vesicle membrane fraction from bovine adrenal medulla that contained no detectable activities of the plasma membrane marker enzymes 5′-nucleotidase and (Na^+-K^+)-ATPase but contained high specific activities of dopamine-β-hydroxylase and adenylate cyclase activities. A rat liver mitochondrial fraction was reported to contain a glucagon-insensitive adenylate cyclase activity, whereas the plasma membrane fractions were enriched in hormone-sensitive adenylate cyclase (Chang et al. 1975). Both basal and hormone stimulated adenylate cyclase activities have been described in isolated nuclei of rat liver (Soifer and Hechter 1971) and ventral prostate gland (Liao et al. 1971). The enzyme present in rat liver Golgi membranes is not stimulated by glucagon (Cheng and Farquhar, 1976).

The hormone-sensitive component of adenylate cyclase activity would appear to serve more accurately as a plasma membrane marker. However, the accurate measurement of hormone-sensitive adenylate cyclase is subject to knowledge of the disposition of the enzyme and hormone receptors in isolated plasma membrane vesicles. There is good evidence that the enzyme is located on the cytoplasmic face of the plasma membrane and this poses in isolated plasma membrane vesicles the usual problems associated with latency (§ 4.1.1), since membranes are relatively impermeable to nucleotides. The activation of the enzyme by polypeptide and adrenergic hormones acting on the outer face of the cell membrane illustrates the importance of also knowing the orientation and the intactness of plasma membrane vesicular profiles seen in sections through the electron microscope.

Some practical difficulties of measuring enzyme activity may be mentioned. First, enzyme activity is labile during fractionation procedures, especially the hormone sensitive component. Pohl et al. (1971) found that the glucagon and epinephrine activated adenylate cyclase activities of partially purified liver plasma membranes exceeded that of the purified membranes. Cheng and Farquhar (1976) studied the loss of basal adenylate cyclase activity in liver homogenates and showed that 2 mM EGTA helped to decrease the decay of enzyme activity. Wisher and Evans (1975) showed that the activation of hepatic adenylate cyclase by a range of glucagon concentrations (10^{-9}–10^{-3} M) was dependent on the topographical origin on the hepatocyte surface of the purified plasma membranes, since a plasma membrane subfraction identified as deriving from the bile canaliculus contained cyclase activity that was slightly inhibited by glucagon. Second, it is difficult to measure accurately a reaction in which usually less than 0.05% of ATP is converted to cAMP and slight contamination of this product leads to serious errors in determining enzyme activity (e.g., see Emmelot and Bos 1974). Thus, it is not surprising that a variety of methods for measuring adenylate cyclase has been described (for detailed reviews see Advances in Cyclic Nucleotide Research, Vol. 2, 1972, and Methods in Enzymology, Vol. 38, 1974). Essentially the methods for assaying adenylate cyclase, or more correctly, for measuring cAMP can be divided into two types. There are methods suitable for in vivo and/or in vitro situations that measure cAMP concentrations, for example, by radioimmunoassay (Steiner et al. 1969) or by its conversion to ATP and its analysis by luciferase luminescence (Johnson et al. 1970). However, by far the most popular in this category are simple and sensitive assays based upon the competition for binding to a protein. The nucleotide protein complex is either absorbed on a cellulose-ester filter (Gilman 1970) or the 'free' nucleotides are absorbed by charcoal and separated by centrifugation (Brown et al. 1971). Standard curves for the estimation of 0.5–16 pmol cAMP are frequently used with commercially available assay 'kits'. Attention has been drawn to factors that affect the saturation assay of cAMP in biological systems (Albano et al. 1974). There are also assays based upon the conversion of radioactive ATP to cAMP which is then separated from other radioactive nucleotides by chromatography and/or adsorption on $BaSO_4$ and $Zn(OH)_2$. Sensitivity of the assay is crucially dependent upon the efficiency with which cAMP is separated from labelled substrate and radioactive contaminants. In other words, at any given level of specific radioactivity of substrate, sensitivity approaches a maximum as the radioactivity in the assay 'blanks' approaches zero. The radiochemical methods frequently followed are those of Salomon et al. (1974) and Krishna et al. (1968) in which an α-labelled [^{32}P]ATP is used. These methods are only suitable for in vitro situations and vary mainly in their claimed efficiencies of cAMP recovery (assessed independently by following the recovery of [^{3}H]cAMP from other nucleotides). These methods have two disadvantages. First, that of the relatively high expense and short half-life of the [^{32}P]ATP and second, the techniques are not directly applicable to the use of [^{3}H] and [^{14}C]ATP due to the poor separation of nucleosides from cAMP.

The determination of nucleotide cyclase activity by measuring free cAMP is subject to two further qualifications. First, cyclic nucleotide phosphodiesterase activity (EC 3.1.4.17) associated with membranes may hydrolyse cAMP. This enzyme is present in low amounts in liver plasma membranes but in other tissues with high activity it may be inhibited by 8–10 mM theophylline or 1 mM EDTA. However, addition of methylxanthines such as theophylline to enzyme reaction mixtures should be done in the knowledge that they may also inhibit other enzymes, e.g., 5′-nucleotidase (Tsuzuki and Newburgh 1975). Second, an ATP generating system (phosphocreatine and creatine phosphokinase) is sometimes included to maintain the level of substrate ATP especially if the membranes used contain high ATP-hydrolysingactivity. Use of synthetic substrates such as adenylyl-imidodiphosphate (AMP-PNP) (Maguire and Gilman 1974) helps to minimise substrate depletion, since this substrate is resistant to hydrolysis by ATPases.

In contrast to information available on the adenylate cyclase, less data have accumulated to assess the suitability of guanylate cyclase (EC 4.6.1.2) as a plasma membrane marker enzyme. This is due mainly to two reasons; first, cells possess lower activities of guanylate cyclase than adenylate cyclase and the substrate GTP is less stable than ATP. Second, there are soluble and membrane bound forms of the enzyme. Assay methods for guanylate cyclase are similar to those for adenylate cyclase and include chromatographic methods for separating the reaction products (Hynie 1974; Nakazawa and Sano 1974). Using an alumina separation column, Krishnan and Krishna (1976) developed a method for measuring cGMP employing [^{32}P]GTP in which 'blank' values of less than 0.0005% of added radioactivity were obtained. However, radioimmunoassay kits for determination of cGMP available from Amersham Searle, Boehringer and Collaborative Research Inc. appear to be becoming the method of choice. These kits include full instructions for use. For information on radioimmunoassay see Chard, this series, vol. VI, 1978.

4.2.1.6. Tissue-specific markers The following enzymic markers have been used to supplement the more conventional markers:

Cytidine monophosphate-N-*acetylneuraminic acid hydrolase* This plasma membrane ectoenzyme has been demonstrated in liver and kidney and catalyses the reaction, CMP-sialic acid → sialic acid + CMP.

Activity is measured by using CMP-[^{14}C]sialic acid and the released sialic acid is separated by ion exchange or paper chromatography (van Dijk et al. 1976). The enzyme is increased in specific activity, relative to the homogenate, 13–18-fold in liver and kidney homogenates but activity is low in brain, spleen, heart (Kean and Bighouse 1974). It has a pH optimum of about 9 and is inhibited by nucleosides, nucleotides, sugar nucleotides and EDTA. The enzyme is localised mainly at the brush border of kidney proximal tubule cells and at the hepatocyte blood sinusoidal plasma membrane.

Triglyceride lipase (EC 3.1.1.3) This enzyme is loosely bound to liver plasma membranes and is released from membranes by heparin. The substrate is glyceryl tri-[1-^{14}C]oleate, and enzyme activity is measured on the basis of fatty acid production (Assmann et al. 1973).

Phospholipase $A_{1,2}$ (EC 3.1.1.4) Phospholipases acting mainly on phosphatidylethanolamine and sphingomyelin are located primarily in liver plasma membrane fractions (Victoria et al. 1971; Newkirk and Waite 1971).

Alkaline phosphatase (EC 3.1.3.1) Highest enrichments of this phosphomonoesterhydrolase activity are obtained in kidney and intestinal brush border membranes, and liver bile canalicular plasma membranes. The enzyme, a glycoprotein, is a useful plasma membrane marker in thymocytes (Ey and Ferber 1977). Activity is determined in a 2 ml reaction mixture by measuring the release of nitrophenol from *p*-nitrophenylphosphate (5 mM) in a Tris buffer, pH 10.9, 5 mM $MgCl_2$ (Pekarthy et al. 1972). Hydrolysis of 0.1 μmol substrate produces an increase in absorbance of 1.20 units. Enzyme activity may also be determined by using a Merckotest®.

γ-Glutamyl transpeptidase (EC 2.3.2.2) This enzyme, an integral membrane glycoprotein (Inoue et al. 1977), catalyses the transfer of the γ-glutamyl moiety of glutathione and other γ-glutamyl compounds to a variety of amino acids and peptides. This membrane-bound enzyme was shown to be a useful plasma membrane marker in lymphoid cells, intestinal mucosa, proximal renal tubules and choroid plexus (Novogrodsky et al. 1976). Transpeptidase activity is measured by using the artificial substrate L-γ-glutamyl *p*-nitroanilide. A typical reaction mixture contains in 2 ml (final concentration): 0.15–1.5 mg protein 5.3 mM L-γ-glutamyl *p*-nitroanilide dissolved in dilute HCl, 110 mM glycylglycine, 90 mM Tris buffer, pH 8.5 (Rosalki and Tarlow 1974). Assay kits (Boehringer 15701; Sigma 415) are available; liberated *p*-nitroanilide is measured at 405 nm. Since hydrolysis is measured actually and transfer of amino acid to acceptor(s) is assumed, the values obtained are probably unreliable when high aminopeptidase activity is present in the fractions.

Sucrase (EC 3.2.1.48) Brush borders of kidney and intestine are highly enriched in disaccharidases. Sucrase, one of the commonest enzymes determined, is measured by its action on sucrose or maltose and the liberated glucose is then estimated by a glucose-oxidase system. Thus, membranes are added to 1 ml containing (final concentration) 30 mM maltose; 0.25 M sodium phosphate buffer, pH 6.0. Liberated glucose is oxidised to gluconolactone by the excess glucose oxidase present in the incubation mixture (approx. 40 μg/ml reaction mixture); subsequently peroxidase and *o*-dianisidine (3,3′-dimethoxybenzidine) are added at 4.5 and 45 μg, respectively, per ml reaction mixture, and the gluconolactone is converted to oxidised *o*-dianisidine which is measured at 530 nm. Membrane fractions should be washed or dialysed against distilled water if they have been prepared in sucrose solutions. Further care should be taken to establish the level of contamination of glucose-oxidase by disaccharidases; in certain circumstances, these are inhibited by the inclusion of Tris salts in a two-stage procedure (Messer and Dahlqvist 1966).

Acetylcholinesterase (EC 3.1.1.7) This enzyme is a neuronal and possibly glial plasma membrane marker (see § 5.2.1), and is measured by the method of Ellman et al. (1961). In a 3-ml volume containing (final concentration) 0.1 M phosphate buffer, pH 8.0, are included membranes, 0.5 mM acetylthiocholine and 0.3 mM dithiobisnitrobenzoic acid (DTNB). Acetylcholinesterase releases thiocholine that interacts with DTNB to liberate a complex that is measured at 412 nm; the molar extinction coefficient is $1.36 \times 10^4\ M^{-1} \times cm^{-1}$.

4.2.2. Endoplasmic reticulum markers

In contrast to plasma membrane markers that are common to many mammalian cells, it is more difficult to generalise about endoplasmic reticular markers, since these may be associated with aspects of cell functioning that vary between cells. A further aspect resulting from advances in subcellular fractionation is that enzymes present in the microsomal fraction can be allocated, for example, to the membranes of the Golgi cisternal and secretory elements also present in this fraction. The following constitutive enzymes are commonly determined for endoplasmic reticulum.

Glucose 6-phosphatase (EC 3.1.3.9) Although frequently used as an endoplasmic reticulum marker, weak activity only is detected in many cells, e.g., adipose cell microsomal fractions, and frequently, recorded activities may reflect non-specific general phosphatases present. The enzyme is latent and unstable and it should be assayed as soon as possible; activity decreases on freezing and thawing, but rapidly frozen samples kept at $-20\,^{\circ}\mathrm{C}$ retain activity. Enzyme activity is measured in a 3-ml volume containing (final concentration) 10 mM glucose 6-phosphate, 0.1 M maleic acid buffer, pH 6.2, 4 mM EDTA, 1 mM KF and the sample (Swanson 1955). After 15 min the reaction is stopped by addition of 25% (w/v) trichloroacetic acid and the released inorganic phosphate is measured. EDTA and KF are included to inhibit acid or alkaline phosphatase activities; divalent cations inhibit activity. Cytochemical and fractionation studies indicate that the enzyme is uniformly distributed in smooth and rough surfaced hepatic endoplasmic reticular membranes. The enzyme is also present in Golgi membranes (Bergeron et al. 1973a) and in nuclear membranes (Kashnig and Kasper 1969).

NADPH-cytochrome c *reductase* (EC 1.6.2.4) This enzyme is frequently used as an alternative marker to glucose 6-phosphatase activity (Phillips and Langdon 1962) but activity is also located in Golgi fractions. Enzyme activity may be determined by the method of Sottocasa et al. (1967) in a 3-ml reaction mixture containing (final concentration) 0.1 mM NADPH, 0.1 mM cytochrome *c*, 0.3 mM KCN, 50 mM phosphate buffer, pH 7.5. Reduction is monitored at 550 nm using molar absorbancy indices of 27.7×10^{3} $cm^{2} \times mol^{-1}$ for reduced cytochrome *c* (Margoliash and Frohwirt 1959). NADPH-linked assays should not be confused with the rotenone insensitive NADH-cytochrome *c* reductase that is mainly located in the outer mitochondrial membrane (Sottocasa et al. 1967).

Subject index p. 261

Other endoplasmic reticular markers Other markers that apply to liver microsomal fractions are cytochromes b_5 and *P*-450, determined spectrophotometrically (see Dallner et al. 1966). Content of RNA (see § 4.3.4.1) does not indicate the level of contaminating smooth microsomal vesicles, Golgi and plasma membranes.

4.2.3. Golgi membrane markers

The extent of contamination of plasma membrane fractions by elements of the Golgi apparatus is rarely determined despite similarities in their densities on sucrose gradients that may result in co-purification (see § 3.1.2). Glycosyl transferases are the major marker enzymes but their determination is not as straightforward as marker enzymes for other membranes and organelles. Transfer of radioactive sugar nucleotides to endogenous or exogenous substrates (prepared by treatment of glycoproteins with sialidases and galactosidases, (Hughes and Jeanloz 1964) is the usual method of determination.

UDP-galactose: N-*acetylglucosamine galactosyl transferase* (EC 2.4.1.38) The method of Fleischer et al. (1969) as modified by Bergeron et al. (1973a) contains in a 100 μl reaction mixture the following (final concentration) 0.3 M cacodylate buffer, pH 6.5, 0.36% Triton X-100, 30 mM $MnCl_2$, 30 mM mercaptoethanol, 1.5 mM UDP[^{14}C]galactose (approx. 2×10^6 cpm), 30 mM *N*-acetylglucosamine and 50–75 μg membrane protein. After 20 min incubation at 37 °C, the reaction is stopped by addition of 20 μl 0.3 M EDTA and cooling on ice. The difference in radioactivity between the experimental samples and the blank in which *N*-acetylglucosamine is omitted gives the amount of [^{14}C]galactose transferred to *N*-acetylglucosamine. Before separation and estimation of *N*-acetyllactosamine by paper chromatography, it is advisable to remove lipid-soluble radioactive compounds in the sample by partitioning in small volumes of chloroform/methanol (2 : 1, v/v) (see § 4.3.2). The upper phase is analysed by addition of a portion to DEAE paper (Whatman No. DE 81) and developing by descending chromatography for 48–72 hr in *n*-butanol/*n*-propanol/water (3 : 1 : 1, v/v). Radioactive lactosamine is located by scanning and estimated by counting 0.5 cm strips of paper in scintillant. Enzyme activity is increased when Triton X-100 is included in the reaction mixture, and this latency is probably due to its location at the inner surface of the Golgi cisternae and vesicles.

Sialyl transferase (CMP-*N*-acetylneuraminate : D-galactosyl-glycoprotein *N*-acetylneuraminyl transferase; EC 2.4.99.1) This activity is measured by following the transfer of CMP-[^{14}C]sialic acid to lactose to form sialyl lactose. The reaction is carried out in a volume of 0.05 ml containing (final concentration) 0.1 mM lactose, 0.6% Triton X-100, 50 mM phosphate buffer, pH 6.9, and cytidine-5′-monophospho-[^{14}C]sialic acid at approx. 1.2 mCi/mmol (New England Nuclear, Radiochemical Centre Amersham; non-radioactive substrate is not yet available commercially). The reaction products are separated by electrophoresis in 1% sodium borate buffer for 4 hr at 100 V/cm, and the distribution of radioactivity, relative to standards, determined by scintillation spectrometry (Carlson et al. 1973).

Other markers for Golgi membranes Fleischer and Zambrano (1974) describe a method for the determination of 3-phosphoadenosine 5′-phosphosulphate-cerebroside sulphotransferase. This enzyme catalyses the conversion of cerebroside to sulphatide and is enriched in Golgi membranes of kidney. Thiamine pyrophosphatase, although sometimes advocated as a Golgi marker enzyme, is fairly widely distributed in hepatic membranes (Cheetham et al. 1971).

Owing to the functional interactions between Golgi and the plasma membrane during secretion, it is likely that low levels of glycosyl transferase activities are measured in plasma membranes deriving from the vascular side of secretory cells.

4.2.4. *Mitochondrial markers*

The size and morphology of intact mitochondria ensures their easy detection in plasma membrane-enriched fractions, but quantitation of the level of contamination requires enzymic determination. In assessing fully the extent of contamination, it is important to select markers present in both inner and outer mitochondrial membranes, since, for example, when hypotonic media or gas cavitation have been used to prepare the fractions, the outer mitochondrial membrane may be stripped off and co-purify with the plasma membranes. Hydrostatic pressure imposed on mitochondria during centrifugation at high *g* forces can also cause damage (Wattiaux et al. 1971).

Succinic dehydrogenase (EC 1.3.99.1) and other inner mitochondrial membrane markers These activities present on inner mitochondrial membranes are determined by measuring the transfer of hydrogen to artificial acceptors. The reaction mixture used to measure succinate dehydrogenase described by Earl and Korner (1965), which is a modification of that of Green et al. (1955) contains in 3 ml the following (final concentration): 50 mM phosphate buffer, pH 7.6; 1 mM KCN; 0.04 mM 2,4-dichlorophenol-indophenol (DCPIP) and 20 mM sodium succinate. After addition of membranes, the reduction of DCPIP at 600 nm is followed; a change in absorbance of 1 unit is produced by the reduction of 0.0524 μmol of indophenol.

Other assay procedures use different artificial acceptors; for example *p*-iodonitroneotetrazolium violet (INT) (Pennington 1961) and phenazine methosulphate (Bernath and Singer 1962). Other marker enzymes, e.g., succinate cytochrome *c* reductase and the rotenone insensitive NADH-cytochrome *c* reductase (Fleischer and Fleischer 1967) are determined by more elaborate methods. Cytochrome oxidase (EC 1.9.3.1) may be determined as described by Sun and Crane (1969). Swick et al (1975) have reviewed mitochondrial enzyme distributions in liver, kidney, heart and ascites cells. The reduction of succinate, when coupled to INT (*p*-iodonitroneotetrazolium violet), results in the formation of a deposit that was exploited by Cotman and Taylor (1972) to make brain mitochondria heavier, thus achieving better separation from synaptosomal plasma membranes. Tissue homogenates may possess other enzymes that will reduce artificial substrates; a blank assay in which, for example, sodium succinate is replaced by sodium malonate which specifically inhibits succinate dehydrogenase corrects for this.

Monoamine oxidase (EC 1.4.3.4) and other outer mitochondrial membrane markers Monoamine oxidase is determined by following the conversion of benzylamine to benzaldehyde. In the method described by Schnaitman et al. (1967) a 2 ml reaction mixture contains (final concentration) 2.5 mM benzylamine, 50 mM phosphate buffer, pH 7.6, and membrane sample. The formation of 0.15 μmol benzaldehyde results in an increase in absorbance at 250 nm of 1 unit.

A further enzyme used as an outer mitochondrial membrane marker in rat liver is kynurenine-3-hydroxylase (EC 1.14.1.2). It may be determined spectrophotometrically by the method of Hayashi (1962) in which the disappearance of NADPH in the presence of L-kynurenine is measured at 340 nm. This enzyme, in contrast to monoamine oxidase, is reported absent from mammary gland mitochondria (Rosano et al. 1976) and hence a caution is raised once more regarding the lack of universality of membrane marker enzymes.

4.2.5. *Lysosomal markers*

A wide range of hydrolytic enzymes with pH optima in the acid range are present in the lysosomal fraction. Extensive analytical fractionation studies have indicated that there may be multiple forms of these enzymes and that they are not confined solely to the lysosomal fraction. For example, β-glucuronidase and aryl sulphatases A and B are distributed approximately two-thirds in the lysosomal fraction and one-third in the endoplasmic reticulum; neuraminidase (sialidase) and α- and β-D-galactosidase are located in lysosomes and plasma membranes and α-D-mannosidase is located in the lysosomes, soluble fraction and Golgi membranes (Touster 1974). Kaulen et al. (1970) have compared lysosomal and plasma membrane enzymes of rat liver and defined conditions for the specific determination of phosphatases.

Acid phosphatase (EC 3.1.3.2) The method described by Gianetto and De Duve (1955) is simple and measures the release of phosphate from (final concentration) 50 mM β-glycerophosphate in a 50 mM acetate buffer, pH 5.0. Approximately 10–15% of liver homogenate activity is located in the supernatant, and the enzyme displays latency properties. Thus, activity should be measured after 3 freezing and thawing cycles or in the presence of 0.1% Triton X-100.

β-Glucuronidase (EC 3.2.1.31) In the method of Gianetto and De Duve (1955) the reaction mixture (2 ml) contains (final concentration) 1 mM phenolphthalein glucuronide, 0.075 M acetate buffer, pH 5.2, and membranes. The reaction is stopped by adding 6 ml of a solution containing 0.133 M glycine, 0.067 M NaCl and 0.083 M Na_2CO_3, pH 10.7. After clarifying the mixture by low-speed centrifugation, released phenolphthalein is measured at 550 nm. *p*-Nitrophenol β-galactoside is another substrate that may be used, and the release of *p*-nitrophenol is determined at 400 nm.

4.2.6. Nuclear membranes (envelopes)

Although rat liver nuclear envelope fractions have been characterised morphologically and chemically (Kashnig and Kasper 1969; Monneron et al. 1972) there are few enzymes that have been identified as specific to the nuclear membrane. Sikstrom et al. (1976) showed, using histochemical and biochemical methods, that a rat liver nuclear envelope fraction contained high levels of glucose 6-phosphatase and 5′-nucleotidase. Microsomal markers, cytochrome b_5 and *P*-450 and NADPH and NADH-cytochrome *c* reductase activities were present at lower levels than found in the microsomal fraction. Rat liver nuclear envelopes are resolved into two subfractions on sucrose gradients (densities 1.16–1.18 and 1.18–1.20 g/cm^3), and the similarities in densities to plasma membrane 'heavy' fractions can result in cross-contamination. Some distinguishing non-enzymic features of the nuclear envelope fractions are a high RNA content (see § 4.3.4) and the nuclear pore complexes (see § 4.5.2). When chicken erythrocyte and liver nuclear membranes were prepared and their enzymic and polypeptide properties compared with those of the corresponding plasma membranes, extensive differences were found (Jackson 1976).

4.2.7. *Cytoplasmic marker enzymes*

Catalase (EC 1.11.1.6) This activity is present in peroxisomes. The enzyme has a broad pH optimum (pH 4–8.5) and is measured as described by Baudhuin et al. (1964). Catalase activity displays considerable latency and freeze-thawing or detergent treatment is required to obtain maximal activity.

Uric acid oxidase (urate-O_2-oxidoreductase, EC 1.7.3.3) is also a 'microbody' marker enzyme and is determined by the method of Henry et al. (1957).

Lactic dehydrogenase (EC 1.1.1.22) is a useful marker for the 'soluble' compartment of cells and is measured by the method of Neilands (1955). Kits for determination of activity are also available commercially.

4.3. *Chemical markers*

Chemical markers are frequently determined when characterising mammalian plasma membranes. The two major positive markers are sialic acid and cholesterol. The latter is usually expressed relative to phospholipid content. Nucleic acids constitute useful negative markers used to determine the absence of nuclear and rough endoplasmic reticular membranes.

4.3.1. *Sialic acids*

The highest concentration of sialic acids is found on the plasma membrane, where they contribute to the overall negative charge of cells. Membrane-bound sialic acids are usually determined by two basically similar methods described by Warren (1959) and Aminoff (1961). Both methods measure free sialic acids released after hydrolysis of membranes (approx. 200–600 μg plasma membrane protein) in

0.5 ml 50 mM H_2SO_4 at 80 °C for 60 min. However, it is stressed that these conditions represent a balance between the release and destruction of sialic acids. Aminoff (1961) showed that the rate of hydrolysis of sialic acids from a range of sialoglycoproteins varied according to temperature and time and according to the source of the sialic acid. The above conditions of hydrolysis are often applied because the hydrolysate is then at the correct pH and volume (0.5 ml) for subsequent dilution with 0.25 ml of 25 mM sodium periodate dissolved in 0.063 N H_2SO_4, after which it is incubated for 30 min at 37 °C. Excess periodate is reduced by addition of 0.2 ml of 2% sodium arsenite in 0.5 M HCl. After the yellow colour of liberated iodine has disappeared (agitate the tube), 2 ml of 0.1 M 2-thiobarbituric acid, pH 9.0, are added and the sample is heated at 100 °C for 7.5 min. The cooled samples are then extracted with 4 ml acid butanol (*n*-butanol containing 5%, w/v, 12 M HCl), and extinction at 532 nm and 549 nm of the extracts are measured. A correction factor is used to calculate the concentration of sialic acid in the sample (Warren 1963):

$$\mu\text{mol sialic acid} = (0.084 \times E_{549}) - (0.031 \times E_{532}).$$

Since membranes are frequently collected from sucrose gradients and are stored in iso-osmotic sucrose, it is important to ensure that the sucrose is removed before measuring sialic acids. This is done either by prolonged dialysis against distilled water or by at least three washings by centrifugation of pellets resuspended by small, tight-fitting homogenisers (see § 2.1.3).

The sialic acid content of mammalian plasma membranes is usually 20–100 μmol/g membrane protein; however a proportion may be attached to lipid (see below). Sialic acid is unevenly distributed on the surfaces of cells exhibiting functional polarity, e.g., hepatocytes, intestinal epithelial cells.

The distribution of sialic acid-containing glycoproteins on SDS-containing polyacrylamide gels was determined by Bretscher (1971) who adapted the Warren (1959) assay to accommodate smaller volumes. A gel slice (approx. 0.1 ml volume) is soaked in 1 ml of

0.05 N H_2SO_4 for 2 hr, heated at 75 °C for 1 hr and shaken overnight at room temperature. The supernatant is then concentrated to 0.2 ml in vacuo and sialic acid determined, as described above using reduced volumes of reagents. For the detection of bound sialic acid, a periodate-resorcinol method, less sensitive than the thiobarbituric acid method, has been described (Jourdian et al. 1971).

Gangliosides Sialic acids attached to lipids are referred to as gangliosides. In thin-layer chromatography systems used to resolve neutral and phospholipids, gangliosides remain near the origin, and more aqueous solvents, e.g., chloroform/methanol/4 M NH_3 (9 : 7 : 2, v/v), can be used to resolve them into mono-, di- and tri-sialogangliosides (Kates 1972). Gangliosides are soluble in chloroform/methanol (2 : 1, v/v) extracts of membranes (proportion of 1 g membrane/ 19 ml) and are then extracted into an aqueous medium by partitioning the organic extract against one-fifth volume of 0.1 M KCl (Folch et al. 1957). After agitating and centrifuging the mixture, the upper phase (aqueous methanol) is collected. The lower $CHCl_3$-rich phase is extracted twice with approx. two to three-fifth volume of $CHCl_3/CH_3OH$/0.1 M KCl (3 : 48 : 47 v/v) (Long and Staples 1959). The combined upper phases are hydrolysed as described above to release sialic acids. Renkonen et al. (1970) showed that plasma membranes of BHK (Baby Hamster Kidney) cells contained 8–13 μg lipid-bound sialic acid/mg protein and that the gangliosides were co-purified in subcellular fractions with (Na^+-K^+)-activated adenosinetriphosphatase and a plasma membrane antigen.

4.3.2. *Phospholipids*

Table 4.2 indicates that there are relatively small differences in the phospholipid composition of subcellular membranes and organelles. Cardiolipin appears to be confined solely to mitochondria and thus serves as an useful indicator of the extent of mitochondrial contamination of plasma membrane fractions. Plasma membranes contain the highest amount of sphingomyelin (between 10–30% of the total phospholipid phosphorus). The presence of high amounts (>10%)

TABLE 4.2

Phospholipid composition (moles %) of some mammalian membranes.

Phospholipid*	Rat liver								L-cells[d] plasma membrane	Milk fat[e] globule plasma membrane
	Nuclear[a] envelope	Rough endoplasmic[b] reticulum	Mitochondria[c]		Golgi[b] apparatus	Plasma[f] membrane				
			inner membrane	outer membrane		*a*	*b*	*c*		
PC	54	61	38	45	45	47	32	33	31	33
PE	22	19	38	30	17	15	20	14	11	25
SP	5	4	1	4	12	13	16	25	24	24
PI	8	9	2	8	9	10	9	7	5	12
PS	7	3	–	–	4	4	8	8	4	4
Others	4	5	1	2	12	8	12	9	13	2
CL	–	–	20	10	–	–	–	–	–	–

Phospholipids are separated by thin-layer chromatography in two dimensions.

* PC, phosphatidyl choline; PE, phosphatidyl ethanolamine; SP, sphingomyelin; PI, phosphoinositides; PS, phosphatidyl serine; Others, predominantly lysophosphatides and phosphatidic acid; CL, cardiolipin.

a, *b*, *c*, plasma membranes from sinusoidal, contiguous and canalicular regions, respectively.

[a] Klenig (1970); [b] Keenan and Morré (1970); [c] Colbeau et al. (1971); [d] Weinstein et al. (1969); [e] Patton and Keenan (1975); [f] Kremmer et al. (1976).

of lysophospholipids suggest lipase action during preparation, e.g., in guinea-pig pancreas (Meldolesi et al. 1971; § 5.7.3).

To determine total phospholipid content of membranes, samples are extracted 3 times with chloroform/methanol (2 : 1, v/v) at room temperature over 24 hr. The use of small grinding homogenisers is helpful and the organic extract can be separated from insoluble residues by centrifugation or filtration. Portions of the organic extract are then dried, hydrolysed in 70% (w/v) $HClO_4$ and phosphate is determined by the method of Bartlett (1959). There are many variations for extracting and partitioning phospholipids (see Kates 1972) and the method of Bligh and Dyer (1959) is useful for small amounts of lyophilised membranes. Dried membranes are wetted in 80 μl of H_2O, and extracted with 300 μl of chloroform/methanol (1 : 2, v/v). After addition of 100 μl chloroform to adjust the chloroform/methanol ratio to 1 : 1, the samples are centrifuged and the supernatants removed. The extraction procedure is repeated 3–4 times and the combined extracts are dried for phospholipid phosphorus or cholesterol analysis. Phospholipid content is calculated by assuming 25 μg of phospholipid/μg lipid phosphorus (Kopaczyk et al. 1966).

4.3.3. Cholesterol

Cholesterol is present at its highest concentration, relative to membrane protein or phospholipid, in plasma membranes. Most (>80%) of the cholesterol is unesterified. Controversy surrounds the idea that cholesterol is mainly confined to plasma membranes; nevertheless, the use of digitonin to perturb specifically the density of plasma membrane components of microsomal fractions by interaction with cholesterol has been described (§ 3.1.9).

Cholesterol content of membranes is usually measured as total cholesterol by colorimetric methods. The most reliable estimations are obtained when the cholesterol is first extracted from membranes. Membrane samples (1–5 mg) are first saponified at 100 °C for 30 min in 15% aqueous KOH. Cholesterol is then extracted in three successive steps with petroleum ether (b.p. 40–60 °C), the extracts are dehydrated by addition of anhydrous Na_2SO_4, filtered and evaporated to dryness. The dry residues are dissolved in acetone, and an alcoholic solution of digitonin is added. After leaving for 12–20 hr, the insoluble cholesterol-digitonin complex is collected by centrifugation and washed successively 3 times with ethanol/acetone/H_2O (3 : 2 : 1, v/v), 3 times with acetone/ethyl ether (1 : 1, v/v) and finally 3 times with ethyl ether. The dry residue is dissolved in 1 ml of warm glacial acetic acid and 1.5 ml of the Liebermann-Burchard reagent (acetic acid/H_2SO_4, 20 : 1, v/v) is added. After 30 min at 25 °C in the dark, extinction at 640 nm is measured. A calibration curve is made by using a solution of cholesterol dissolved in acetone/ethanol (1 : 1, v/v).

Plasma membranes are characterised by cholesterol/phospholipid molar ratios that vary between 0.5 to 1.1. Values for intracellular membranes vary from 0.3 for Golgi membranes, to less than 0.1 for mitochondrial fractions. Cells with functional polarity appear to have an uneven distribution of cholesterol in plasma membranes, e.g., liver bile canalicular plasma membranes contain the highest cholesterol content (Kremmer et al. 1976). This fact, coupled with the different methods used to estimate cholesterol, may explain the wide variation in the reported cholesterol contents of mammalian plasma membranes.

A number of colorimetric methods, using ferric salts in concentrated acids, have been described for determining mainly serum cholesterol levels. Cholesterol can also be measured by gas liquid chromatography.

4.3.4. Ribonucleic acid

The RNA content of membranes is measured by the Schmidt–Thannhäuser method as described by Fleck and Munro (1962). Depending on the amount of ribosomal contamination (this may be estimated by electron microscopy), membranes containing about 5 mg of protein are precipitated by 2 ml of cold 0.3 M perchloric acid. After 15 min, the solution is centrifuged and the pellets washed by centrifugation in 2 ml, 0.3 M perchloric acid and then digested in 1 ml of 0.3 M KOH at 37 °C. After 1 hr, 1 ml of 0.6 M perchloric acid is added and the precipitate removed by centrifugation. The extinction of the supernatant at E_{260} and E_{275} is determined and the concentration of RNA calculated using the following formula:

$$\mu g\ \text{RNA/ml} = (125 \times E_{260}) - (109 \times E_{275}).$$

The final precipitate is kept and analysed for DNA.

4.3.5. Deoxyribonucleic acid

The method of Giles and Myers (1965) is generally used. To the final precipitate obtained in the RNA determination (§ 4.3.4) is added 1.5 ml 0.5 M perchloric acid and the solution is heated at 70 °C for 30 min. The remaining insoluble material is removed by centrifugation and 1 ml of the supernatant is added to 3 ml of Burton's reagent (1.5 g diphenylamine dissolved in 100 ml glacial acetic acid containing 0.5 ml of 16 mg/ml acetaldehyde) and the sample incubated at 30 °C for 16 hr. The E_{600} is then determined and the DNA content calculated by reference to a standard curve constructed for 0–80 μg DNA.

4.3.6. Estimation of membrane protein

Membrane protein is usually measured by the Lowry et al. (1951) technique, with crystalline bovine serum albumin, fraction IV (care must be taken to keep it dehydrated) as standard.

More sensitive fluorometric techniques are available. The method of Hiraoka and Glick (1963) was modified by Peters et al. (1972). The subcellular fraction (0.25 ml) is mixed with 1 ml of 0.1 M sodium citrate buffer, pH 3.1. 1 ml of 33 μM eosin Y dissolved in 0.05 M sodium citrate buffer, pH 3.1, is added, and fluorescence determined (excitation wavelength, 519 nm; emission wavelength, 540 nm). A standard curve constructed with 1–100 μg/ml bovine serum albumin is used, and quenching of eosin Y is closely proportional to protein concentration. This technique is claimed to be 5 times more sensitive than the Lowry procedure, and is not influenced by the presence of up to 60% w/w sucrose. However, the protein concentration is 20% lower than that obtained by the Lowry technique.

Subject index p. 261

4.3.7. Estimation of phosphate

Phosphate, released by the action of phosphohydrolases on organic substrates is determined spectrophotometrically. The methods used are based on the original method of Bell and Doisy (1920) and depend on the development of a blue colour of a phosphomolybdate complex in the presence of reducing agents, such as hydroquinone, 1,2,4-aminonaphthol-sulphonic acid (Fiske and Subbarow 1925; King 1932; Barlett 1959), ascorbic acid (Lowry and Lopez 1946; Chen et al. 1956), or stannous chloride (Martin and Doty 1949). The method of Bartlett (1959) by heating at 100 °C, increases the sensitivity of the Fiske and Subbarow (1925) method from $3.6 \times 10^3\ M^{-1} \times cm^{-1}$ to $2.6 \times 10^4\ M^{-1} \times cm^{-1}$ at 660 nm. However, although this increased sensitivity is useful in determining total phosphorus, e.g., in phospholipid phosphorus determinations (see § 4.3.2), it can result in hydrolysis of organic phosphate esters. Thus, methods of Fiske and Subbarow (1925) and Martin and Doty (1949) feature mainly in the measurement of enzymically released phosphate. In the presence of detergents, especially 0.1% Triton X-100, the analytical procedures may prove unreliable (Wattiaux and De Duve 1956; Snoke and Nordlie 1956). However, the increased turbidity produced by Triton X-100 is proportional to the concentration of the phosphomolybdate complex, and this forms the basis of a further method for determining phosphate (Eibl and Lands 1969; Carson 1976). Inorganic phosphate may also be measured by using the Merckotest Kit No. 3331 (Merck).

4.3.8. Analysis of membranes by polyacrylamide gel electrophoresis

Electrophoresis in polyacrylamide gel has proved an useful adjunct to other techniques in the characterisation of plasma membrane fractions. Using this technique, Schnaitman (1969) showed that the membranes of various subcellular fractions of rat liver contained different and characteristic polypeptide profiles. Glossmann and Neville (1971) made a thorough comparison of the plasma membranes of different rat tissues and came to the same conclusion. In comparing the authentic polypeptides of isolated membrane fractions (i.e., after removal of luminal and/or absorbed soluble proteins, see § 3.7), electrophoresis carried out in buffers containing sodium dodecyl sulphate (SDS) has completely superceded systems using urea-acetic acid as a membrane-solubilising agent, and only this system will be described.

First, it is necessary to choose between a variety of electrophoretic systems using various types of apparatus, buffer systems and gel compositions. In the present account, systems that give the best resolution of membrane polypeptides are described, and more comprehensive accounts of the theory and practice of the technique are found in Gordon (1975), Maizel (1971) and Fish (1975).

4.3.8.1. Solubilisation of membranes In general, approximately 50–200 μg of membrane proteins are required to be resolved to enable dyes to stain 1–20 μg of the separated polypeptides. For visualisation of glycoproteins by Schiff-periodate staining, the amount of protein applied to the gel should be increased by a factor of 2–3.

Membranes are solubilised by addition to the smallest possible volume of the electrophoretic buffer containing a final concentration of 1–2% SDS, 3% mercaptoethanol, 0.001% bromophenol blue (tracking dye) and 10% glycerol or sucrose. When membranes have aggregated or are in pellet form, their dispersion and solubilisation is aided by use of small (0.1 ml) homogenisers (§ 2.1.3). For further protection of free sulphydryl groups (produced by inclusion of 1 mM dithiothreitol in the solubilising solution), they can be alkylated by reaction with 100 mM iodoacetic acid or iodoacetamide (Pitt–Rivers and Impiombato 1968; Williams and Gratzer 1971). Reducing agents may be omitted to establish the profiles of unreduced proteins; however this may result in a larger proportion of the sample applied failing to enter the gel and the apparent molecular weights of the components are less reliable (Tung and Knight 1972). In some cases, 1 M urea may be included in addition to SDS in the sample-solubilising medium to encourage recalcitrant proteins to dissolve more completely and to enter the gel. To accelerate the solubilisation of membranes, the samples are heated at 100 °C for

3–5 min to obtain clear solutions. With membranes, a further important reason for heating in SDS is to inactivate proteolytic enzymes which can maintain activity in SDS solutions at high temperature. Peptidases retaining activity after this avidly hydrolyse susceptible peptide bonds of membrane proteins unfolded by interaction with SDS during micelle formation. An example of low molecular weight artifacts induced in membrane characterisation resulting from a failure to denature proteolytic enzymes is provided by 'mini-proteins' described by Laico et al. (1970) and discussed in retrospect by Dreyer et al. (1972). When possible, it is advisable to dissolve freshly prepared membranes immediately in SDS; membranes stored frozen may not solubilise completely, as indicated by the proportion of material failing to enter the resolving gel. Attempts to remove membrane lipids prior to solubilisation (e.g., by extraction with chloroform/methanol, ether, etc.) can make proteins recalcitrant to full solubilisation in SDS solutions. Lopez and Siekevitz (1973) have reported a procedure for membrane delipidisation that subsequently yielded good resolution of polypeptides in gels. With lipid-rich membranes, e.g., myelin, partial delipidation with acetone or ether/ethanol mixtures may prove useful (Agrawal et al. 1972; Greenfield et al. 1971), for although lipids migrate ahead of proteins in SDS gels, the presence of excessive amounts of lipids can distort the separation patterns of proteins.

4.3.8.2. Operation of various gel systems Phosphate buffers, pH 7.4, were used originally to effect separation of membrane proteins (see Weber and Osborn 1969), but the best separations are obtained by using the discontinuous buffer systems and their variants described by Laemmli (1970), Maizel (1971). These are constructed of a high porosity 4.5% acrylamide 'stacking' gel (1–2 cm) made in 0.125 M Tris–HCl buffer, pH 6.7, overlaying a separating gel (10–20 cm long) made in 0.37 M Tris–HCl, pH 8.9. The composition of various porosities of polyacrylamide gels is given in Table 4.3. The acrylamide concentration is varied according to the polypeptides to be resolved. For optimal separation of polypeptides of 35,000–300,000

TABLE 4.3

Composition of polyacrylamide gels used in the separation of membrane proteins.

Components (ml)	Acrylamide (%, w/v)				
	Stacking gel	Resolving gel			
	4.5	7.5	10	12.5	15
Total volume	10	30	30	30	30
Acrylamide/bis[a,d] (30 : 0.8, w/w)	1.5	7.5	10	12.5	15
10% SDS	0.1	0.3	0.3	0.3	0.3
3 M Tris–HCl, pH 8.9	–	3.75	3.75	3.75	3.75
1 M Tris–HCl, pH 6.7	1.25	–	–	–	–
Glass distilled water	7.1	18.3	15.8	13.3	10.8
Temed[b]	0.007	0.015	0.015	0.015	0.015
Ammonium persulphate (10%)[c,d]	0.06	0.15	0.15	0.15	0.15

[a] *N,N*-bismethylene acrylamide.

[b] *N,N,N′,N′*-tetramethylethylene-diamine.

[c] 2% solution used for constructing gradient gels.

[d] Solutions to be made up on day of use.

Electrode buffer concentration is 0.05 M Tris–HCl, 0.3 M glycine, 0.1% SDS, pH 8.3.

daltons, 7.5% acrylamide is suggested; for 20,000–250,000 daltons, 10% acrylamide; for 20,000–50,000 daltons, 12.5% acrylamide; for 7000–40,000, 15% acrylamide.

Two types of gel systems are currently in use; cylindrical tube gels and flat slabs of various dimensions and gel thickness. In general, slab gels are gaining in popularity for the following reasons: (a) several samples can be run side by side under identical electrophoretic conditions. Hence, determination of the R_f and apparent molecular weight of components can be made directly from markers. In tube gels, the relative electrophoretic mobility for a polypeptide needs to be determined as follows (Weber and Osborn 1969):

$$R_f = \frac{\text{length of gel before staining}}{\text{length of gel after staining, destaining}} \times \frac{\text{distance of stained band from origin}}{\text{distance of tracking dye from origin}};$$

(b) there is better heat dissipation. Gel thickness may be varied from 1 mm (useful for small amounts of protein and autoradiography) to 3–4 mm when semi-preparative separation is possible; (c) densitometry, photography and gel drying are relatively easier; (d) slab gels are especially useful when large numbers of samples are to be analysed, but care must be exercised to ensure that no cross-contamination of adjacent sample slots occurs; (e) two-dimensional separations are possible (see § 4.3.5.6).

Two-dimensional gel electrophoresis This technique is being used increasingly to analyse membrane proteins and glycoproteins. Proteins are dissolved in SDS and electrophoresed in polyacrylamide gels containing SDS in the first direction and in the second dimension in detergent-free gel. When this technique was applied to erythrocyte membranes, band III, a major glycoprotein was resolved into five components (Anselstetter and Horstmann 1975). However, in general, a complete resolution of membrane proteins is frequently limited by inadequate solubilisation in the second dimension. Conrad and Penniston (1976) have reported a method in which resolution in the first dimension in SDS-containing buffers is followed by a second dimension run in phenol/8 M aqueous urea/acetic acid (50/25/25 w/v/v). A further variant using chloral hydrate in the first dimension and dodecyl sulphates of Tris and 4-picoline in the second dimension has been described (Booth 1977). Further developments of this technique should allow resolutions complementing those described by O'Farrell (1975) that use isoelectric focussing gels containing 8 M urea/Nonidex P-40 in one direction followed by conventional electrophoresis in SDS-containing gels in the second direction.

Acrylamide is available in a form specially purified for electrophoresis; however, further purification by recrystallisation may be carried out (Maizel 1971). Unpolymerised acrylamide is carcinogenic and neurotoxic, and gloves should be worn when handling. Acrylamide solutions should be prepared on the day of use as a stock solution of 30 g acrylamide and 0.8 g *N*,*N*-bismethylene acrylamide dissolved in 100 ml distilled water (see Table 4.3). Temed is used undiluted. Ammonium persulphate should be made up on the day of use.

Apparatus for tube or slab polyacrylamide gel electrophoresis are now available from a number of manufacturers. Less sophisticated, but equally effective apparatus can be constructed from Perspex in laboratory workshops (Fig. 4.4). Electrophoresis may be carried out at a low current for 10–16 hr (overnight). The usual currents used for tube gels are approx. 4 mA/tube 10 mm long, 4 mm diameter, and for slab gels 10–20 mA/gel 12–20 cm long, 1 mm thick. The electrophoresis time will vary according to the amount of polyacrylamide gel, its concentration and the buffer concentration. Electrophoresis at higher currents can result in localised heating in the gel, and this may result in the stained bands appearing 'moon'-shaped. To speed up the electrophoresis in slab gels for short periods, e.g., near the end of a separation, the current may be increased and a blast of cold air from a hair-dryer may be directed at the gel.

Discontinuous gels (as described in Table 4.3) use a spacer gel to concentrate the sample and the discontinuous pH aids in band sharpening, an important consideration when analysing the complex polypeptide composition of mammalian cell membranes. Further resolution of complex membrane systems may be obtained by using gradient polyacrylamide gel systems. The superior resolution of a Tris-glycine buffered gradient gel system over a phosphate buffered uniform polyacrylamide system was demonstrated by Hubbard and Cohn (1975) and an example of the resolution of complex membranes is shown in Fig. 4.5. Continuous gradient gels may be constructed by mixing two concentrations of acrylamide as shown in Fig. 4.4. To ensure gradient stability, sucrose or glycerol (15%, v/v) may be added to the stronger acrylamide. Gradient gels need to be formed quickly to avoid polymerisation in the narrow tubes leading from

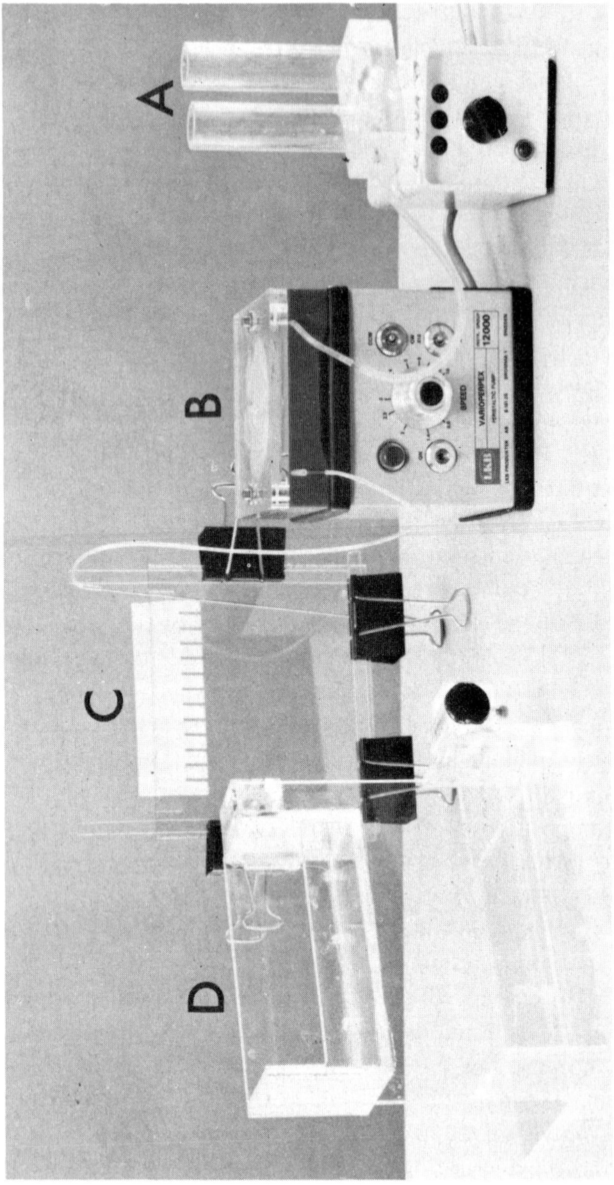

Fig. 4.4. Construction of a gradient polyacrylamide slab gel. Two concentrations of acrylamide (containing 3 and 12% sucrose) are mixed magnetically in gradient maker A (a similar one is used to make sucrose gradients) and the gradient is pumped (B) carefully between the glass plates C where the acrylamide polymerises. The long needle is held immediately above the rising level of the acrylamide, so minimising 'wavyness'. The gel slab is then attached to the gel electrophoresis apparatus D. For further details, see text.

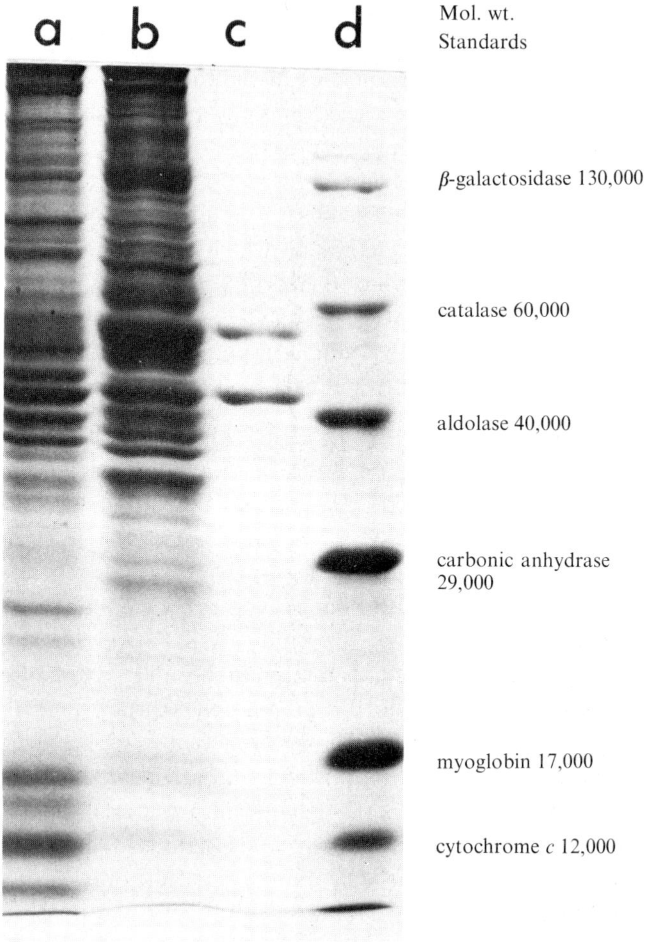

Fig. 4.5. Polyacrylamide gradient gel electrophoresis of plasma membranes. A 10–15% polyacrylamide gradient was run in the discontinuous buffer system of Laemmli (1970) in the apparatus shown in Fig. 4.4. Samples of plasma membranes (≈100 μg) are dissolved in 2% SDS, 5% mercaptoethanol, 10% glycerol, 40 mM Tris–HCl, pH 6.7. Liver plasma membranes track (a) (Evans 1970), prepared as described in § 5.1, and synaptosomal plasma membranes track (b) from rat forebrains were prepared according to Jones and Matus (1974) (§ 5.2.1). Track (c) contained tubulin and actin. Molecular weight markers are in track (d).

the mixing chambers. Speed of polymerisation is reduced by lowering the ammonium persulphate concentration to one-fifth. Two methods for making gel gradients are in use. In the first, the acrylamide of higher concentration is introduced into the polymerising chamber first, followed by acrylamide of continuously decreasing concentration (Fig. 4.4). This method of construction can lead to 'waves' forming on flat-bed gels. Problems of obtaining uniform gradients are overcome in a second method in which the solution of lower acrylamide concentration is introduced first and this is 'pushed' upwards by acrylamide solutions of gradually increasing concentration. Preformed gels containing varying acrylamide concentrations are available commercially (e.g., from Pharmacia and Gradient Pty. Ltd.).

4.3.8.3. Choice of detergents Although SDS is the most widely used detergent to solubilise and separate membrane proteins by electrophoresis in polyacrylamide gels, other detergents have been investigated and found, in general, to be inferior. However, under certain conditions, e.g., to investigate acyl phosphate bonds stable only at acid pH, the use of other detergents is warranted. Fairbanks and Avruch (1972) compared the utility of SDS and the cationic detergent CTAB (cetyl trimethyl ammonium bromide; available from Eastman Kodak, Cat. No. 5650) to separate erythrocyte membrane proteins at pH 8.3 and 2.4, and reported that CTAB gave satisfactory resolution at pH 2.4, but poor resolution at the alkaline pH. Marjanen and Ryrie (1974) working with mitochondrial membranes reported that CTAB produced fewer bands than SDS and suggested that this was due to the exclusion from the gel of the protein subunits of greatest hydrophobicity. Urea and neutral detergent mixtures are used in isoelectric focussing gels.

4.3.8.4. Post-electrophoretic staining of constituents After electrophoresis, gels are immediately stained for protein by immersing in 0.2% Coomassie Blue in methanol/acetic acid/water (8 : 1 : 11, v/v); other dyes used include Amido Black and Fast Green FCF. Thin gels (1 mm) stain in about 15 min, whereas up to 3 hr are recommended for thicker flat gels and cylindrical gels. Destaining may be speeded up by warming the destaining solution (i.e., the solution the dye is dissolved in); increasing the methanol concentration also speeds up destaining, but gel shrinkage occurs. The insertion of pieces of felt sheets or porous rubber into the destaining solution also speeds up the process by absorbing excess dye from the gels thus minimising the number of changes of destainer necessary. The use of electrophoretic destaining methods is not recommended for gels containing SDS. The estimation of the relative amounts of polypeptides in membranes on the basis of staining intensity is not accurate. For example, in examining the subcellular distribution of actin in renal medulla, Phillips and Slater (1975) showed that actin binds less dye than serum albumin, and Bray and Thomas (1975) estimated that muscle actin was stained by Coomassie Blue at 67% the intensity of bovine serum albumin. Coomassie Blue staining of protein deviates from Beer's law at high protein concentrations (Chrambach et al. 1967) and Amido Black may stain metachromatically (Gorovsky et al. 1970).

After the background has fully destained, gels may be photographed and/or traced in spectrophotometers equipped with gel-tracing attachments. Coomassie Blue stained gels are traced at 595 nm.

'Stains-all' Differential staining of phosphoproteins is claimed for the cationic carbocyanine dye 'Stains-all' [1-ethyl-2-[3-(1-ethyl-naphthol[1,2*d*]thiazolin-2-ylidene)-2-methylpropenyl]-naphthol-[1,2*d*-thiazolium bromide] (Kodak Organic chemical No. 2718, British Drug Houses 442272F) by Green et al. (1973). Phosphoproteins stain blue, whereas a red-coloration is obtained with non-phosphorylated proteins. King and Morrison (1976) using 'Stains-all' show that, additionally, erythrocyte membrane sialoglycoproteins stained blue, the proteins red and the lipids yellow orange. The gels are washed extensively in isopropanol to remove SDS (see below) and placed in staining solution overnight in the dark (Stains-all (5 ml of a 0.1% stock solution) is dissolved in 20 ml formamide, 100 ml isopropanol and 275 ml of 15 mM Tris adjusted to pH 8.5). Gels are destained in distilled water.

Subject index p. 261

Location of carbohydrates Since glycoproteins are more distinctive of plasma membranes than proteins, identification of their profiles can provide important characteristic features. Two methods are now described.

Schiff-periodate staining Procedures for staining gels for carbohydrates by the Schiff-periodate procedure have been described by Zacharias et al. (1969) and Glossman and Neville (1971). As for protein staining, it is first necessary to immobilise the bands in the gel, and this is done by immersing the gels in 5% trichloroacetic acid or acid methanol for 30 min with frequent changes of liquid; this, together with subsequent H_2O-washes, also serves to help remove excess SDS. Gels are then immersed in 2.5% (v/v) periodic acid dissolved in 3% (w/v) acetic acid, for 1 hr. Excess periodic acid is removed by extensive washing in distilled water until the addition of a drop of concentrated $AgNO_3$ to the eluate shows the absence of I^-. Vicinal OH groups in sugars are oxidised to the aldehydes and these are visualised by covalent reaction with Schiff's reagent (available from various chemical manufactures) for 40 min in the dark. Excess Schiff's reagent is removed by washing repeatedly in 1% sodium metabisulphite. Gels may be densitometrically traced at 560 nm. Control gels may be subjected to the same procedure, but omitting the periodate step. The Schiff-periodate procedure is prone to positive artifacts, since proteins can also interact electrostatically with the Schiff base. An additional control to identify artifactual staining of non-glycosylated protein is to show that bovine serum albumin or trypsin is not stained under the conditions adopted. A modification of this method involves coupling the aldehyde group to dansyl hydrazine (Eckhardt et al. 1976).

Lectin-affinity binding Techniques based on the specific binding of labelled lectins to glycoproteins in gels have been described (Gurd and Evans 1976; Tanner and Anstee 1976). Following electrophoresis, gel strips (gels may be sliced longitudinally with a sharp blade or with a wire cutter and each slice treated separately) are washed successively at 4 °C with 25% isopropanol/10% acetic acid (one change, 8 hr) and 10% isopropanol/10% acetic acid/5 mM Tris–HCl, pH 7.6 (4 changes, 4 hr) to remove SDS. The gel is then incubated for 60 min at 24 °C in 7.5 ml 5 mM Tris–HCl, pH 7.6, containing 4 μg ^{125}I-labelled Concanavalin A (see § 4.6.3.2 for comments on iodination of lectins). After labelling, gels are washed with 3 changes in 5 mM Tris–HCl, pH 7.6 (250 ml per gel per wash) at 4 °C over a 3-hr period. They may then be stained with Coomassie Blue. After cutting into equal pieces (1–2 mm) the ^{125}I-radioactivity distribution is determined, either by using a γ-counter or by spectrometry by placing each slice in 10 ml of scintillation mixture containing 0.16 g *p*-bis(*O*-methylstyryl)benzene, 7.84 g 2,5-diphenyloxazole and 30 ml of Protosol (New England Nuclear) in 1 l of toluene and leaving it to swell overnight at 37 °C before counting. A control gel is incubated with ^{125}I-labelled Concanavalin A in the presence of 2.5% α-methyl-D-glucopyranoside, and no peaks of radioactivity should be observed in gel slices. This procedure is effective with several other lectins, e.g., wheat germ agglutinin, *Ricinus comminus* agglutinin and *Lens culinaris* agglutinin thus allowing fine discrimination between classes of carbohydrate moieties that may be present as receptors and enzymes on plasma membranes. Using immunoglobulins conjugated to horse-radish peroxidase, this technique detects antigens in SDS-polyacrylamide gels (Olden and Yamada 1977). Radioactivity may also be determined by autoradiography (§ 4.3.8.5).

Location of enzymes Most enzymes are denatured in SDS solutions (especially after heating at 90 °C). However, Dulaney and Touster (1970), Nelson (1971) and Mather and Keenan (1974) have shown that some classes of enzymes can be detected in SDS-polyacrylamide gels. Alkaline phosphatase and alkaline phosphodiesterase activities are detected as follows. It is imperative that the membranes suspended in SDS are not boiled and that no disulphide reducing agents are present. Gel strips are dipped into a test tube at 37 °C containing suitable substrates; (a) alkaline phosphatase staining: 1 mM *p*-nitrophenyl phosphate, 0.5 mM $MgCl_2$-Tris buffer, pH 10.0; a yellow band (scan at 400 nm) becomes visible at the position in the gel where the enzyme is located; (b) alkaline phosphodiesterase: as above, but the substrate is 5 mM thymidine *p*-nitrophenylphosphate, pH 8.5.

4.3.8.5. Autoradiography of gels – direct or by fluorography Gels, usually thin slab gels, can be subjected to autoradiography directly, or following staining. Gels are first dehydrated by placing flat on drying plates in a rectangular box that is evacuated using a water or vacuum pump. The drying surface of the evacuated box is covered by a porous polypropylene sheet and the flat gel is positioned on this and covered with cellophane that is sealed around the box. After drying the gel has become attached to the cellophane. A number of closely similar methods for dehydrating gels are in use, and these are described in greater detail by Gordon (1975) and Maizel (1971). Commercially available gradient gels that contain a high bis concentration (see Table 4.3) have a tendancy to crack on dehydration. Membrane polypeptides labelled with ^{125}I or ^{131}I are usually visualised by autoradiography. As an alternative to autoradiography, intact gels can be sliced into equal portions and peak heights quantitated, but this is less sensitive and may not discriminate as well as between closely migrating peaks. Equally spaced blades or more complex apparatus (e.g., Gilson) are available that partition slices directly into scintillation vials for determination of radioactivity.

^{3}H- or ^{14}C-labelled membrane proteins may also be detected in situ after gel dehydration by soaking in a solution of 2,5-diphenyloxazole in dimethylsulphoxide, drying and exposing to X-ray film (Bonner and Laskey 1974; Laskey and Mills 1975). β-Particles from ^{3}H interact with the 2,5-diphenyloxazole emitting light which causes blackening of X-ray films. The practical details are as follows. After electrophoresis and staining, gels are soaked in 2 changes of 20 volumes of dimethylsulphoxide for 30 min each time. The gel is then immersed in 4 volumes of 20% (w/w) PPO (2,5-diphenyloxazole) in dimethylsulphoxide (22.2%, w/v) for 3 hr and then washed in 20 volumes H_2O before being dried under vacuum. Royal X-Omat or equivalent medical X-ray film (Kodak; Ilford) briefly pre-exposed to light is placed in contact with the dried gel in the dark and exposed at −70 °C. Using this method, 3000 disintegrations ^{3}H/min in a band 1 × 0.1 cm is detected in 24 hr at −70 °C. Detection of ^{14}C radioactivity is approximately 10 times more sensitive. This method thus supplements other methods in which gel slices are solubilised in H_2O_2 or proprietary gel solubilisers (New England Nuclear; Amersham Searle) followed by scintillation counting.

4.3.8.6. Molecular weights of membrane proteins Gel electrophoresis in SDS not only reveals characteristic profiles of membrane polypeptides but also provides an indication of their apparent molecular weight, by reference to a series of 'marker' proteins (Table 4.4) run under identical electrophoretic conditions. However, it should be stressed that whereas the marker proteins commonly used are mainly 'soluble' proteins demonstrated to bind approximately constant amounts of SDS (Pitt–Rivers and Impiombata 1968; Reynolds and Tanford 1970; Steinhardt et al. 1977) the amount of SDS bound by the various constituent polypeptides of membranes is largely unknown. Indeed, membrane proteins may differ greatly from most soluble proteins which bind 1.0–2.0 g SDS/g protein; an erythrocyte glycoprotein bound 5–7 g SDS/g protein (Grefrath and Reynolds 1976) and cytochrome b_5 of liver microsomes bound 3 g SDS/g protein (Robinson and Tanford 1974). Proteins without disulphide bonds bind increased amounts of SDS.

Membrane polypeptides vary in molecular weight from about 10,000 to 250,000. When the molecular weights of marker proteins are plotted semi-logarithmically against relative mobility, a straight line relationship (irrespective of the % acrylamide in the gel) from 20,000 to 80,000 is obtained. Above this range, the relationship becomes hyperbolic (Neville 1971; Neville and Glossman 1971) and since many membrane proteins are found in the molecular weight range 80,000–150,000, determination of a reliable molecular weight becomes difficult. Proteins of molecular weight below 15,000 also show aberrant behaviour on polyacrylamide gels.

Glycoproteins migrate anomalously in SDS gels and usually a

TABLE 4.4

Protein molecular weight markers useful in membrane characterisation by polyacrylamide gel electrophoresis.

Marker	Location	Mol. wt.	Other comments
Carbamoyl-phosphate synthase (ammonia) (EC 2.7.2.5)	Liver mitochondria	165,000	191,000 mol. wt. in Triton; Clarke (1976)
Albumin	Marker for serum and liver soluble proteins occluded inside or attached to membranes	65,000–67,000	Useful in determining purity of plasma membranes and liver Golgi fractions
Tubulin, actin	Components of contractile elements that remain attached to plasma membrane of non-muscle cells	55,000 43,000	CNS rich in tubulin. 'Arrowheads' formed on treatment of actin with heavy meromyosin. Actin may be a universal plasma membrane component
Myosin	Plasma membrane marker?	200,000	Present in rat kidney fibroblast membranes. Amount related to transformation (Shizuta et al. 1976)

Collagen	A frequent contaminant of plasma membranes from adult tissue	α_1 124,000[a] α_2 114,000	Values for chick collagen
Myelin proteins	Myelin markers	23,000 18,400 (A_1) 12,000 (P_2)	Peripheral (P_2) and CNS myelin
Haemoglobin	Marker on erythrocyte membranes	16,700	

[a] Runs anomalously. Apparent mol. wt. 130,000–140,000.

The following soluble proteins are useful mol. wt. markers. β-Galactosidase (130,000), Worthington 4090; rabbit muscle phosphorylase *a* (94,000), Sigma P1261; catalase (60,000), Sigma C-40; aldolase (40,000), Sigma A-5511; carbonic anhydrase (29,000), Sigma C-7500; myoglobin (17,000), Sigma M-0380; cytochrome *c* (12,300), British Drug Houses 44025. They give single bands under reducing conditions. Manufacturers (e.g., Gallard–Schlesinger, BDH and Boehringer) also supply 'kits' containing marker proteins of similar mol. wt. range to that indicated above. For an extensive list of proteins and authentication of their mol. wt., see Weber and Osborn (1969).

higher apparent molecular weight is obtained than by other physical methods, or by amino acid sequence analysis. The extent of deviation may vary considerably, being quite unrelated to the amount of carbohydrate attached to protein (Glossman and Neville 1971). Examples of anomalous molecular weights of membrane glycoproteins are provided by erythrocyte components (Segrest and Jackson 1972) and liver plasma membrane 5′-nucleotidase (Evans and Gurd 1973). The apparent molecular weight of a membrane protein may decrease also as the % polyacrylamide in the gel is increased (Ferguson 1964); for example, the apparent molecular weight of rhodopsin decreased from 40,000 to 34,500 as the acrylamide concentration was increased from 4–11% (Frank and Rodbard 1975).

4.3.8.7. Isoelectrofocussing of membranes Although isoelectrofocussing in thin polyacrylamide gels or sucrose gradients incorporating ampholines and Triton X-100 has proved useful in analysing membrane components that appear as single components when electrophoresed in polyacrylamide gels, e.g., transplantation antigens (Parham et al. 1975) erythrocyte components (Bahkdi et al. 1975), the technique shows several shortcomings for membrane analysis and characterisation. First, the separation is carried out in neutral detergents, e.g., Triton X-100, Lubrol, and, as indicated in § 4.3.8.3, these detergents are not as effective as SDS in dissolving and resolving membranes containing many polypeptides. Second, resolution by isoelectrofocussing is limited by the tendancy of membrane proteins to band mainly in the pH range 5.5 to 6.5 (Zalik and Cook 1976). Third, the molecular heterogeneity of membrane components, real or artifactual (due to deamidation occurring during preparation), results in multiple components (electrophoretic heterogeneity); profiles can be simplified by treatment of membranes with neuraminidase. Fourth, glycoproteins that are rich in sialic acid residues are difficult to resolve because of current shortcomings of ampholine pH gradients below pH 3. Focusing of membrane proteins in neutral detergents followed by SDS gel electrophoresis in a second dimension gives improved resolution (O'Farrell 1975).

4.4. Antigenic markers

Cell-surface antigens have not been extensively used in plasma membrane characterisation. Kamat and Wallach (1965), Allan and Crumpton (1970) and Gahmberg and Simons (1970) provide examples of the use of antigenic markers to help characterise plasma membrane fractions of Ehrlich ascites cells, lymphocytes and baby hamster kidney cells, respectively.

Histocompatibility antigens These are biologically important immunological markers found on all mammalian cells (except erythrocytes (Herberman and Stetson 1965). They have not been extensively used in plasma membrane characterisation owing to the elaborate assay systems required for their estimation. Two major assay systems exist for measuring histocompatibility activity. The chromium-release method (Sanderson 1964; Wigzell 1965) measures, semi-quantitatively, the inhibitory activity on the cytotoxic reaction of an allo-antisera against lymph node lymphocytes prelabelled with ^{51}Cr. In the fluorochromasia assay of Bodmer et al. (1967), histocompatibility antigen levels are measured by inhibition of a complement mediated cytotoxicity.

The H-2 and HLA histocompatibility antigens are present at high concentrations on the plasma membrane fraction of rodent liver (Herberman and Stetson 1965; Evans and Bruning 1970) and in a human diploid lymphoblastoid cell line in which HLA-2 activity was determined (Snary et al. 1974).

Anti (plasma membrane) antisera in membrane characterisation A widely-used technique for exploring antigenic differences between subcellular fractions is to raise an antiserum (usually in rabbits) against a saline-washed plasma membrane fraction. This antiserum (or more specifically, an IgG fraction prepared by ammonium sulphate precipitation and exclusion chromatography of the plasma (see Clausen 1969)) may then be used to, (a) resolve the number of antigenic specificities present in the fraction (Gurd et al. 1973; Blomberg and Perlmann 1971); (b) to investigate enzymic polymorphism by using, for example, two-dimensional immunoelectrophoresis (Deheer et al. 1974; Blomberg and Raftell 1974); and (c) to assess cross-reactivity between plasma membrane and other subcellular fractions by absorption experiments, using iodinated antisera (Gurd et al. 1972). A full description of the techniques used in determining the specificity of the antibody-antigen reaction and in its quantitation falls outside the scope of the present account (for an extended account of the use of immunological techniques in membrane research, see Axelsen et al. 1975).

The following practical points are apposite when using antisera to characterise membrane fractions. Membrane proteins solubilised in deoxycholate, Triton X-100, Nonidet P-40, Brij 58, Tween 80, can be immunoprecipitated (Crumpton and Parkhouse 1972). However, immunological results obtained in the presence of detergents must be interpreted carefully. For example, the presence of small amounts of SDS can give rise to non-immune precipitation lines (Green et al. 1975). The enzymic complexity of plasma membrane fractions can be assessed by two-dimensional immunoelectrophoresis in neutral detergents; for example, antigens displaying nucleoside diphosphatase, triphosphatase, aminopeptidase and epinephrine-binding activities may be compared in plasma membrane and microsomal fractions (Blomberg and Raftell 1974; Blomberg and Berzius 1975). Schmidt-Ullrich et al. (1975) and Nielsen and Bjerrum (1977) have extended this approach by using crossed immunoelectrophoresis techniques to investigate Concanavalin A-reactive proteins of thymocyte and milk fat globule membranes. Antisera raised against plasma membrane fractions may, when incubated with plasma membranes prepared from the same or different tissues and organs, or from other species, specifically inhibit membrane enzymes, e.g., 5′-nucleotidase (Gurd and Evans 1974; Newby et al. 1975). Antisera with inhibitory activity towards membrane enzymes show potential as tools in determining interelationships between plasma membrane enzymes and in assessing their topographical position on the membrane.

4.5. Morphological markers

Intact organelles (e.g., nuclei, mitochondria, Golgi dictyosomes) and plasma membrane specialisations (e.g., intercellular junctions, brush border complexes) have proved useful in providing estimates of the overall quality of plasma membrane fractions. Strictly, a thorough morphological examination in the electron microscope of stained sections of the membrane pellets should follow and largely support the quantitative enzymic and chemical analyses of the putative plasma membrane fraction, especially to assess contamination by nuclei, mitochondria and rough endoplasmic reticular vesicles.

Most of the plasma membranes of eukaryotic cells fragment during homogenisation into vesicular profiles that, when observed in isolated fractions, will provide little clue as to their origin in the intact cell. Furthermore, similar vesicular profiles can also be created from other cell components under certain conditions. Methods of plasma membrane preparation that entail vigorous cell disruption procedures and/or media of low ionic strength can destroy the normally distinctive morphological features of nuclei and mitochondria. Hence, the demonstration of vesicular profiles, unsupported by enzymic and chemical results, does not constitute a demonstration of a 'clean' plasma membrane fraction. Pitfalls and some features of isolated fractions will now be briefly discussed.

4.5.1. Outer mitochondrial vesicles

Vesicles, indistinguishable from plasma membrane vesicles are formed when fractions containing mitochondria are exposed to hypotonic media, e.g., 1 mM bicarbonate, Tris–HCl buffers of low ionic strength. They are probably formed by osmotic rupture of mitochondria and have a similar density to plasma membrane fragments. Therefore, enzymic markers for the outer mitochondrial membrane (see § 4.2.4) should be assayed, especially when hypotonic media are employed during isolation.

4.5.2. Nuclear envelope components

It is difficult to quantitate the extent of contamination of a plasma membrane fraction by nuclear membranes which have been released by mechanical or osmotic damage to nuclei during processing. The paucity of specific nuclear membrane markers, together with the common use of 'low speed' or 'nuclear' fractions in hypotonic media suggest that the extent of contamination of plasma membrane fractions may be underestimated. The outer nuclear membrane appears to be in direct contact with the endoplasmic reticulum (Watson 1965). Thus, the presence of ribosomes attached to membrane vesicles (usually the outer surface) may indicate contamination by outer nuclear or rough endoplasmic reticulum membranes. In some instances, it is possible to recognise the nuclear pore complex, the appearance of which is shown in Fig. 4.6. The attachment of chromatin to the inner nuclear membrane indicates that estimation of DNA (see § 4.3.5) can also provide clues of the level of contamination of the fraction by nuclear membranes.

4.5.3. Golgi apparatus components

Most plasma membrane preparations have not been assessed enzymically for Golgi contamination. Although Golgi dictyosomes become unstacked during cell breakage and subcellular fractionation, it is still possible to identify them in thin stained sections and also by negative staining. Fig. 4.7 shows that examination of sections of a liver 'plasma membrane' subfraction indicates the presence of cisternal elements and also secretory vesicles with their densely-stained contents (probably very low-density lipoproteins). Subsequent estimation of Golgi marker enzymes (galactosyl and sialyl transferases) confirmed the morphological evidence (Wisher and Evans 1975). In non-secretory cells, morphological identification of Golgi membranes in isolated fractions is difficult, but Warley and Cook (1976) have shown that this is possible in leukemic cells.

4.5.4. Filamentous structures attached to membranes

A specific association between polymeric forms of actin, i.e., microfilaments, and the cytoplasmic face (IO) of the plasma membrane of non-muscle cells is suggested by an increasing number of reports

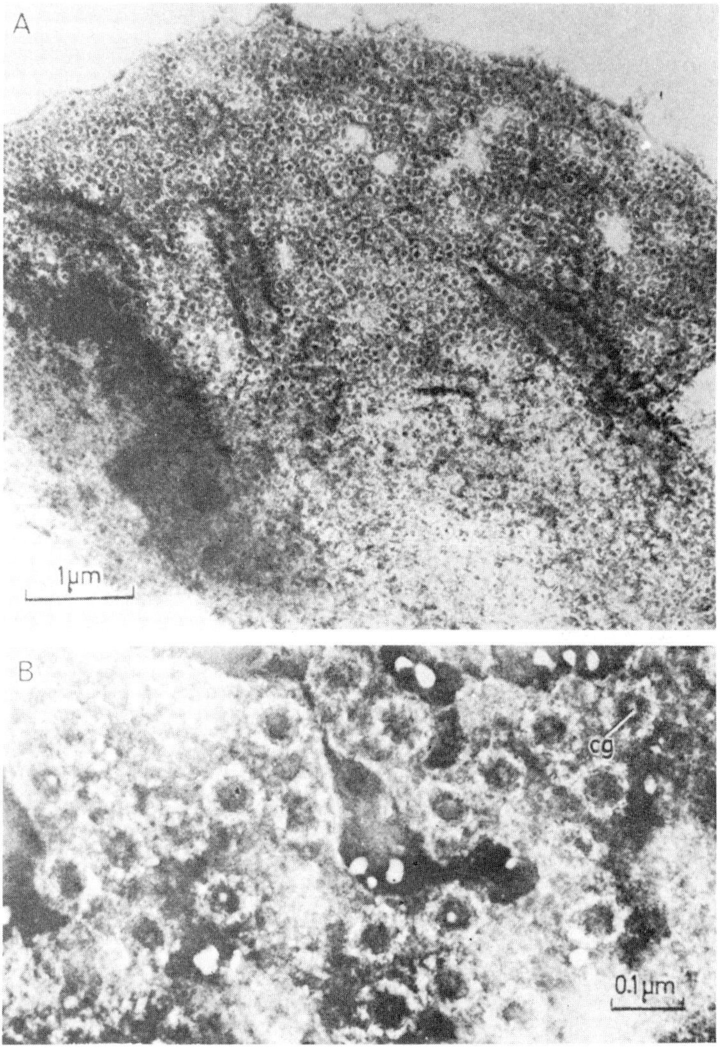

Fig. 4.6. Rat liver nuclear membranes. Negatively-stained electron micrograph, (A) magnification × 18,000. (B) × 120,000, shows nuclear pores; annular subunits and an occasional central granule (cg arrowed) are well preserved (Kay et al. 1972). Reproduced by permission of the Federation of European Biochemical Societies.

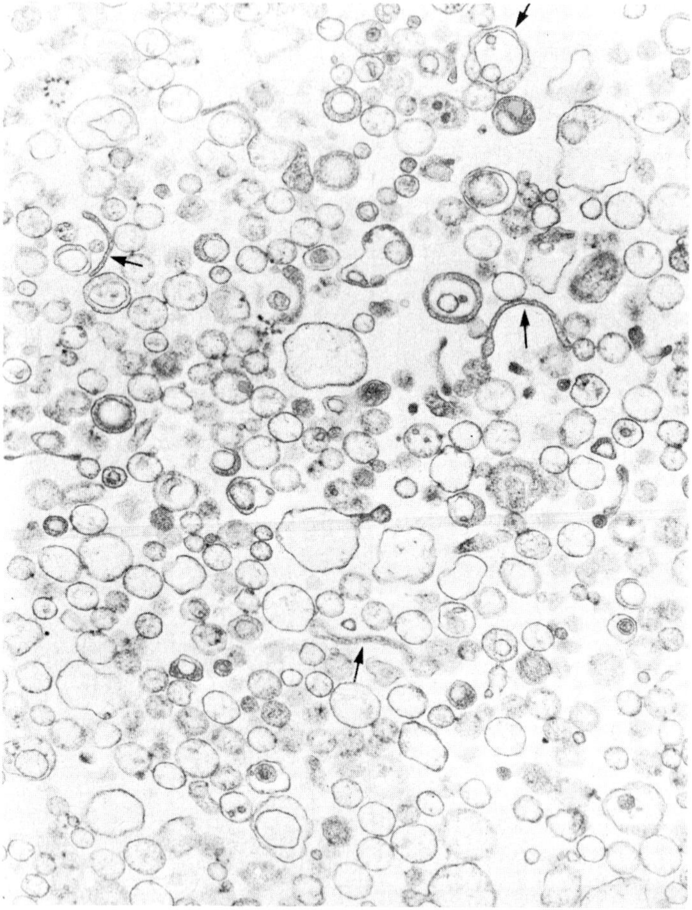

Fig. 4.7. A liver plasma membrane fraction contaminated by Golgi elements. A liver 'blood sinusoidal' plasma membrane fraction was prepared as described by Wisher and Evans (1975). Arrows point to Golgi-cisternal-element-derived membranes; Golgi secretory element membranes containing densely-stained material are also present. Magnification × 33,000.

(Pollard and Weihing 1974). Isolated plasma membranes from a variety of mammalian tissues and cell lines contain filaments attached to the inner surface of vesicular elements. These filaments persist through several washing steps in iso-osmotic media, but may be lost after hypotonic washing. Membrane adherent filamentous material similar in appearance to microfilaments correspond to those found underlying the cell surface membrane and microvillar protrusions of intact cells (Yahara and Edelman 1975). Figure 4.8 illustrates the appearance of microfilamentous material attached to plasma membranes of HeLa cells and human lymphoblastoid cells. Although the unequivocal identification of this material as an actin-polymer must await labelling of the filaments by heavy meromyosin (a proteolytic fragment of myosin) (Pollard and Weihing 1974), the demonstration by polyacrylamide gel electrophoresis of a major band co-electrophoresising with, e.g., pig actin will suggest that actin is at least associated with the membranes. Further proof of identification that the membrane-associated component is actin is obtained after its extraction and the two-dimensional mapping of tryptic digests (e.g., Barber and Crumpton 1976). Although thought to be a highly conserved plasma membrane-associated protein in mammalian cells, comparison of the tryptic peptides and amino acid sequence studies of actin extracted from various sources indicates that multiple forms exist (Gordon et al. 1977). Hence, association of microfilamentous material with isolated fractions can provide morphological clues as to their origin, and importantly, provides an indicator of the orientation (usually right-side-out) of the plasma membrane vesicular profiles.

4.5.5. Plasma membrane specialisations

These have played an important role in the isolation and identification of plasma membrane fractions from tissues, guiding subsequently to the introduction of enzymic and chemical markers. Tissue cells form juctional complexes with neighbouring cells, and provided that tissues are disrupted by mild methods, these complexes remain intact and are easily identified in subcellular fractions. Farquhar and Palade (1963) classified epithelial plasma membrane junctional complexes into, (a) tight junctions (*zona occludens*) in which the intercellular space is completely obliterated, (b) intermediate junctions (*zona adherens*) characterised by the presence of a 2 nm inter-

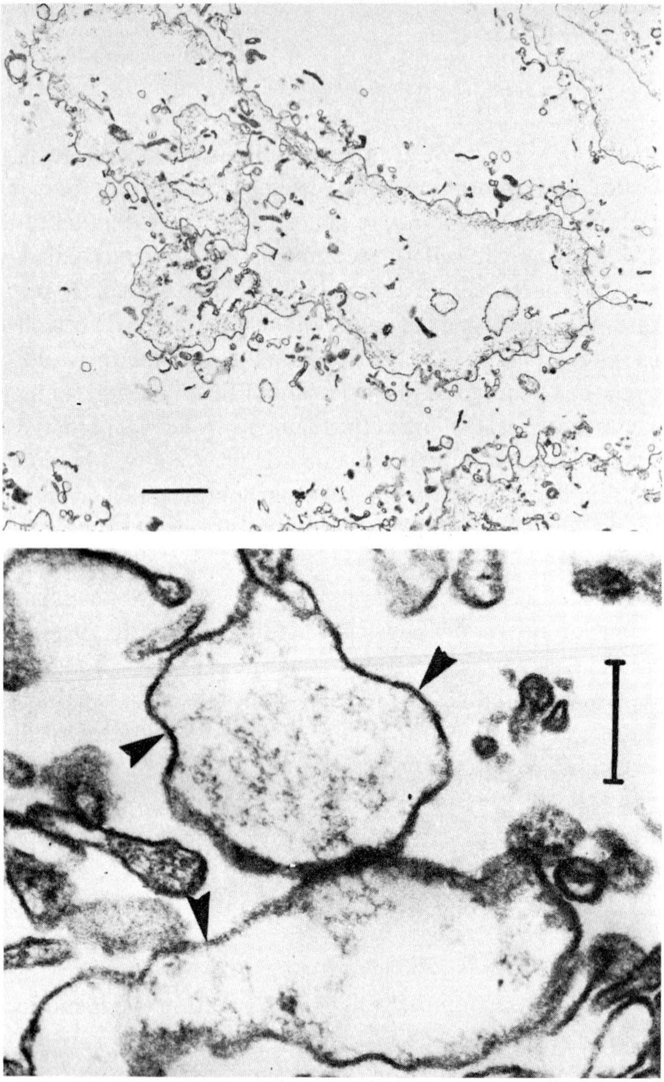

Fig. 4.8. Filamentous material attached to isolated plasma membranes. (A) Electron micrograph of HeLa plasma membrane ghosts. Bar = 1 μm. Note presence of microvilli and filamentous material probably corresponding to microfilaments attached to the cytoplasmic side of the plasma membrane. Photograph by courtesy of Dr. P.H. Atkinson and by permission of Academic Press. (B) Electron micrograph of a cultured human lymphocyte plasma membrane fraction. Arrowhead indicates point of contact between microfilaments and the cytoplasmic side of the plasma membrane. Photograph by courtesy of Dr. M.J. Crumpton. Diameter of microfilaments is 5–7 nm, of inter mediate filaments 10 nm and of microtubules 25 nm.

cellular space, and (c) desmosomes (*macula adherens*) which possess a central disc of dense material to which is attached filamentous material. Revel and Karnovsky (1967) defined a further category of intercellular contact – the gap junction in which the two adjacent plasma membranes run closely parallel maintaining an intercellular gap of approximately 2 nm permeable to lanthanum salts. Gap junctions (sometimes referred to as communicating junctions) also display a polygonal substructure in negatively stained preparations (see Fig. 4.9C).

When isolating plasma membranes from tissues, the desmosomes and gap junctions have proved to be important morphological markers, especially in identifying membrane fragments from the lateral (contiguous) regions of the cell (Fig. 4.9A and B). Tight junctions are not usually seen in isolated fractions. Differences in the arrangement of intramembranous particles of gap and tight junctions are seen by freeze-fracturing these regions of the plasma membrane. Brush border complexes isolated from intestinal or kidney cells, synaptosomes isolated from brain, and bile canaliculi from liver are further examples of useful morphological markers for specific regions of the plasma membrane.

4.5.6. Morphometric techniques

Quantitative morphological information may be obtained from electron microscopy of particulate components present in isolated fractions by applying stereological techniques (see Bauhduin et al. 1967). Before these techniques can be applied to assess with reasonable accuracy the purity of isolated fractions, it is necessary to recognise clearly the components present in the fields observed. Since vesicles of unknown origin usually preponderate in plasma membrane fractions, the usefulness of this technique to assess plasma membrane purity is limited (see Loud 1962; Elder and Morré 1976). Even when there are characteristic cell surface specialisations present the approach is fraught with difficulties. For example, the identification of synaptosomal plasma membrane fraction (§ 5.9) relies on the identification of post-synaptic densities. Yet, the majority of sections in many

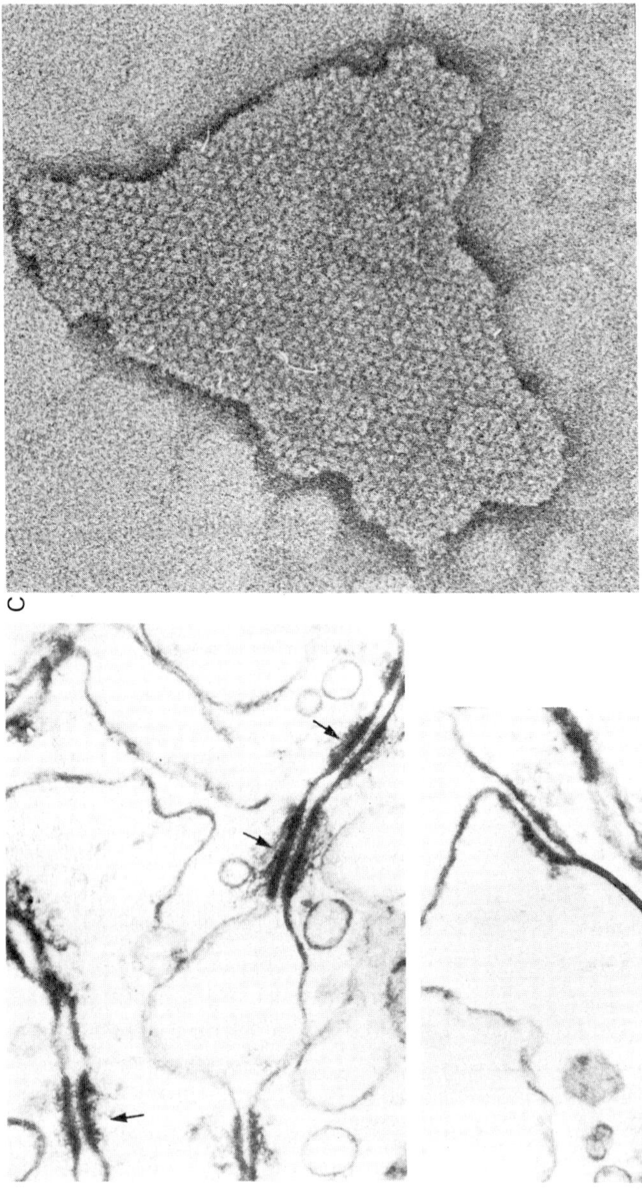

Fig. 4.9. Plasma membrane specialisations. (A) Section through a battery of desmosomes on contiguous liver plasma membranes (arrows) (× 51,000). (B) A hepatocyte gap junction (× 51,000). (C) Isolated gap junction negatively-stained with sodium silicotungstate (× 250,000).

planes through an isolated synaptosomal structure or remnant is unlikely to indicate the presence of the characteristic post-synaptic thickening.

4.6. *Radioactivity as a plasma membrane marker*

An increasing number of techniques is becoming available for attaching, to the surfaces of the intact cells, markers that remain bound to the plasma membrane during subsequent fractionation and that are detectable at low concentrations. These approaches, using isotopic labelling, complement the enzymic and other indigenous markers described in § 4.2–4.5. The most extensively used radio-label is ^{125}I or ^{131}I and this is either attached directly and covalently to the plasma membrane of intact cells by means of an enzymically-catalysed reaction or is attached to a ligand that then interacts strongly with the plasma membrane of cells. The wide range of biologically specific ligands available, e.g., lectins, polypeptide hormones, toxins, immunoglobulins results in increasing use of the second procedure. In addition, a number of low molecular weight chemical reagents, reactive towards the NH_2 or SH groups of membrane proteins have been described (Carraway 1975), but these have not been used as extensively to label the plasma membrane for subcellular distribution studies mainly because of doubts about the extent of their permeation into cells and consequent labelling of intracellular components.

4.6.1. *Enzymically catalysed radio-labelling of plasma membranes*

4.6.1.1. Catalysis by galactose oxidase This enzyme has featured in cell surface carbohydrate labelling studies, since it oxidises the C-6 of galactosyl and *N*-acetylgalactosaminyl groups to the aldehyde. The radioactive marker is introduced into the plasma membrane when the aldehyde is reduced by addition of tritiated borohydride (Gahmberg and Hakomori 1973; Steck and Dawson 1974). An useful adjunct to the technique involves pretreatment of cells with a sialidase that removes sialic acid thereby exposing the penultimate sugar that is frequently galactose (Juliano and Behar–Bannelier

1975a,b). The galactose-oxidase catalysis method appears to label similar glycoprotein species (on a molecular weight basis determined by polyacrylamide gel electrophoresis) as the lactoperoxidase catalysis method.

4.6.1.2. Catalysis by lactoperoxidase Radio-iodination, catalysed by lactoperoxidase, was initially devised as a probe for determining the vectoral arrangement of the membrane proteins of erythrocytes (Phillips and Morrison 1971; Hubbard and Cohn 1972) and for labelling lymphocyte surface immunoglobulins (Marchalonis 1971). The technique is now used increasingly to monitor the distribution of plasma membranes in subcellular fractionation schemes.

The technique depends on excluding the enzyme (mol. wt. 78,600) from cells, thus ensuring that iodination is confined to the outer aspect of the plasma membrane. The reaction is:

$$X^- + H_2O_2 + \text{membrane acceptor-H} \xrightarrow{\text{lactoperoxidase}} \text{membrane acceptor-X} + OH^- + H_2O$$

X^- = oxidisable halogen ion (usually $^{125}I^-$).
The H_2O_2 is either added in small portions or is continually generated by a glucose oxidase system incorporated in the reaction mixture (Hubbard and Cohn 1972).

There is a great deal of variation in the relative amounts of the reactants used by investigators working with a variety of cell types. Lactoperoxidase (EC 1.11.1.7) is available from commercial sources; however it may be purified from raw skimmed milk by extraction with a weak cation-exchange resin followed by ion-exchange and gel chromatography (Morrison and Hultquist 1963). Procedures used to identify the iodinated species in membranes (mainly tyrosine residues) are described by Hubbard and Cohn (1972), who also describe methods for ascertaining the activity of the lactoperoxidase and glucose oxidase (EC 1.1.3.4) components of the system.

The following suggestions may be helpful in designing iodinations:

(1) Cells to be iodinated should be intact. Thus, with cell sus-

pensions, it is essential that viability is >95% before and after iodination. Viability is usually assessed on the basis of the exclusion of eosin or trypan blue (0.2% dissolved in medium). Non-viable cells are permeable to lactoperoxidase and their internal components are thus liable to extensive iodination (Podulso et al. 1972, Schmidt–Ullrich et al. 1974). Damaged cells were shown, by autoradiography, to incorporate on the basis of grain density, up to 4 times the radioactivity of intact cells (Gonatas et al. 1976).

(2) Soluble proteins are iodinated more easily than membrane proteins; presumably because they contain more accessible tyrosine residues. Hence, cells collected from body cavities or maintained in culture media should first be washed by repeated centrifugation at low speeds in phosphate-buffered saline.

(3) Peroxidases can generate I_2 and this will enter cells reacting extensively with lipids containing unsaturated fatty acids. It is thus advisable to use low concentrations of iodide ions, usually by using carrier-free $Na^{125}I$ (Fig. 4.10A).

(4) The method depends on the availability of exposed tyrosine residues on the cell surface.

(5) To obtain reproducible levels of incorporated radioactivity (especially with suspension cultures) steps should be taken to keep constant the number of cells per ml in relation to other parameters. Conditions applicable to one cell type may not be optimal for others.

(6) It is advantageous to generate H_2O_2 enzymically by including glucose and glucose-oxidase in the reaction mixture, since direct addition of H_2O_2 may cause cell lysis.

(7) Differences are observed in the amount of radio-iodine incorporated into fibroblasts depending on whether they are maintained in monolayer on suspension cultures (Hunt and Brown 1975), and on the position in the cell cycle. For example, mouse L-cells were most amenable to iodination at the G_1 phase of the cell cycle (Hunt et al. 1975).

Iodination of cells in suspension Cells (about 1×10^7), first washed by low speed centrifugation in phosphate-buffered saline are suspended in 1 ml of phosphate-buffered saline, pH 7.2, containing 4 mU lactoperoxidase (approx. 5 μg), 3.6 mU glucose oxidase (approx 0.5 μg), 5 μmol glucose and 100 μCi carrier-free $Na^{125}I$. Cells are normally left for 10–30 min at temperatures varying between 4 and 37 °C with gentle agitation. To minimise membrane activity such as pinocytosis, lateral diffusion of membrane proteins and proteolysis, it is advisable to expose cells for the minimal time that results in acceptable incorporation of ^{125}I (Fig. 4.10B). The reaction is terminated by addition of a 10-volume excess of phosphate-buffered saline that may also contain 10^{-5} M $Na_2S_2O_3$. That the incorporation of ^{125}I into membrane macromolecules precipitated by addition of trichloroacetic acid depends on the presence of lactoperoxidase and H_2O_2 should be ascertained in control tubes lacking these components. Usually, $<5\%$ of iodine incorporation should occur in the absence of enzymes. Trichloroacetic acid precipitates are washed several times with phosphate-buffered saline, pH 7.4, containing $K^{127}I$ before determining protein-bound radioactivity. When many samples are to be determined, this can be done by pipetting aliquots onto Whatman GF/C discs held on a Millipore filtration apparatus, and the discs are then washed in 10% trichloroacetic acid, $K^{127}I$, dried and radioactivity determined.

Iodination of cells in monolayer cultures Hynes (1973) used iodination catalysed by lactoperoxidase to label the surface membranes of hamster fibroblasts and their virally-transformed variants. Cells (approx. 5×10^6/per 9 cm diameter Petri dish) are first washed 3 times in phosphate-buffered saline, pH 7.2, to remove serum components. Phosphate-buffered saline, 5 mM glucose is then added to cover the cells, followed by carrier-free $Na^{125}I$ to a final concentration of 400 μCi/ml. Addition of lactoperoxidase and glucose oxidase to final concentrations of 20 μg/ml and 0.1 units/ml, respectively, starts the reaction which is continued for 10 min at room temperature with occasional swirling. The reaction is terminated by addition of phosphate-buffered iodide (150 mM $Na^{127}I$) containing 2 mM phenylmethyl sulfonylfluoride to inhibit protease activity (§ 2.3.5). The cells

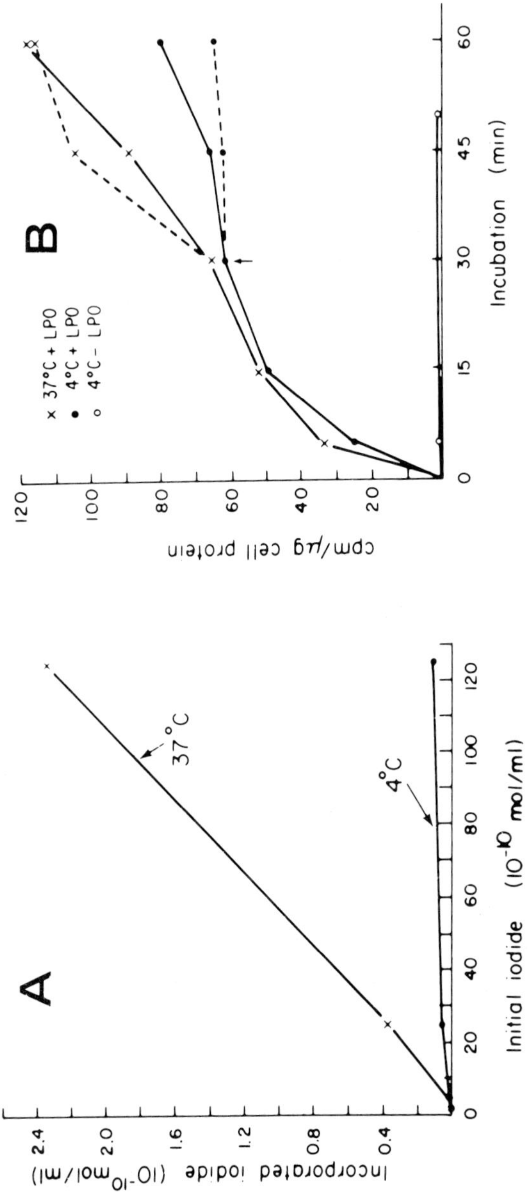

Fig. 4.10. (A) Effects of temperature and iodide concentration on the iodination of L-cells. 1 ml of reaction mixture contained 4×10^6 cells in buffered saline, 5 μmol glucose, 5 mU galactose oxidase, 3 mU lactoperoxidase, 50 μCi Na^{125}I (0.02 nmol/ml iodide) and varying amounts of K^{127}I. Cells were iodinated for 30 min at 4 °C or 37 °C, rinsed 5 times in buffered saline, lysed in 0.5 ml H_2O and aliquots analysed for trichloroacetic-acid-insoluble radioactivity.

(B) Effects of temperature on the time course of iodination of L-cells. 1 ml of reaction mixture contained 9×10^6 cells in buffered saline, 5 μmol glucose, 6 mU galactose oxidase, 3 mU lactoperoxidase (LPO), 50 μCi Na^{125}I. At the indicated times, 1 ml was removed, added to 10 ml 10^{-5} M $Na_2S_2O_3$ in buffered saline, the cells were rinsed and lysed and aliquots analysed for trichloroacetic-acid-insoluble material. Arrow at 30 min indicates addition of extra 3 milliunits of lactoperoxidase; (——) additional lactoperoxidase added; (– – –) no lactoperoxidase added. Cells incubated at 4 °C without lactoperoxidase incorporated less than 1% the radioactivity of cells exposed to the complete labelling system (bottom curve). 1 μg L-cell protein ≃ 6×10^3 cells.

Figures from Hubbard and Cohn (1975) by courtesy of Journal of Cell Biology.

are then washed several times in phosphate-buffered iodide until the washings are low in ^{125}I-radioactivity before being removed for homogenisation.

Examples of the use of lactoperoxidase catalysis to iodinate and identify plasma membranes in subcellular fractionation

(a) *Hamster fibroblasts* In view of the difficulties in preparing plasma membranes from fibroblast homogenates (see § 5.6), lacto-peroxidase-catalysed iodination of the cells has been used to provide an additional marker so as to assess their recovery and purity. Graham et al. (1975) used hamster fibroblasts prelabelled with ^{125}I (see § 4.6.1.2) in 0.25 M sucrose, 5 mM $MgCl_2$, 5 mM Tris–HCl (pH 7.4) and homogenised by nitrogen cavitation (54 atm for 15 min). The post-nuclear supernatant was introduced into a B-XIV zonal rotor filled with a sucrose gradient (Fig. 4.11). Analysis of the fractions collected from the gradient showed a coincidence in the peaks of ^{125}I-radioactivity and the two plasma membrane marker enzymes 5′-nucleotidase and (Na^+-K^+)-ATPase. The use of cells with surface membranes prelabelled with ^{125}I-radioactivity thus corroborates the validity of the selected enzymic markers and their subcellular distribution. The method also provides information on the heterogeneity of the plasma membrane fragments. A second 'heavy' peak of ^{125}I-radioactivity that did not coincide with enzymic markers helped identify remnants of a surface coat or extracellular micro-exudate that contained increased amounts of an iodinated high-molecular weight protein that was reduced in intensity in virally-transformed derivatives of this cell line. This example is representative of a general approach applicable to a number of cell types. The technique not only allows an assessment of the behaviour of plasma membrane fragments in subcellular fractionation but also indicates the extent of the heterogeneity of plasma membrane fragments as also shown in the next section.

(b) *Baso-lateral and brush border plasma membranes of intestinal cells* Cells that exhibit gross functional polarity, e.g., kidney and intestinal cells, contain two major plasma membrane regions. Hence,

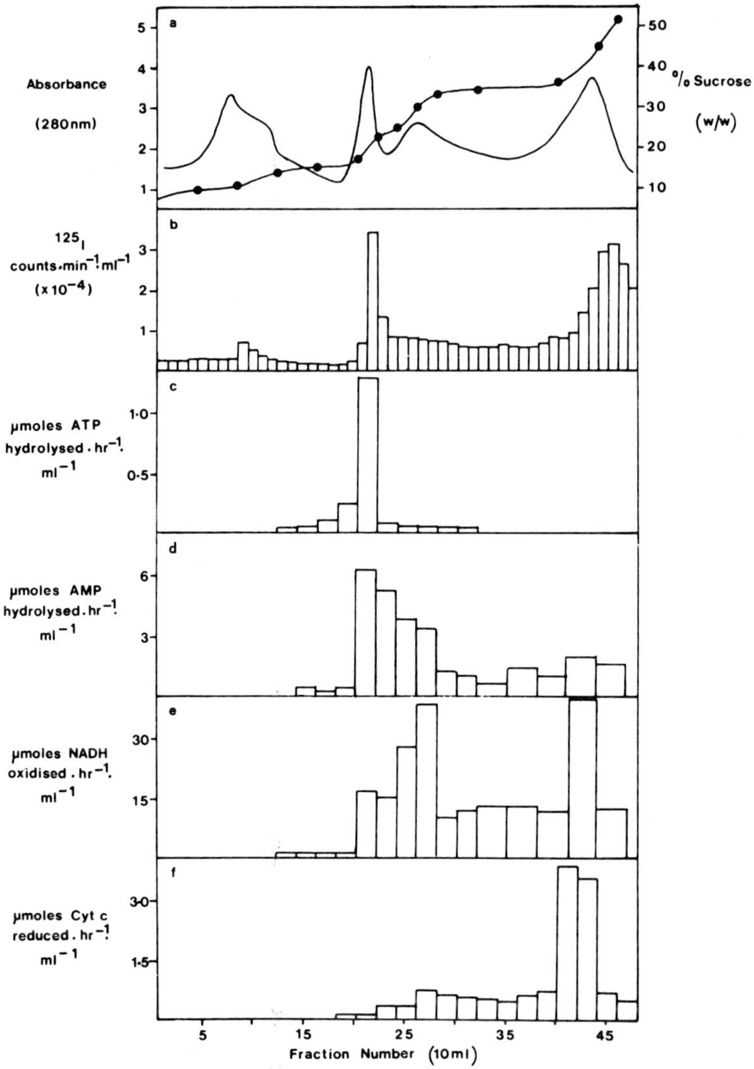

Fig. 4.11. Distribution of marker enzymes and ^{125}I radioactivity in a post-nuclear supernatant of NIL8 hamster fibroblasts. Samples were centrifuged at 110,000 $g \times 90$ min in a B-XIV zonal rotor containing the following discontinuous gradient: 90 ml 15% (w/w) sucrose, 40 ml 19% (w/w) sucrose, 50 ml 22.5% (w/w) dextran (mol. wt. 40,000), 200 ml 35% (w/w) sucrose and approx. 175 ml 60% (w/w) sucrose. The sample was overlayed with 50 ml of 5% (w/w) sucrose.
(a) E_{280} absorption (———) and sucrose density profile (●——●) of gradient; (b) Trichloroacetic acid precipitated ^{125}I-radioactivity; (c–f) (Na^+-K^+)-ATPase, 5′-nucleotidase, NADH-diaphorase and succinate cytochrome c reductase activities, respectively, across the gradient. Reproduced from Graham et al. (1975) by permission of M.I.T. Press.

a chosen plasma membrane marker may also indicate the subcellular distribution of membranes originating from one or other of these regions. Lewis et al. (1975) investigated the subcellular distribution of intestinal cell plasma membranes and this work provides a useful example of the use of lactoperoxidase-catalysed iodination in locating functionally specialised plasma membrane fragments. A suspension (3.4 ml) of collagenase-dissociated and washed cells or sheets of cells ($>90\%$ viable) of guinea-pig intestine are incubated at 25 °C for 15 min in polypropylene tubes containing in 0.2 ml lactoperoxidase (1 mg/ml), 12 units of glucose-oxidase, 3.6 mg glucose and 0.5 mCi of carrier-free ^{125}I. The cells were then homogenised in a Potter–Elvehjem homogeniser and subjected to zonal centrifugation (Fig. 4.12). Fractions were analysed for the distribution of marker enzymes and ^{125}I-radioactivity. The figure allows the position of the two populations of plasma membrane fragments derived from the brush border or baso-lateral regions to be identified and shows the positions of markers for intracellular membranes and organelles. The ^{125}I-radioactivity peak at higher density coincided with the brush border plasma membrane markers alkaline phosphatase and leucylnaphthylamidase, while a lower density radioactive peak banded with (Na^+-K^+)-ATPase, a baso-lateral plasma membrane marker. As expected, the labelling system used was not as effective in iodinating the baso-lateral region of the plasma membranes.

4.6.2. Biosynthetically incorporated plasma membrane markers

Radioactive amino acids are rapidly incorporated into all cell proteins and therefore, they offer no advantages as markers for the preparation of plasma membranes. However, the presence of high concentrations of glycoproteins and glycolipids in the plasma membranes creates the potential for the use of sugar precursors to label mainly the plasma membrane. The incorporation of carbohydrate precursors into membrane glycoproteins, although used mainly as tools to study their biogenesis and structure, has also been used in conjunction with other parameters to follow the position of plasma membranes in subcellular fractionation.

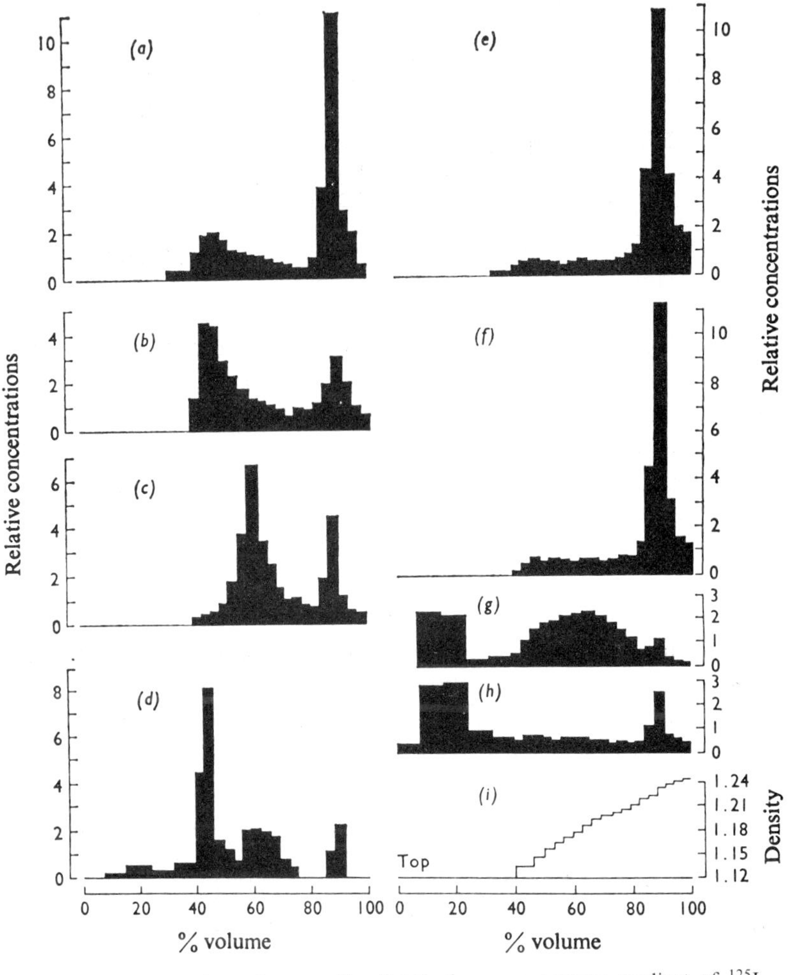

Fig. 4.12. Correspondence between the distribution on a sucrose gradient of ^{125}I-radioactivity and the marker enzymes of iodinated guinea pig intestinal brush border and baso-lateral plasma membranes. The components, whose distributions are shown, are (a) particulate ^{125}I, (b) (Na^+-K^+)-ATPase, (c) succinate dehydrogenase, (d) monoamine oxidase, (e) alkaline phosphatase, (f) leucylnaphthylamidase, (g) esterase, (h) protein, (i) density. Relative concentration is defined as the concentration in the fraction divided by the concentration that would be present if the recovered material had been evenly distributed in all fractions. After iodination and washing of intestinal cell sheets, homogenates were loaded onto the sucrose gradient (30–50%, w/w) in a B-XIV zonal rotor and centrifuged at 200,000 *g* for 1.5 hr. See text for further details.

The majority of studies directed at plasma membrane carbohydrate labelling has used glucosamine or fucose, and these are incorporated, after a delay of 15–30 min into membrane glycoproteins, although a minor but variable proportion appears in glycolipids. The use of sugar nucleotides as plasma membrane precursors (Patt and Grimes 1974) depends on the activity of cell surface ectoenzymes that can directly transfer sugars to incompletely glysolysated polypeptide chains. Whether glycosyltransferases are located on the surfaces of cells is a matter of dispute (Keenan and Morré 1975) but the low efficiency of sugar-nucleotides as direct precursors of membrane carbohydrates is due to the presence in most mammalian cells of an ecto-nucleotide pyrophosphatase that, for example, hydrolyses UDP-galactose to UMP and galactose 1-phosphate and a phosphatase that then releases galactose (Sela et al. 1972; Deppert et al. 1974; Evans 1974; Munro et al. 1975; Abney et al. 1976). Autoradiography of a variety of cell types has shown that labelled sugars are mainly incorporated into membranes in the saccules of the Golgi apparatus, and then redistribute to other parts of the cell, especially the plasma membrane (Bennett et al. 1974). Thus, the uptake of radiolabelled membrane precursors into intracellular pools that then redistribute the radioactivity to various membrane compartments and exported components is a complex problem that impinges on the mechanism of uptake (active or passive), metabolism of precursors, reutilisation of radioactive labels, and the rate of cell growth and division. Clearly, the results obtained with one category of cells can only be extrapolated to other cell types with caution.

Cultured cells Radioactive glucosamine and fucose have been used to label the plasma membranes of cultured cell lines (Bosmann et al. 1969; Kaufman and Ginsburg 1968). Examples of the amounts of radioactive sugars used and the length of exposure of cells are shown in Table 4.5. Cells are either grown to confluency in the presence of radioactive sugars, or are exposed to sugars for shorter periods. In HeLa cells, Atkinson (1973) surmised that it takes 15–20 min for fucose-containing marker molecules to arrive at the plasma membrane, and a further 20–30 min to dilute the internal radioactive pool of precursors. The rate of glycoprotein synthesis may vary according to the growth phase, since maximal incorporation occurs during the S phase, when growth is maximal (Nowakowski et al. 1972).

Tissue cells Conditions for incorporating sugars into tissue glycoprotein are summarised in Table 4.5. Radioactive sugars may be injected either intraperitoneally or intravenously when more efficient incorporation is obtained. In rodents injections are conveniently made into the caudal vein by using a syringe with a narrow needle (approx. 50 μl per animal); entry into the caudal vein, previously expanded by warming the animals under a lamp, is indicated by the paling of the vein. Various cage designs are available for confining the activity of the animal during manipulation of the tail, e.g., a wide test tube with a 5 mm slit down one side through which the projecting tail is held is effective for mice. For shorter term incorporation experiments, perfused organs may be used connected up to recirculating systems. Intravenous injection of [^{14}C]glucosamine into mice resulted in approx. 10% of radioactivity being incorporated into the liver; approx. 5% of liver homogenate radioactivity was recovered in a plasma membrane fraction (Evans and Gurd 1971). Brain tissue is labelled by intracerebral injection of 10 μl of radioactive glucosamine or fucose through holes in the skull that extend 3 mm into the cerebral cortices (Barondes 1968; Zatz and Barondes 1970).

The period during which maximal incorporation of plasma membrane precursors is attained will vary according to a number of factors; in intestine and liver it is 4–6 hr, whereas in brain it is 16 hr (Table 4.5). In intestinal and liver plasma membranes, maximal levels of radio-labelled glycoproteins are found initially in the brush border and sinusoidal plasma membranes respectively. In brain tissue, the specific activity in synaptic plasma membrane fractions is no greater than in other subcellular fractions (Zatz and Barondes 1970).

Extraction of labelled plasma membrane fractions with chloroform/methanol followed by partitioning against aqueous media (see

TABLE 4.5

Examples of conditions for incorporation of radioactive sugars into plasma membrane fractions.

Cell line or tissue	Conditions of incorporation of radioactivity	Other comments	Reference
HeLa	4×10^5 cells in 1 ml incubated at 37 °C for 60–160 min in 5 μCi/ml of L-[^{3}H]fucose (4.3 Ci/mmol).	Plasma membrane ghosts prepared. Plasma membrane proteins also labelled.	Atkinson and Summers (1971)
KB	Cells grown to confluency in media containing D-6-[H^3]glucosamine or D-[^{14}C]glucosamine (1–3 μCi/ml; 0.05–5 Ci/mmol).	Plasma membrane ghosts prepared. Radioactivity coincided on sucrose gradients with 5′-nucleotidase activity.	Butters and Hughes (1974)
3T3 Mouse fibroblasts	Cells grown to confluency in media containing ^{14}C-glucosamine (1.5–5 mCi in 3.8–8 mg glucosamine HCl or [^{3}H]glucosamine 0.5–1.0 mCi in 3.8–8 mg glucosamine HCl).	Plasma membranes prepared from microsomal fraction. 1–2% radio-activity incorporated by cells. Glucosamine/galactosamine ratios in normal and virally-transformed cells determined.	Wu et al. (1969)

Intestinal epithelial cells Rat (100–120 g)[a]	Intraperitoneal injection of 1-[^{14}C]-glucosamine 50 μCi/mg in 0.2 ml 0.15 M NaCl; 5 μCi per rat. After 4 hr animals were killed and brush border plasma membranes prepared.	Max. incorporation of ^{14}C into brush border occurred in 90–180 min and remained so for 12 hr. 14-fold enrichment over homogenate ^{14}C activity.	Forstner (1971); Weiser (1973)
Brain Mice (30 g)[a] Rats (150 g)[a]	250–500 μCi ^{3}H-fucose (13.4 Ci/mmol) in 50 μl buffered saline, pH 7.4, injected intracerebrally into 5 rat brains. Animals killed 16 hr later.	General incorporation of label into all membrane fractions.	Zatz and Barondes (1970); Gurd and Mahler (1974)
Liver Mouse (25 g)[a]	Injection of 60 μCi [^{14}C]glucosamine (0.3 Ci/mmol) into caudal veins of 10 mice. A further 30 mice livers were added to make a homogenate for plasma membrane preparation.	Radioactivity recovered mainly in plasma membrane glycoproteins and glycolipids. Maximal incorporation into plasma membrane fraction occurred in 4–6 hr.	Evans and Gurd (1971)

[a] Weight of animal

§ 4.3.2) can indicate the extent of incorporation into lipid components. To establish the identity of the labelled sugars, membranes or their extracts are hydrolysed in 4 M HCl at 100 °C for 4 hr. The hydrolysate is dried, dissolved in H_2O and chromatographed on Whatman 3 MM paper in *n*-butanol/pyridine/H_2O (6 : 4 : 3, v/v) with the appropriate sugar standards. Sugars are located by spraying with alkaline $AgNO_3$ and the distribution of radioactivity in the spots is then determined by scintillation spectrometry. The incorporation of radioactive sugars into plasma membrane fractions is a function of the extent of existing metabolic interconversions and degradations. At short intervals (up to about 4–6 hr with tissues) glucosamine incorporated into plasma membranes is subject to less than 20% breakdown; fucose is metabolised even less, but may not be incorporated into glycoprotein as efficiently as glucosamine. Galactose is mainly incorporated into membrane glycolipids. The use of radioactive sugars (^{14}C or ^{3}H labelled) constitutes an expensive method of monitoring the preparation of plasma membranes, and their main use is in the chemical and functional characterisation of the membrane glycoproteins.

4.6.3. Radio-labelled ligands as plasma membrane markers

Affinity ligands that bind specifically with high dissociation constants to plasma membrane receptors show desirable features as markers in preparative studies. Ligands used in these approaches include heteroantibodies to surface antigens and viral receptors (Allan and Crumpton 1970; Philipson et al. 1968) (see also § 3.3 and 4.5), and a range of toxins, hormones and plant lectins. The basic approach entails the labelling of isolated cells or tissues by a brief exposure to low concentrations (10^{-9} to 10^{-11} M) of a high affinity, highly radioactive ^{125}I-ligand. After homogenisation, the subcellular distribution and recovery of the specifically bound label is followed and equated with the location of the endogenous marker enzymes.

Examples of the use of a range of iodinated ligands are given in Table 4.6.

TABLE 4.[illegible]

Examples of the use of various ligands to label cells before preparation of plasma membranes.

Ligand	Conditions for binding	Cell type	Specificity	Reference
^{125}I-insulin	30 min at 24 °C, 8×10^{-10} M, 7×10^6 cpm in Krebs–Ringer 1% serum albumin	Fat cells[a]	Possible with 'target' organs; slight dissociation.	Chang et al. (1975)
^{125}I-WGA[b]	30 min 24 °C, 5×10^6 cpm, 50 μg/ml in Krebs–Ringer 1% serum albumin	Fat cells Lymphocyte Liver	Binds strongly to α-D-*N*-acetyl-glucosamine groups. Very little dissociation.	Chang et al. (1975); Carey and Evans (1977)
^{125}I-glucagon	2 min at 24 °C, 1.5×10^6 cpm, 1×10^{-9} M in Krebs–Ringer 0.1% serum albumin	Liver	Highly specific binding 'slight dissociation' Enhances adenylate cyclase activity.	Carey and Evans (1977)
^{125}I-cholera toxin	5 min at 24 °C, 7×10^6 cpm (0.14 μg/ml)	Fat cells	Binds to G_1-ganglioside[c] Negligible dissociation.	Chang et al. (1975)
^{125}I-α-bungarotoxin	10 min at 37 °C in 30 pmol (8×10^6 cpm) in culture medium (Eagle's solution)	Cultured muscle cells	Strong binding to acetyl-choline receptor. Reversible by addition of 1 mM carbonylcholine.	Schimmel et al. (1973)
^{125}I-Concanavalin A	0.1 μg/ml and 10 μCi/μg	Fat cells[a]	Binds to glucose and mannose groups. Dissociates rapidly, especially in presence of sucrose.	Chang et al. (1975)

[a] A 120–150 g rat will contain in its epididymal pads approx. 8×10^6 fat cells (Greenwood and Hirsch 1974).
[b] Wheat germ agglutinin.
[c] Monosialoganglioside.

Subject index p. 261

4.6.3.1. Practical considerations in the selection and use of radioactively labelled ligands

(1) The corresponding receptors or binding sites should be present on the cell surface membrane to bind specifically the radioactive ligand. Hence, the use of iodinated polypeptide hormones is confined to those tissues that are hormonally-sensitive; thus insulin and glucagon are used to label hepatocytes and adipocytes, corticotropin with adrenal cortex cells, TSH (thyroid stimulating hormone) with thyroid tissue, etc. Plant lectins, e.g. wheat germ agglutinin, Concanavalin A, ricin, lentil lectin, etc. (see Table 4.7), have a relatively broader specificity towards the carbohydrate groups of glycoproteins and glycolipids on the cell surface; hence, lectins react with a wide range of animal cells.

(2) The interaction between the radioactive ligand and the binding site on the cell surface membrane should remain relatively stable during the various manipulations involved in preparing plasma membranes. Thus, the ligand-membrane combination must withstand, for example, several dilution steps, membrane resuspension and exposure to various concentrations of sucrose solutions. Since the dissociation constant of the ligand-receptor interactions is probably temperature dependant, the temperature should not rise above 4 °C during fractionation. Hence, it is essential to maintain all rotors and solutions at 0–4 °C. An extreme example of rapid dissociation of the ligand occurring during membrane preparation is provided by Concanavalin A, since sucrose solutions compete with the ligand for the fructofuranoside groups on the plasma membrane. In contrast, the interaction of wheat germ agglutinin with *N*-acetyl-D-glucosamine residues on membranes is relatively stable.

(3) Care should be taken to ensure that the majority of the ligand is specifically bound. Non-specific binding is more labile. Preparation of iodinated ligands that retain their conformational properties helps to ensure specific binding (see § 4.6.3.2).

(4) The exposure time of cells to ligands should be as brief as possible in order to confine the ligand to its original reaction site. Ligands remaining on cells for about 30 min or longer may be

TABLE 4.7
Physical and chemical properties of some lectins used in membrane research.

Lectin	Source	Molecular wt.	No. of subunits	Specificity	No. of binding sites
Concanavalin A	*Canavalia ensiformis*	102,000	4	glucose, fructose, mannose	4
Soya Bean Agglutinin	Glycine max.	110,000–120,000	4	*N*-acetyl-D-galactosamine D-galactose	2
Lentil Lectin	*Lens culinaris*	42,000–69,000	2	mannose, glucose	2
Ricin (RCA II)[a]	*Ricinus communis*	60,000	2[a]	D-galactose *N*-acetyl-galactose	1
Wheat Germ Agglutinin	*Triticum vulgaris*	36,000	2	*N*-acetyl-D-glucosamine	4
Phyto-Hemaglutinin	*Phaseolus vulgaris*	128,000	4	*N*-acetyl-D-galactosamine	4
Pea Nut Agglutinin	*Arachis hypogaea*	110,000	4	*N*-acetyl-D-galactosamine D-galactose	–

Data compiled from Sharon and Lis (1972); Lis and Sharon (1973); Lotan et al. (1975) and Nicolson and Blaustein (1972).
[a] Linked by disulphide bonds.

'interiorised' or move laterally in the plane of the membrane, thus complicating the interpretation of subcellular fractionation results.

(5) The combination of ligands with plasma membrane receptors may modify marker enzymic activities used to monitor the distribution of plasma membranes, e.g., glucagon stimulates adenylate cyclase and various effects of lectins on plasma membrane enzymes have also been described (Luly and Emmelot 1975; Edelson and Cohn 1976).

(6) Polypeptide ligands that attach to the surface membrane may be subject to rapid proteolysis, e.g., insulin or glucagon after attachment to hepatic plasma membranes are rapidly inactivated.

(7) The degree of specific binding of ligand to the isolated plasma membranes should be ascertained by displacement studies involving the addition of an excess of unlabelled ligand. Most of the [^{125}I]-glucagon present in isolated plasma membranes after interaction of the radio-labelled ligand with intact cells, is rapidly displaced by an excess of glucagon. Similarly, specifically bound radioactively-labelled lectins should be rapidly displaced by addition of an excess of sugar with which the lectin is believed to bind to on the surface membrane, e.g., wheat germ agglutinin is displaced by *N*-acetyl-D-glucosamine (Table 4.7).

4.6.3.2. Iodination of ligands; comparison of enzymic and chemical methods Iodination is the most widely used technique for labelling ligands since it allows their preparation at very high specific activity. Two methods are generally used, enzymic and chloramine-T iodination; an additional chemical method has recently been described in which an ^{125}I-containing acylating agent is used (Bolton and Hunter 1973).

Both the enzymic and chloramine-T methods directly substitute ^{125}I or or ^{131}I into tyrosyl groups. Generally, the methods using lactoperoxidase or horseradish peroxidase are preferred since they operate under conditions of controlled release of H_2O_2 from a glucose-oxidase-glucose system (Hubbard and Cohn 1972; Thorell 1972) (see § 4.6.1.2). The chloramine-T method of iodination (Hunter and

Greenwood 1962) involves the release of 'active' iodine and hypochlorous acid at slightly alkaline pH (optimum pH 7.4), and the disadvantage of mild oxidising conditions compared to enzymic methods is compensated by the ease of separation, mainly by dialysis, of the radioactively labelled ligand from the other reactants. The high molecular weight of the reactants in enzymic iodination can cause subsequent separation problems; these are exacerbated by strong and sometimes irreversible adsorption of ligands to the gel filtration columns used in purification. Iodination of small polypeptide hormones, e.g., insulin, glucagon, can modify their biological activity (Bromer et al. 1973) and up to about 1 Ci/μmol can be incorporated with retention of specific binding (Cuatrecasas 1971).

Both enzymic and chloramine-T methods are unsuitable for peptides devoid of tyrosyl residues, e.g., secretin. They may cause deleterious biological modification during contact between reagents and the ligand. No such direct contact is involved by using the Bolton and Hunter (1973) reagent, where the ligand is exposed to the ^{125}I-labelled acylating conjugate, 3-(4-hydroxyphenyl)propionic acid-*N*-hydroxysuccinimide ester which reacts with free amino groups in the polypeptide/protein so attaching the ^{125}I-labelled groups by amide bonds. The acylating conjugate is previously iodinated by the chloramine-T method and purified by solvent extraction; the iodinated derivative is available commercially (Amersham Searle). The immunoreactivities of polypeptide hormones labelled by this method are identical with those optimally iodinated at tyrosine residues.

Chloramine-T iodination is carried out as follows. To the tube containing ^{125}I (100–200 μCi) are added in order 50 μl phosphate buffer, pH 7.4, 6–20 μg of protein (dissolved in 20 μl buffer) and chloramine-T (6–20 μg) also dissolved in 20 μl buffer. After 40–90 s at room temperature, the reaction is terminated by addition of 2 volumes of sodium metabisulphite (in a 2-fold molar excess to chloramine-T) followed by a similar volume of cold phosphate. The protein is then dialysed against H_2O to remove the low molecular weight reactants (Greenwood et al. 1963). Lactoperoxidase-catalysed iodination is carried out as described in § 4.6.1.2.

The iodinated ligand is usually purified by gel filtration. A Sephadex G-25 or G-50 column (other grades, depending on the molecular weights of the reactants, may be used) 10 × 1 cm, equilibrated with 50 mM phosphate buffer, pH 7.4, containing 0.1% bovine serum albumin, is used; however, longer columns (up to 60 cm) may be necessary when damaged enzymes and hormones are also present. The use of lactoperoxidase covalently coupled to Sepharose beads (available from, e.g., Boehringer) offers obvious advantages in the subsequent purification of iodinated ligands (David 1972).

4.6.3.3. Comparison of the binding properties of various ligands Chang et al. (1975) compared the suitability of ^{125}I-labelled insulin, Concanavalin A, wheat germ agglutinin and cholera toxin as plasma membrane markers. The conditions adopted to label fat cells or lymphocytes are applicable generally for cell suspension and are as follows: 1 × 10^8 cells are incubated for 5–30 min at 24 °C with the iodinated ligand (see Table 4.6 for radioactivity and concentration of ligands). The incubation volume is about 1 ml and contains a Krebs–Ringer buffer and 1% albumin. After incubation, cells are washed by centrifuging at low speed in iso-osmotic sucrose/Tris–HCl buffer (10 mM, pH 7.4), 1 mM EDTA before homogenisation. The subcellular fractionation scheme is then closely monitored for correspondence between the distribution of radioactivity and plasma membrane marker enzyme (see § 4.6.2.5).

4.6.3.4. Domain-specific plasma membrane labelling In contrast to free-living cells in which the cell surface membrane appears morphologically uniform and is fully exposed to the ligand-containing environment, the surface membrane of differentiated cells organised into tissues may interface with two or more environments. Domain-specific labelling techniques involve the confinement of the ligand to one environment, thus resulting in the labelling and identification by subcellular fractionation of the appropriate plasma membrane fragments. This approach thus allows an assessment of the contribution of the various surface membrane domains to the plasma membrane fraction. For example, the hepatocyte cell surface is bathed by regions containing blood or bile; introduction of labelled ligands into the portal vein results in the confinement of ligand-receptor interactions to the blood-sinusoidal and possibly lateral plasma membrane (Carey and Evans 1977). Similarly the kidney epithelial cell surface is specialised into microvillar and baso-lateral regions and since the latter region interfaces with the kidney's blood circulation, perfusion through the renal vein with radioactively labelled ligands should help identify the baso-lateral plasma membranes (a similar approach using lactoperoxide-catalysed covalent iodination of intestinal brush border membranes is described in § 4.6.1.2). A further example entails the use of [^{3}H]phlorizin to label the brush border surface of the proximal tubule of dog kidney (Silverman 1974), since high affinity phlorizin receptors are concentrated at this region. [^{3}H]Phlorizin (New England Nuclear) is administered in vivo by perfusion into the renal vein and is filtered into the lumina before becoming attached to the brush border membranes.

The adoption of domain-specific labelling techniques requires knowledge of the permeability properties of the tissue's intercellular junctions, since confinement of the ligand to the correct surface membrane region is essential. In liver tissue, gap and tight junctions (see § 4.4.5) should prevent the passage of molecules >10,000 MW from the blood to the bile sinusoids; the intervening lateral surface areas of neighbouring hepatocytes will be labelled to varying extents. In general, the tight junctions of tissues constitute a permeability barrier that will ensure the restriction of ligands to the cell surface to be radioactively labelled. Perfusion time of organs or tissues should be short so as to avoid increasing intercellular permeability.

4.7. Fluorescent probes as plasma membrane markers

Maddy (1964) was the first to examine the combination of fluorescent probes with erythrocyte membranes. Maddy's work was extended by Marinetti and Gray (1967) who examined the binding of 4-

acetamido-4′-isothiocyanate-stilbene-2,2′-disulphonic acid (SITS) (British Drug Houses) to isolated liver cells and subcellular fractions from liver. SITS is claimed not to penetrate cells and appears to bind mainly to plasma membrane sulphydryl groups. Its location and quantitation in subcellular fractions is determined by measurement of its fluorescence in strongly alkaline conditions. A further series of fluorescent probes and markers for cell membrane studies have been described by Pincus et al. (1975). Dansyl cadaverine and other dansylated compounds (Cyclo Chemical Corporation, Pierce Chemical Company) are non-covalently incorporated into the surface membrane of cultured mouse lymphoma cells and human fibroblasts, and have potential as plasma membrane markers in subcellular fractionation work, and as probes in exploring membrane fluidity, etc.

Hawkes et al. (1976) have described the use of fluorescamine (Hoffman–La Roche, Inc.) to label cell surfaces. Cultured fibroblasts were exposed for 30 s to fluorescamine (dissolved in acetone and then added to a buffer). The distribution of fluorescent-labelled proteins of cell homogenates was then analysed by electrophoresis in SDS polyacrylamide gels.

Preparation of plasma membranes from certain tissues and cell lines

In this chapter, procedures for the preparation of cell or tissue homogenates (described in Chapter 2), coupled with the general methods available for separating and characterising plasma membranes (Chapters 3 and 4) are amplified in the physiological context of a range of tissues and free and cultured cells. Methods for preparing plasma membrane specialisations are also described. Little attempt will be made to describe comprehensively various aspects of the experimental details in each case, but general preparative guidelines and some problems applicable to the tissues and cell types will be discussed. Methods for plasma membrane preparation should be applicable to small and large amounts of starting homogenates; they should be rapid, not merely for operator convenience but also to minimise membrane breakdown or modification; they should yield plasma membranes demonstrably free from contamination by cytoplasmic material, and if they do not produce a product representative of most of the cell's surface membrane, then the surface region selectively purified should be known.

5.1. Liver plasma membranes

There are probably more papers describing methods for the preparation of liver plasma membranes than for any other mammalian tissue. Most of these miss a major point, namely that the plasma membrane of the hepatocyte is strictly a complex organelle that is differentiated into at least three major functional regions. Since complete plasma membrane 'ghosts', incorporating all three areas are not obtained, the stage is set for the isolation of a permutation of plasma membrane fractions containing one or more of these regions. The relative cellular homogeneity of liver tissue makes it an attractive candidate for subcellular fractionation. However, it should be noted that although hepatocytes account for more than 90% of adult rodent liver weight, they account for only 50–60% of the cell population (Greengard et al. 1972; Blouin et al. 1977). Other cells present in adult liver tissue include erythrocytes (removed by prior perfusion), reticulo-endothelial cells, Kupffer cells (a class of macrophages), fibroblasts (synthesising basement membrane and collagen), the bile ductule cells and fat-storing (stellate) cells. Ruminants have a more well developed reticulo-endothelial network with extensive basal lamina (Gemmel and Heath 1972), a factor worth noting when comparing bovine and rodent subcellular fractions.

The three major functional surface regions of the hepatocyte are the blood sinusoidal, the lateral and the bile canalicular regions, accounting in rodents for about 50, 40 and 10% of the total surface area, respectively. Preparative procedures for plasma membranes from each of these areas are now described.

5.1.1. Lateral and bile canalicular plasma membranes

The method of Neville (1960) and its variants (e.g., Emmelot et al. 1964; Song et al. 1969; Ray 1970) isolates from a low speed 'nuclear' fraction plasma membrane fragments, deriving mainly from the contiguous area of hepatocytes with attached bile canalicular complexes. Some blood sinusoidal derived plasma membranes are also present as judged by the persistence of hormonally activated adenylate cyclase activity (Pohl et al. 1971), but this microvillar area has largely vesicularised during the homogenisation of tissue and is recovered from the 'microsomal' fraction (§ 5.1.2).

Livers, whole (mouse) or in small segments (rat), are homogenised in 1 mM $NaHCO_3$, pH 7.4, in a loose fitting Dounce homogeniser (clearance 0.127 mm, 5–15 up and down strokes) at a tisue to medium ratio of approximately 100 ml/rat liver and 20 ml/mouse liver. The mechanical stability of the bile canalicular complex and the junction containing lateral plasma membranes is enhanced by standing for 15–30 min in the alkaline hypotonic medium; the membranes are sedimented at low speed in the 'nuclear' fraction (Fig. 5.1A; see § 2.4.3). The low speed pellet is multilaminate and is best separated from the supernatant and a loose covering of mitochondria by careful aspiration using polycarbonate centrifugation pots. The liver plasma membrane fragments are resolved from nuclei, mitochondria, vesicles and other components present in the low speed pellet, either by repeated low speed centrifugation (Neville 1960), or by rate-zonal centrifugation in zonal rotors (see § 3.1.8). Separation by floatation (Neville 1968) is a further alternative obtained in swinging bucket rotors by resuspending the low speed pellet in sucrose of density 1.22 g/cm^3 and overlaying with sucrose solutions of density 1.16 and 1.18 g/cm^3. After centrifugation at 50,000 *g* for 1–2 hr, plasma membranes are found at the 1.16/1.18 g/cm^3 density interface, and the nuclei and mitochondria are found in the higher density sucrose. These plasma membranes have been used for a range of studies, but can be subfractionated further as follows. Plasma membranes suspended in, for example, iso-osmotic sucrose, are subfractioned by mechanical treatment (10–20 strokes of a tight-fitting Dounce homogeniser or 10 s in an 'Ultraturrax' homogeniser); on density gradient centrifugation (3 hr at 100,000 *g*) three bands (Z-L, Z-H_A and Z-H_B) are obtained at sucrose density interfaces of 1.12–1.13, 1.15–1.17 and 1.17–1.18 g/cm^3, respectively. The 'light' plasma membrane subfraction Z-L contains mainly bile canalicular membranes, whereas the two 'heavy' subfractions contain membrane strips, intercellular junctions and large vesicles, derived mainly from the lateral plasma membrane region (Wisher and Evans 1975; Evans et al. 1976). The distribution of typical plasma membrane marker enzymes among the liver subfractions prepared as described in Fig. 5.1 is shown in Fig. 5.2.

5.1.2. Blood sinusoidal plasma membranes

The blood sinusoidal microvillar region of the hepatocyte plasma membrane is converted into vesicles during tissue homogenisation, and these are recovered in the 'microsomal' fraction. In the method

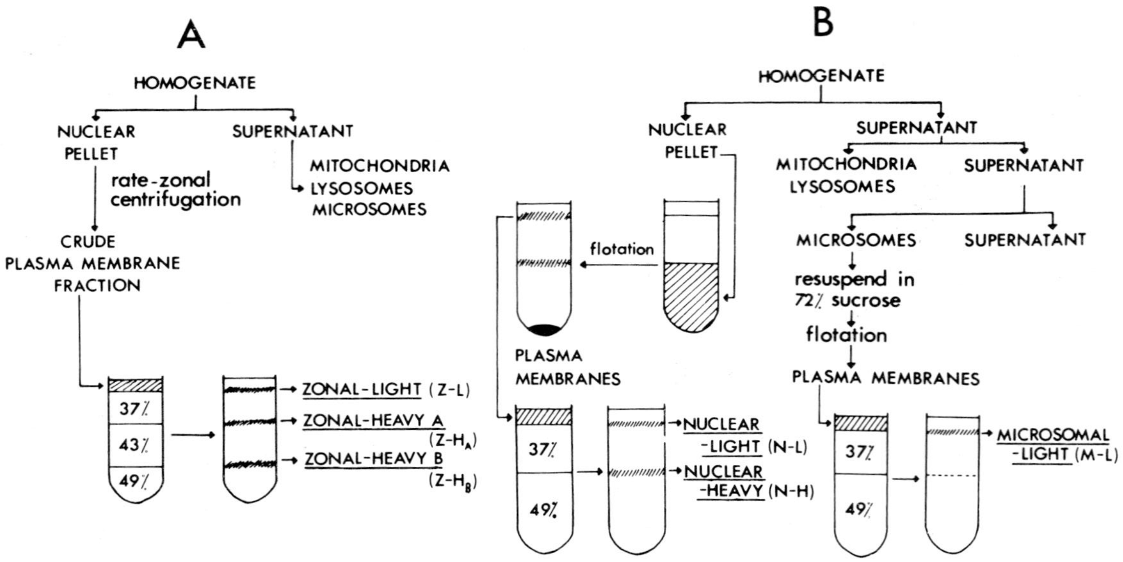

Fig. 5.1. General scheme for the preparation of 6 liver plasma membrane subfractions. In procedure A, livers are homogenised in 1 mM $NaHCO_3$ and the plasma membrane fraction, prepared by a rate-zonal procedure (see Fig. 3.5A) is resolved to yield 3 subfractions (Z-light, Z-Heavy-A and Z-Heavy-B). In procedure B, livers are homogenised in iso-osmotic sucrose and two subfractions (Nuc-light, Nuc-heavy) are prepared from the nuclear pellet, and a Mcs-light subfraction is prepared from the microsomal fraction (Wisher and Evans 1975). For assignment of these fractions to the blood-sinusoidal, contiguous or bile-canalicular areas of the hepatocyte plasma membrane see text, and for details of their enzymic properties see Fig. 5.2.

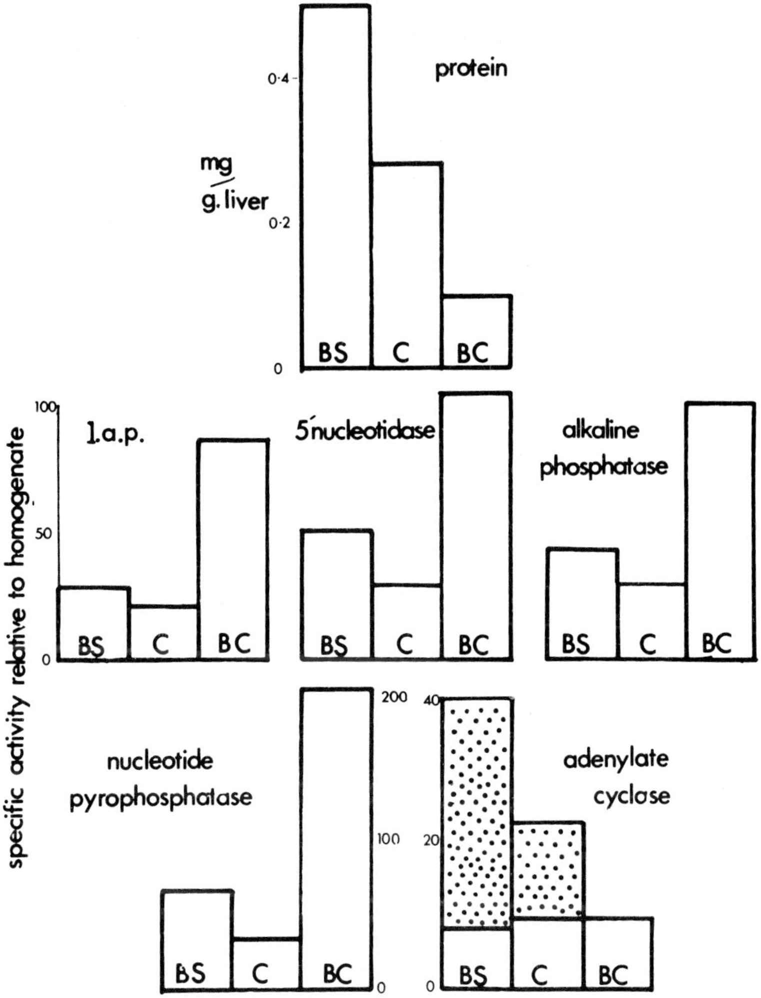

Fig. 5.2. Distribution of 5 plasma membrane marker enzymes among the fractions derived from the three major surface areas of the hepatocyte. Note that the 4 ecto-enzymes (see Table 4.1) have a different distribution to basal (clear histogram) and glucagon-activated (dotted histogram) adenylate cyclase. BS, blood sinusoidal (corresponding to M-L in Fig. 5.1B); C, contiguous (corresponding to Zonal-Heavy-A and Zonal-Heavy-B in Fig. 5.1A), and BC, bile canalicular fraction (corresponding to Z-Light in Fig. 5.1A), l.a.p. = leucine-aminopeptidase. See text for further details.

of Touster et al. (1970) the plasma membranes are prepared by floatation from a 'microsomal' fraction of liver tissue dispersed in iso-osmotic sucrose by using a Potter–Elvehjem homogeniser. This 'microsomally-derived' plasma membrane fraction can be subfractionated by using a tight-fitting Dounce homogeniser followed by sucrose density gradient centrifugation to yield a 'microsomal-light' (M-L) fraction of density 1.12–1.13 g/cm^3 (Fig. 5.1B). The 'microsomal-light' plasma membrane subfraction is enriched 20–30-fold in plasma membrane markers (see fig. 5.2) and it shows an adenylate cyclase activity that is highly stimulated by glucagon. Morphological and enzymic (galactosyl and sialyl transferase activities) analyses (Wisher and Evans 1975) indicate that this plasma membrane subfraction, unlike other subfractions in Fig. 5.1 is contaminated by Golgi apparatus components (§ 4.5.3).

5.1.3. Plasma membranes from hepatoma

To prepare hepatoma plasma membranes, Emmelot and Bos (1966) modified their procedure, as applied to normal liver, by including in the homogenising medium either dilute citric acid (2.8 mM, pH 6.0) or 2 mM $CaCl_2$ (cf. Ray 1970) to stabilise the nuclei against disruption. Hepatoma plasma membranes are then prepared from homogenates in the same way as liver tissue membranes, but they have a lower density on sucrose gradients (1.14 g/cm^3), suggesting that junctional complexes are reduced in number or absent. The biochemical and morphological complexity of the liver plasma membrane fraction isolated at different densities stresses that direct comparison between membranes of normal liver and hepatoma is full of pitfalls.

5.1.4. Plasma membranes from isolated hepatocytes

The increasing use made of isolated hepatocytes for metabolic studies makes it desirable to prepare plasma membranes from these cells. The inclusion of a soybean trypsin inhibitor in the collagenase-containing perfusion mixture introduced into the liver via the portal vein to dissociate the tissue into its constituent cells helps to increase the recovery of plasma membrane marker ectoenzymes (e.g., 5′-nucleotidase and alkaline phosphodiesterase) in the three plasma membrane subfractions prepared from hepatocytes according to the scheme shown in Fig. 5.1B. To disrupt isolated liver cells, a tight-fitting Dounce homogeniser is required (Wisher and Evans 1977).

5.2. *Central nervous system plasma membranes*

The brain is a complex regionalised organ composed of many cell types. The most important cells are the neurons (i.e., excitable cells capable of transmitting an impulse) and the glial cells; these are present in approximately equal amounts although this varies according to the age of the animal. Functionally, neurons are complex and in the cerebellar cortex, for example, there are five types, i.e., pyramidal, purkinje, granular, stellate and basket cells. Glial cells are divided into two major classes, astrocytes and oligodendrocytes; the latter class is involved in myelination. Superimposed on the complex regionally-variable cellular heterogeneity of brain tissue, is the topographical specialisation of the neuronal plasma membrane into synaptic, dendritic, axonal and perikaryal surface areas.

The neuronal synapse and its plasma membranes have attracted a great deal of attention in view of their involvement in intercellular communication, neurotransmitter release, etc. A neuron in the central nervous system may have over the surface area up to 80,000 synapses. The major efforts have been directed to the preparation and characterisation of the synaptic area of the neuronal plasma membrane. However, the functional heterogeneity of the nerve-ending (synaptosome) fractions used to prepare synaptosomal plasma membranes is again illustrated by variation in the neurotransmitter content of synaptic vesicles (Snyder et al. 1973; Bretz et al. 1974). Plasma membranes of varying degrees of purity have been prepared from different regions of the brain (summarised in Table 5.1), and procedures that result in separation of neuronal and glial cells for preparation of membranes have been described, e.g., Bloomstrand

TABLE 5.1
Preparation of subcellular fractions including plasma membranes from areas of the nervous system.

Region	Fraction prepared	Source	Reference
Hypothalamus	Nerve-ending	Rat	Bradford et al. (1973)
Cerebellum	Glomerular	Rat	Tapia et al. (1974)
Olfactory nerve	Axonal plasma membrane	Garfish	Chacko et al. (1974)
Optic tectum and neural retina	Plasma membrane	Chick embryo	Gottlieb et al. (1974)
Cerebral cortex	Perikaryal plasma membrane	Young rats	Hemminki and Suovaniemi (1973)
Neuroblastoma cultures	Neuronal plasma membrane	Cultured cells	Truding et al. (1974)
Cerebral cortex	Glial plasma membrane	Rat glial cells	Levitan et al. (1972)
Cerebral cortex	Neuronal plasma membrane	Rabbit neuronal cells	Henn et al. (1972)
Embryonic and neonatal cerebral cortex	Nerve-ending	Chick embryo Rat (5-day-old)	Den et al. (1975) Norman and Howard (1977)
Anterior pituitary	Plasma membrane	Bovine pituitaries	Poirer et al. (1974)
Neural glands	Synaptosomal plasma membrane	Bovine pituitaries	Vilhardt et al. (1975)

and Hamberger (1970). Cohen et al. (1974), Sellinger and Azcurra (1974), Chao and Rumsby (1977).

5.2.1. Synaptosomal plasma membranes

When brain tissue, especially the cerebral cortex, is subjected to controlled mild liquid shearing forces in iso-osmotic sucrose (§ 2.4.1), synapses become detached from their terminal axons and seal up to form fragments of characteristic morphology termed synaptosomes or nerve ending particles. During subcellular fractionation, the synaptosomes sediment mainly in the conventional mitochondrial fraction and they can then be separated from mitochondria, myelin fragments, etc., by density gradient centrifugation (Gray and Whittaker 1962; De Robertis et al. 1963). Isolated synaptosomes prepared by these methods retain a complex morphology, consisting of a plasma membrane with a localised thickening (with an attached post-synaptic fragment) enclosing intraterminal mitochondria and synaptic vesicles (Fig. 5.5). Synaptosomes have been observed to form after homogenisation of the cerebral neocortex, hippocampus, caudate nucleus, spinal cord, hypothalamus and hypophysis (Bradford 1974). The synaptosome fraction, after exposure to controlled hypotonic shock that releases the constitutive membranes and organelles, can be resolved further by density gradient centrifugation on multi-step gradients constructed of either sucrose, Ficoll or sodium diatrizoate (Tamir et al. 1976). Synaptosomal plasma membranes prepared by these procedures, when initially analysed, are impure by a variety of criteria (see below). Furthermore, the synaptosomal plasma membrane can be subjected to a variety of detergent treatments, yielding fractions enriched in post-synaptic densities (§ 5.9.3).

Preparation of purified synaptosomal plasma membranes The flow sheet (Fig. 5.3) summarises the methods described by Morgan et al. (1971) and Gurd et al. (1974) to prepare synaptosomal plasma membranes. Essentially, these methods are developments of the procedures originally described by Whittaker et al. (1964) and Rodriguez de Lores Arnaiz et al. (1967). Commencing with cerebral cortices trimmed free of white matter from up to 120 adult rats, a synaptosomal plasma membrane fraction claimed to be more than 80% pure judged by enzymic, chemical and morphological criteria, and containing approx. 15 mg of membrane protein, was finally recovered. Cerebral cortices are homogenized at a ratio of 10 g tissue per 100 ml of 0.32 M sucrose containing 0.1 mM EDTA and 1 mM phosphate

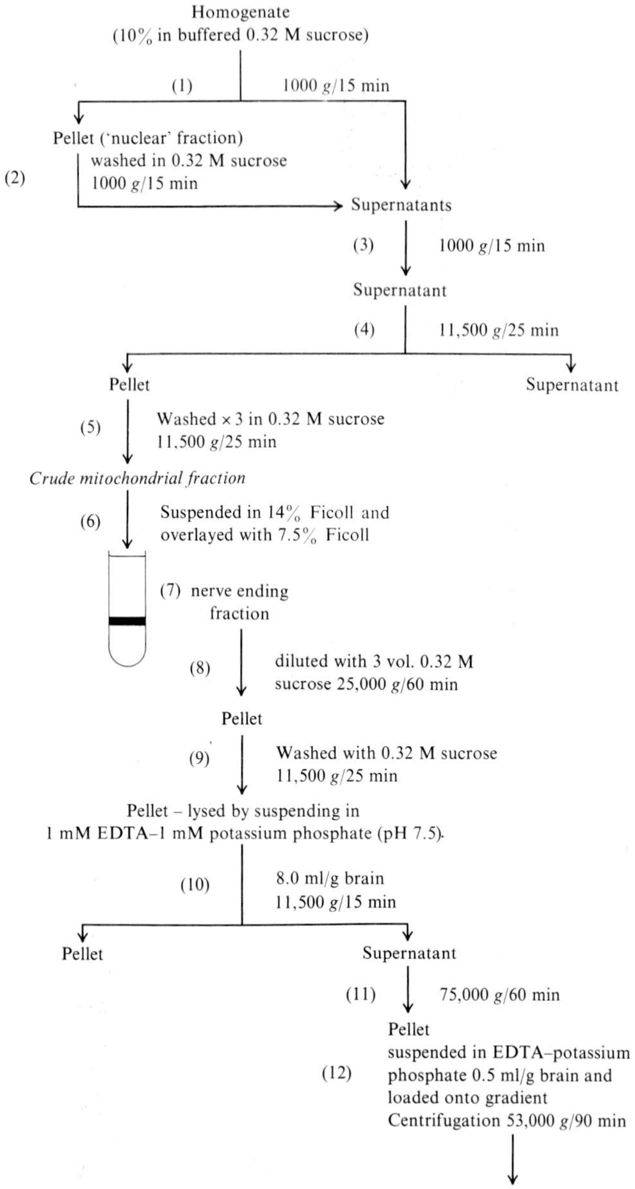
Homogenate
(10% in buffered 0.32 M sucrose)
(1)
1000 g/15 min
Pellet ('nuclear' fraction)
washed in 0.32 M sucrose
1000 g/15 min
(2)
Supernatants
(3)
1000 g/15 min
Supernatant
(4)
11,500 g/25 min
Pellet
Supernatant
(5)
Washed × 3 in 0.32 M sucrose
11,500 g/25 min
Crude mitochondrial fraction
(6)
Suspended in 14% Ficoll and
overlayed with 7.5% Ficoll
(7) nerve ending
fraction
(8)
diluted with 3 vol. 0.32 M
sucrose 25,000 g/60 min
Pellet
(9)
Washed with 0.32 M sucrose
11,500 g/25 min
Pellet – lysed by suspending in
1 mM EDTA–1 mM potassium phosphate (pH 7.5).
(10)
8.0 ml/g brain
11,500 g/15 min
Pellet
Supernatant
(11)
75,000 g/60 min
Pellet
suspended in EDTA–potassium
phosphate 0.5 ml/g brain and
loaded onto gradient
Centrifugation 53,000 g/90 min
(12)

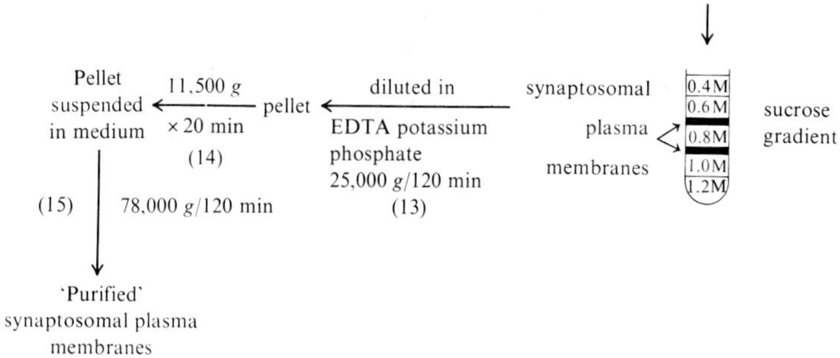

Fig. 5.3. Preparation of rat brain synaptosomal plasma membranes. See text for further details.

buffer (pH 7.6), using a motor-driven Potter–Elvehjem homogeniser (clearance between Teflon pestle and glass 0.12 mm) rotating at about 800 revs/min. Approximately 12 up and down strokes are used. The 1000 $g \times 15$ min supernatant is subjected to a further low speed centrifugation (step 3) before a crude mitochondrial pellet is prepared (step 4). Extensive washing of the mitochondrial pellet is carried out (step 5) to reduce the contamination of this fraction by vesicles that normally sediment at higher gravitational forces in the microsomal fraction. The gravitational force used to sediment the mitochondrial fraction is also lower than that used in earlier schemes, a further modification introduced to reduce the contamination of this fraction by light membranes and myelin fragments. The nerve-ending particles (synaptosomes) present in the mitochondrial fraction are separated by floatation from a 14% Ficoll solution into an overlay of 7.5% Ficoll (steps 6 and 7). Ficoll solutions dissolved in isotonic sucrose are first prepared by making up a stock 40–50% Ficoll solution in water and dialysing this extensively against distilled water before dilution to the desired concentration. It is often convenient to float the synaptosomal fraction by overnight centrifugation, but at least 3–4 hr at 97,000 g is necessary. The time necessary to effect separation can also be reduced by centrifuging the crude mitochondrial fraction into a discontinuous Ficoll gradient. Many advantages are claimed over sucrose gradients for Ficoll/iso-osmotic sucrose gradients in the preparation of nerve-ending particles (Morgan et al. 1971).

The nerve-ending (synaptosmal) fraction is first washed in iso-osmotic sucrose (steps 8 and 9) and then lysed by suspending in EDTA phosphate buffer. The conditions for optimal lysis of nerve-endings are poorly defined, but approx. 15 min at room temperature with occasional mild disruption by passage up and down a narrow syringe needle is effective. Cotman and Matthews (1971) claim that plasma membrane markers separate better from mitochondrial markers if the synaptosomes are lysed at pH 8.4. After further low and high speed centrifugations (steps 10 and 11), the disrupted synaptosomal fraction is loaded onto a discontinuous sucrose gradient, and the synaptosomal plasma membranes are collected at the 0.6/0.8 and 0.8/1.0 M sucrose interfaces. The amount of unlysed synaptosomes at the 1.0/1.2 M interface indicates the efficiency of lysis. Further purification, albeit at the cost of yield, may be obtained by centrifuging the synaptosomal plasma membranes repeatedly at low speed, to sediment residual mito-chondria leaving the synaptosomal plasma membrane fragments in the supernatant (steps 13 to 15).

An abbreviated method for the preparation of rat synaptosomal plasma membranes is described by Jones and Matus (1974). The membranes are prepared from the crude lysed mitochondrial pellet by a one-step floatation-sedimentation method (Fig. 5.4). These synaptic plasma membranes, in contrast to those produced by the method shown in Fig. 5.3, are characterised morphologically by synaptic and post-synaptic thickenings (Fig. 5.5) and are thus suitable for the preparation of post-synaptic densities (§ 5.9.3). Their enzymic marker content is summarised in Table 5.2. Polyacrylamide gel analysis of a rat brain synaptosomal plasma membrane fraction prepared by this method is shown in Fig. 4.5. Babitch et al. (1976) describe modifications of these procedures suitable for chicken brain homogenates.

Assessing the purity of neuronal and synaptosomal plasma membranes

There are few exclusive synaptosomal or neuronal plasma membrane

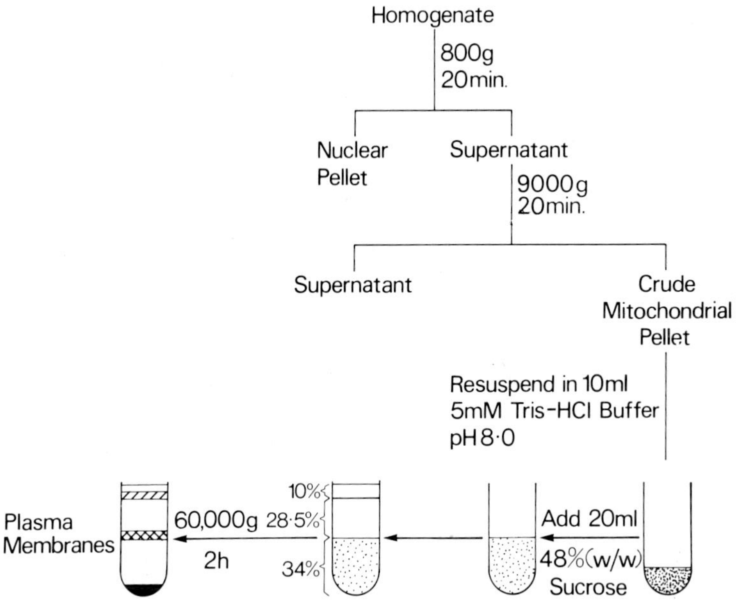

Fig. 5.4. A rapid method for preparation of rat brain synaptosomal plasma membranes. Volumes apply to six rat forebrains homogenised in 9 volumes of 10% (w/w) sucrose using a Potter–Elvehjem homogeniser. Lysis of the crude mitochondrial fraction is effected by resuspension in 5 mM Tris–HCl buffer, pH 8.1, for 30 min at 0 °C followed by 6 complete strokes in the homogeniser. (Jones and Matus 1974)

enzymic markers and some general plasma membrane markers are used. This, of course assumes that there is little or no distinction between the various areas of the neuronal surface in the distribution of these markers. By analogy with other cells showing differentiated function, e.g., intestinal and renal columnar epithelia, hepatocytes, this is unlikely to be a valid assumption.

The most widely used neuronal plasma membrane marker is (Na^+-K^+)-ATPase, an enzyme believed to participate in transporting Na^+ and K^+ across the cell membrane. Supporting its role as a general neuronal plasma membrane marker is the histochemical demonstration that it is present at least on the synaptosomal plasma membrane (Daniel and Guth 1975). However, the enzyme is also

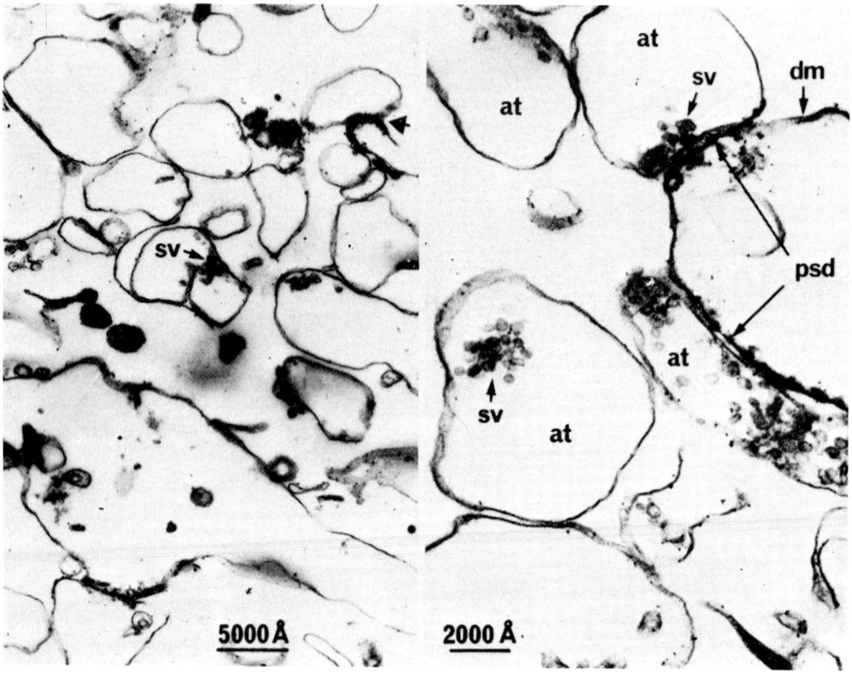

Fig. 5.5. Synaptosomal plasma membrane fraction prepared from rat forebrain. sv, synaptic vesicles trapped inside presynaptic axonal terminal boutons (at). dm, dendritic membrane bearing post-synaptic densities (psd). Electron micrograph by courtesy of Dr. A. Matus and reproduced from Matus and Walters (1975) by permission of Chapman and Hall.

present on glial membranes (Henn et al. 1972) and there is evidence that glial membranes contaminate synaptosome derived fractions (Cotman et al. 1971; Henn et al. 1976). Values for the level of enrichment of the (Na^+-K^+)-ATPase in different preparations of synaptosomal plasma membranes over the homogenates vary from 2.5 to 14 (Table 5.2). Such differences in the level of enrichment of (Na^+-K^+)-ATPase in putative neuronal or synaptosomal plasma membranes prompt the question: what is the maximum theoretical yield of such a marker enzyme? The neuronal surface area in 1 g rat

TABLE 5.2
A survey of synaptosomal plasma membrane marker enzymes.

Authors	Source	Enrichment over homogenate			Yield of tissue (mg/g wet weight of tissue)
		$(Na^{+}$-$K^{+})$-ATPase (ouabain-sensitive)	Acetyl-cholinesterase	5′-Nucleotidase	
Cotmaṇ and Matthews (1971)	rat forebrain	2.5–3.5	1.36	1.24	1–2
Morgan et al. (1971)	rat brain	9–14	0.5–1.0	1–4	0.1–0.2
Levitan et al. (1972)	rat brain	7.6	n.d.	n.d.	5
Gurd et al. (1974)	rat cerebral cortices	8–10	n.d.	n.d.	0.25
Jones and Matus (1974)	rat forebrain	1.85	1.2	n.d.	1

n.d. = not determined

brain neocortex is estimated to be 10^4 cm^2, and assuming an average membrane density of 1.2 g/cm^3, a membrane thickness of 100 Å and a protein/lipid ratio of 2 : 1, it is calculated that 1 g rat neocortex should yield maximally 8 mg of neuronal plasma membrane protein (Morgan et al. 1971). It follows that if a putative marker is exclusively present on the neuronal plasma membrane, the maximum enrichment in such a fraction over the homogenate would be about 12. However, if the marker is present at higher concentrations at the synaptic area of the surface relative to the non-synaptic area, this enrichment figure should increase and the protein yield correspondingly decreases if the synaptosomal plasma membrane fraction is obtained. Gurd et al. (1974), by assuming, (a) average volume of a nerve terminal of 0.1 μm^3 of which 8% is taken up by the external membrane, (b) the presence of 10^{10} synapses per g cortex (Whittaker 1969), (c) an average membrane density of 1.2 g/cm^3 and (d) 50% of membrane is protein, calculated that 1 g cortex should yield maximally 0.5 mg of synaptosomal plasma membrane protein. Variations in the yield of protein per g wet tissue are summarised in Table 5.2.

Acetylcholinesterase is not an ideal neuronal plasma membrane marker, for it may be unequally distributed among various categories of neurons and thus lower enrichment values for this enzyme in neuronal or synaptosomal plasma membranes are to be expected (Table 5.2). Although 5′-nucleotidase has proved a useful plasma membrane marker in most mammalian tissues, this does not appear to be so in the mammalian brain where variable degrees of enrichment for this enzyme have been reported. Such variation may be indicative of an uneven distribution of the 5′-nucleotidase on plasma membrane fragments of varying densities reflecting an uneven distribution on the cell surface. For example, Van Leeuwen et al. (1976) found in chick brain a population of membranes with high 5′-nucleotidase activity that co-sedimented with 'light' mitochondria. Further complicating the issue are reports (e.g., Podulso 1975; Henn and Hamberger 1976) that glial cell membranes of rat brain contain high 5′-nucleotidase activity.

The problem of the contamination of neuronal plasma membranes by endoplasmic reticular membranes is not helped by the paucity of our knowledge about the biochemical properties of these intracellular membranes. Cotman et al. (1971) reported that endoplasmic reticulum markers were present in a synaptosomal plasma membrane fraction at one-third their levels in a 'microsomal' fraction. The chemical and biochemical composition of mammalian cerebral cortex membranes is made more uncertain by the report that rat synaptosomal and microsomal membranes possessed similar polypeptide profiles when electrophoresed in polyacrylamide gels (Gurd et al. 1974). It may be concluded that the microsomal fraction of mammalian brain consists of a heterogeneous population of fragments of endoplasmic reticulum, nerve endings, small axons and plasma membranes. The status and subcellular distribution of Golgi organelles in the brain is as yet unknown, thus further complicating this issue. There are contradictory reports of the presence of sugar transferase enzymes in nerve-ending fractions.

Myelin contamination of neuronal plasma membrane fractions is usually assessed by electron microscopy of stained sections. Although present in myelin (despite the growing realisation of the importance of cyclic nucleotides in isolated or cultured neurons and the modulation of cyclic nucleotide levels by drugs, e.g., analgesics) the enzyme activity is also present in considerable amounts in glial membranes (Zanetta et al. 1972). Since glial cells generate the myelin sheath, the presence of the cyclic AMP phosphohydrolase activity may be an indicator of the extent of contamination of neuronal by glial plasma membranes.

To assess the presence of lysosomes in cerebral fractions, acid glycerophosphatase, β-glucosidase and acid phosphatase have been measured in brain subfractions. Cotman and Matthews (1971), who reported considerable amounts of lysosomal activities in synaptic membranes, washed away nearly all activity with 0.1% Triton X-100. Gurd et al. (1974) also reported high activities of lysosomal activities in their synaptosomal plasma membranes. Synaptic vesicles are not a major source of contamination of plasma membranes, in view of their lower buoyant density in sucrose; thus acetylcholine-containing vesicles are of density 1.07 g/cm^3; this contrasts with the higher densities of the neurosecretory granules of the pituitary (1.22 g/cm^3) and the chromatin granules of adrenal medulla (1.28 g/cm^3).

The conclusion emerges from studies on the properties of plasma membrane preparations classified as neuronal or synaptosomal from mammalian brain namely that the more comprehensive the enzymic characterisation of the 'purified' putative membrane fraction, the more heterogeneous it appears to be. Morgan et al. (1971) in their multi-step procedure for preparing synaptosomal plasma membranes from rat cerebral cortices (Fig. 5.3) claimed a higher than 80% purity for this fraction; this has now been lowered to 50–75% on the basis of further intensive characterisation (Morgan 1976). In summary, insufficient markers exist to allow accurate determination of the purity of neuronal plasma membranes, and the extent of the contribution of synaptosomal, dendritic and glial plasma membranes has yet to be elucidated. Separation of neurons and glia prior to membrane preparation is a possible approach, but the use of antigens specific to one or other classes of cells that may prove useful as markers has not yet yielded clear-cut answers.

5.3. Plasma membranes from kidney

Membranes prepared from kidneys derive mainly from the epithelial cells of the uriniferous tubules, although some contribution is made by the endothelial and mesangial (phagocytic) cells of the glomerulus. The proximal convoluted tubule that leads from the glomerular capsule consists of columnar and cubical epithelial cells with highly differentiated borders on the luminal side. The brush border consists of a parallel array of microvilli extending into the tubule lumen. The distal convoluted tubules also consist of cubical epithelial cells, but the cells lack a well-developed brush border. The proximally located cells are concerned with transepithelial transport of molecules to be reabsorbed from the glomerular filtrate. The renal medullary region is involved in the concentration of urine, and these membranes are used in studying the effects of vasopressin and antidiuretic hormones. The effects of parathyroid hormone are usually studied by using renal cortex membranes (Marx et al. 1972).

The above indicates that plasma membrane fragments with varying properties should be obtained from the various epithelial cells constituting the nephron unit. Initially, methods were directed at isolation of the brush border region of the plasma membrane, followed by the establishment of enzymic and chemical markers. Methods are now available also for the preparation of the epithelial baso-lateral membranes and their separation from brush border membranes, and this has helped to clarify the distribution of sites of hormone action and extent of enrichment of enzymic markers in plasma membrane fractions (Schwartz et al. 1974). In view of the different target cells of various hormones (§ 5.3.4) it is preferable to dissect the medullary and cortical regions before preparing subcellular fractions.

The isolation and biochemical characterisation of brush border (microvillar) and the baso-lateral membranes of renal cortices is an important achievement in view of the different functions of these plasma membranes. The extent of contamination of kidney plasma membranes by Golgi membranes is rarely assessed, but the availability of a specific marker enzyme, e.g., cerebroside sulphotransferase (Fleischer and Zambrano 1974) (see § 4.2.3) will help to correct this. Methods for the separation of glomerular and tubular segments of kidney that utilise collagenase treatment and separation on sucrose gradients have been described (Helwig et al. 1974, 1975) and preparation of plasma membranes from more defined areas of kidney should ensue.

5.3.1. Brush border plasma membranes

The preparation of brush borders from rat and rabbit kidneys was described by Thuneberg and Rostgaard (1968), Fitzpatrick et al. (1967) and Burger and Sacktor (1970) who used methods analagous to that of Miller and Crane (1961) for intestinal brush borders. Mild homogenisation of kidneys releases the brush borders as large fragments, and these are separated from mitochondria and vesicular membranes by repeated low speed centrifugation. A range of homogenisation procedures is summarised in Table 2.7, and in general a loose-fitting Dounce or Potter–Elvehjem homogeniser is used. Gentle treatment of liberated brush borders at 0–4 °C during the resuspension of pellets produced by repeated centrifugation helps to keep these structures intact. Wilfong and Neville (1970) elaborated the general procedure by introducing a sucrose floatation step that helped minimise the number of differential centrifugation steps and yielded a purer product. To prepare brush borders in bulk, zonal rotors have been used (§ 3.1.8) (e.g., Price et al. 1972; Quirk and Robinson 1972; Fleischer and Zambrano 1974). The two-phase dextran/polyethyleneglycol separation system (§ 3.2.1) was used by Glossman and Gips (1974) to speed up the Wilfong and Neville (1970) procedure.

In detail, rabbit brush borders are prepared rapidly as follows. Rabbit cortex is dispersed by forcing through a tissue press equipped with a wire gauze (20 mesh per cm) or using a loose-fitting Dounce homogeniser. The homogenate is diluted to 10% (w/v) with 0.5 M sucrose and samples (22 ml) are layered over 17 ml of 1.5 M sucrose and centrifuged at 50,000 *g* for 40 min in a Beckman SW27 rotor. The supernatant and the material at the interface are combined, diluted and centrifuged at 7000 *g* for 15 min. The fluffy layer at the top of the pellet is carefully collected and may be further purified by centrifugation at 4000 *g* for 10 min (Thuneberg and Rostgaard 1968; Booth and Kenny 1976).

5.3.2. *Brush border microvilli*

Microvilli are prepared from isolated brush borders by homogenisation for 20 s in the Ultraturrax machine (§ 2.1.8) followed by their collection from the supernatant of a low speed centrifugation step. Microvilli are used to study peptidases and disaccharidases present in this plasma membrane region concerned with absorption. Commencing with rabbit kidneys, the procedure shown in Fig. 5.6 is used. Approx. 20 g rabbit cortex is homogenised in 200 ml of 10 mM mannitol, 2 mM Tris–HCl, pH 7.4, in a Waring blender at full speed

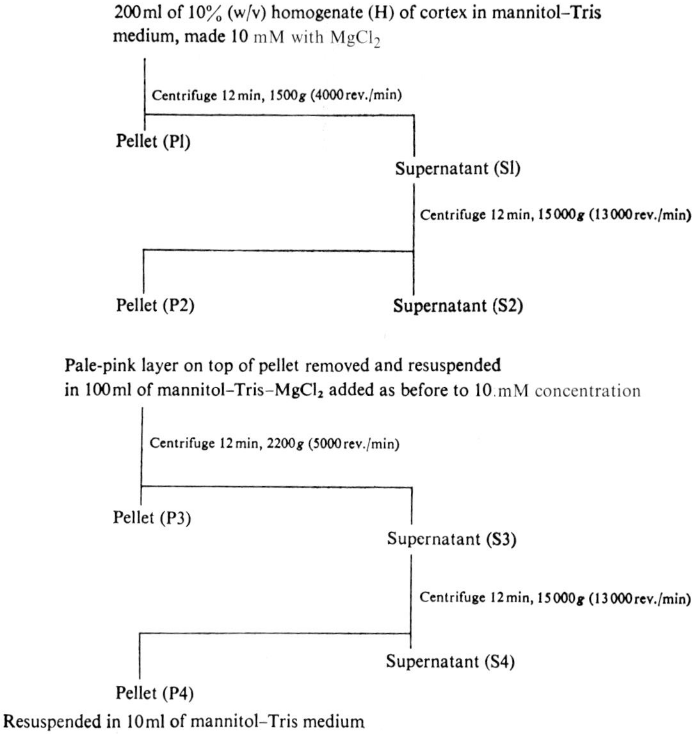

Fig. 5.6. Preparation of a kidney cortex microvillus fraction. (Booth and Kenny 1974.)

for 2 min. The homogenate is adjusted to 10 mM $MgCl_2$ and stirred for 15 min at 4 °C and fractionated as shown above.

5.3.3. Baso-lateral plasma membranes

Renal plasma membrane fractions described as brush border membranes usually consist of varying proportions of brush borders and baso-lateral derived membranes. Separation of membranes from both areas of the renal epithelial cells was obtained by conventional subcellular fractionation (Liang and Sacktor 1977) and by free-flow electrophoresis (Heidrich et al. 1972; § 3.4.1) and this helped establish, for example, that alkaline phosphatase, 5′-nucleotidase, peptidases and disaccharidases were primarily located at the brush border plasma membrane, whereas (Na^+-K^+)-ATPase and adenylate cyclases were baso-lateral plasma membrane markers. The preparation of baso-lateral membranes from rat kidneys is as follows. Rat kidney cortices are homogenised in 0.25 M sucrose, 1 mM EDTA, 1 mM Tris–HCl, pH 7.6, in a Potter–Elvehjem homogeniser (clearance 0.15 mm, 5 strokes at 45 revs/min and 25 strokes at 90 revs/min) at a tissue to medium ratio of 1 : 5. After filtration through a nylon mesh, the baso-lateral membranes in the homogenate are separated from nuclei, large mitochondria and intact brush borders by two centrifugation steps at 1350 *g* for 10 min, when they remain in the supernatant, followed by repeated centrifugation at 2500 *g*/20 min to separate them as a pellet from light mitochondria and vesicular elements that remain in the supernatant. The final pellet is then resolved by a density gradient centrifugation step on sucrose steps of density 1.14, 1.15, 1.18, 1.20 g/cm^3. The membranes collect mainly at the 1.14–1.15 g/cm^3 interface and are washed by centrifugation (Ebel et al. 1976). The baso-lateral enriched plasma membrane fractions have a (Na^+-K^+)-ATPase activity approx. 20-fold that of the homogenate, whereas alkaline phosphatase was increased 2-fold. The marker enzyme distribution is approximately the reverse of that obtained with brush border and microvillar preparations. The fraction, although shown morphologically to contain some brush border elements, contains numerous junctional complexes (gap junctions, desmosomes).

5.3.4. Kidney medulla plasma membranes

The medulla is the best source of plasma membranes for the study of antidiuretic hormone-stimulated adenylate cyclase. While the parathyroid-sensitive adenylate cyclase is found mainly in plasma membranes prepared from the cortex, the vasopressin-sensitive enzyme is concentrated in the medulla (Neer 1973), but the calcitonin and isoproterenol-sensitive adenylate cyclase is found in both zones. Marx et al. (1972) describe a method for preparing medullary plasma membranes similar to that described for baso-lateral plasma membranes (§ 5.3.3). The biochemical and hormonal sensitivity differences between apical (brush border) and baso-lateral plasma membranes are not so clear-cut in the distal tubules present in the medulla, and the separation of the topographical areas has not been achieved in such a clear-cut fashion as with the cortically derived membranes.

5.4. Intestinal epithelial brush border and baso-lateral plasma membranes

Epithelial cells (enterocytes) scraped from mucosa of saline washed everted small intestines have a similar surface morphology to the proximal cells of the kidney tubule (§ 5.3). Hence, methods described for the preparation of kidney and intestinal brush borders and baso-lateral plasma membrane fractions correspond closely, as also do many of the biochemical and morphological properties of the isolated fractions. Microvillar membranes are prepared by the disruption of isolated brush borders as described for kidney cells, and these have featured in studies of the role of enzymes, e.g., disaccharidases, alkaline phosphatase and aminopeptidases in the absorptive function of these plasma membranes.

The methods generally followed for preparation of intestinal 'scrape' brush borders are those of Miller and Crane (1961), Eichholz and Crane (1965), Forstner et al. (1968) and Schmitz et al. (1973). The basic rationale involves the gentle disruption (see Table 2.6 for methods) of intestinal cell sheets (Evans et al. 1971) followed by repeated low speed differential centrifugation to sediment the brush borders from other components; brush borders are subsequently separated from nuclei by centrifugation on sucrose cushions. The brush border preparations may then be fractionated to separate microvilli from a fibrillar residue (Forstner et al. 1968; Mooseker 1976). The isolation of a purified baso-lateral membrane fraction has proved a more difficult task since they are not readily separated from mitochondria due to similar sedimentation coefficients and densities (see Peters 1976 for modal equilibrium densities in sucrose of human jejunal epithelial cell components). The use by Lewis et al.

(1975) of indigenous enzymic markers in combination with covalently attached radio-active iodine to identify intestinal microvillar and baso-lateral plasma membranes is described in § 4.6.1.2, and the application of density perturbation techniques to modify selectively the density of intestinal plasma membranes is described in § 3.1.9. Satisfactory methods for preparation of predominantly baso-lateral plasma membranes are described by Quigley and Gotterer (1972), Fujita et al. (1972) and Murer et al. (1976). Techniques for labelling the intestinal plasma membrane glycoproteins and glycolipids are described in § 4.6.3.

Assessing the purity of brush border and baso-lateral plasma membranes The baso-lateral plasma membrane of intestine has a specific activity of (Na^+-K^+)-ATPase up to 30-fold higher than that of the homogenate. The modal equilibrium density in sucrose of baso-lateral membranes is 1.12 g/cm^3. Mitochondrial contamination is checked by determining marker enzyme activities (inner and outer membranes). Adenylate cyclase also co-purifies, as in kidney membranes, with the (Na^+-K^+)-ATPase (Murer et al. 1976). The brush border fraction is enriched to a similar extent in sucrase, trehalase, alkaline phosphatase and leucylnaphthylamidase activities. Brush border (modal density in sucrose gradients 1.21 g/cm^3) and derived fractions should be checked for contamination by endoplasmic reticular membranes (density 1.16 g/cm^3) and Golgi membranes. Continuous free-flow electrophoresis (§ 3.4), successfully used to separate baso-lateral and brush border plasma membranes of kidney, has not been applied to intestinal plasma membrane fractions. Chemical analysis of the lipids and proteins of the membranes indicates that brush border microvillar membranes contained higher amounts of cholesterol and glycolipids than the baso-lateral membranes (Kawai et al. 1974).

5.5. Plasma membranes from lymphocytes

A range of methods for preparing lymphocyte plasma membranes is discussed and the properties of the membranes are compared in

§ 2.4.2. A method for preparing lymphocyte plasma membrane from lymphoid organs and cultured lymphocytes in which the cells are broken by passage through a pump (§ 2.1.5) is shown in Fig. 5.7. It is preferable to separate lymphocytes (e.g., teased from mouse spleens by chopping in a parsley chopper or lymph nodes by passage through a wire mesh) from other cells and 'dead' lymphocytes by gradient centrifugation as follows. Cell suspensions from spleen or thymus in saline are added to a cushion of Isopaque/Ficoll solution (14%, w/v) Ficoll and 32.8% (w/v) metrizoate (Isopaque) solution (Nyegaard and Co., 12 : 5 v/v) and centrifuged in plastic tubes at 2000 $g \times 15$ min at 20 °C in a swinging bucket rotor. Lymphocytes are collected from the interface and are washed in 0.15 M saline, 10 mM Tris–HCl, pH 7.4 (Davidson and Parish 1975; Abney et al. 1976). A large scale isolation method described by Kornfield and Siemens (1974) is suitable for frozen calf thymus or human spleen and the tissue homogenate, prepared by using a Potter–Elvehjem homogeniser, is also fractionated as shown in Fig. 5.7. The plasma membrane marker enzymes 5′-nucleotidase, alkaline phosphodiesterase, γ-glutamyl transpeptidase, were enriched 15-fold over the homogenate values and between 15–45% of the homogenate marker enzyme activity was recovered (see Table 2.4). The isolation of two plasma membrane fractions from lymphoid tissue by Demus (1973) and Jett et al. (1977) from a human lymphoma cell line indicates that the lymphocyte plasma membrane is probably complex. In addition, the functional heterogeneity of lymphocyte populations (e.g., T and B subpopulations) combined with variations in surface topography ('bald' or 'hairy') with environmental factors such as temperature and position in the cell cycle, etc. (Alexander and Wetzel 1975) may also result in heterogeneity of isolated plasma membranes. Contamination of lymphocyte plasma membranes by membranes of the Golgi apparatus is rarely determined despite an anticipated similar isopycnic density.

In the isolation of leukemic lymphocyte plasma membranes similar methods using tight-fitting Dounce homogenisers (Marique and Hildebrand 1973) and, in contrast, gentle homogenisation through

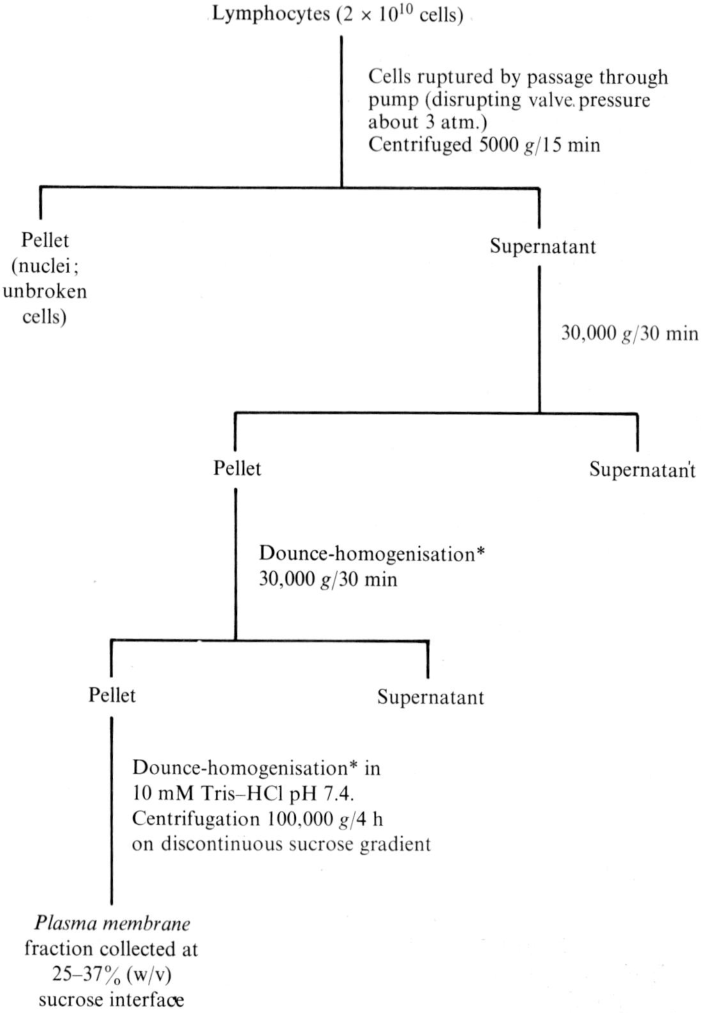

Fig. 5.7. Preparation of lymphocyte plasma membranes. A mouse spleen yields approx. 1.5×10^9 cells. Cell rupture and density gradient techniques elaborated in the relevant sections. From 1×10^9 cells, approx. 1 mg.membrane protein is recovered from cultured lymphocytes and 0.5 mg from lymphocytes teased from lymphoid organs (Crumpton and Snary 1974).

* A tight-fitting Dounce homogeniser is used.

syringe needles (cf. fat cells; § 2.3.4; Dods et al. 1972) and by stirring cell suspensions in hypotonic borate buffers (Warley and Cook 1973) have been described.

5.6. *Plasma membranes from cultured cells*

Cultured mammalian cells have proved useful in investigating the role of the plasma membrane in growth control, metabolism, metastasis, etc. One of the main advantages of using cultured cells is the ease with which they divide rapidly to provide large uniform cell populations under conditions that can be closely controlled. Cultured fibroblastic cells and their virally or chemically 'transformed' counterparts have served as models to investigate cell surface modifications that may be related to malignancy. Most studies have used rapidly dividing fibroblastic cell lines derived from tumours. Epitheloid lines have been used to a lesser extent to investigate biochemical modifications of the cell surface that correlate with tumour formation in carcinomas (for review see Tooze 1973). Properties of cultured cells that can influence the quality and yield of the membranes include the following. A population of cultured cells may consist of three major categories of cells – growing, non-growing and dead cells. The number of dead cells can account for up to 5% of the cell population, and following the release of cells from monolayer cultures, this may increase. Cultured cells will be at various stages of the cell cycle, and this is reflected in a range of cell volumes, from large polyploid cells ready to undergo mitosis, to small daughter cells. Changes in surface morphology (e.g., number of microvilli) during the cell cycle have been reported (Porter et al. 1973). Large cells tend to swell easier in hypotonic buffers (as evidenced in metaphase spread techniques) and will be disrupted more efficiently than the smaller cells. Large rounded cells are also more easily disrupted than long thin fibroblastic cells for obvious geometric reasons and, to date, it appears easier to prepare surface membranes from rounded epithelioid cells, e.g., HeLa, KB lines, ascites cells, than from fibroblastic cell lines. Variation in cell volume can be minimised by use of synchronised cell lines; the use of monolayers approaching confluency is an approximation to this ideal. The use of synchronous cell populations is also important from the viewpoint of the cell surface properties, since cell surface membrane constituents have been shown to vary in composition and accessibility at different stages of the cell cycle (Johnsen et al. 1975; Mastro et al. 1974; Hunt et al. 1975).

Cultured cells display the full complement of subcellular organelles. Cultured rat embryo fibroblasts for example possess in their cytoplasm an extensive endoplasmic reticulum network heavily studded with ribosomes. The Golgi apparatus is also well developed in the perinuclear zone (Tulkens et al. 1974), although in subcellular fractionation studies assessment of the distribution of Golgi marker enzymes is rarely made (Chandrabose et al. 1976).

5.6.1. Plasma membranes from various cultured cell lines

Despite the large number of reports describing the isolation and properties of plasma membranes using a variety of cultured cells (a selected number is analysed in Table 5.3), there is no satisfactory general method that gives adequate consideration to yield, representativeness and a full analysis of the properties of the isolated membranes; also, a number of anomalies regarding density and enzyme distribution is evident. Further, the behaviour during subcellular fractionation by the same protocol of various cell types, or even the same cell type at different points in its growth cycle, can be different. Since most methods involve density gradients for isopycnic separation of plasma membranes, it is helpful to know the isopycnic density of a wide range of subcellular marker enzymes and chemical constituents. The median densities described by Tulkens et al. (1974; see Table 3.3) are similar to those obtained with other cell types, e.g., Ehrlich ascites tumour cells (Horvat and Touster 1967) and Chinese hamster ovary fibroblasts (Milson and Wynn 1973). The plasma membrane median density in sucrose is usually 1.13–1.15 g/cm^3.

Pretreatment of cells Cells grown in suspension (Spinner cultures) usually have serum proteins loosely attached to their outer surface, so it is necessary to remove these proteins by washing the cells 2 or 3 times by low speed centrifugation (usually 500 $g \times 10$ min using a refrigerated centrifuge) in an iso-osmotic buffer, e.g., a Krebs–Ringer phosphate buffered saline (pH 7.4); Johnsen et al. (1974) recommend inclusion of 1 mM Ca^{2+} in the saline. To obtain sufficient plasma membranes for biochemical analysis, approx. 1×10^8 or more cells are required. To harvest cells from monolayers, the use of trypsin or collagenase, even at low concentrations, can result in damage to plasma membrane components facing the medium and gentle scraping using a rubber policeman to release cells and cell aggregates (from confluent monolayers) is the best method. Material remaining on the glass surface after cell removal has been analysed (Culp 1976). Washed cells may be suspended in 10 mM Tris buffer, pH 7.0–7.8 (approx. 20-fold the volume of the pellet, approx. 5×10^7 cells/ml) and allowed to swell for 10–15 min. Prolonged swelling in hypotonic buffer, especially in the absence of Mg^{2+}, can result in lysis of nuclei. A range of other buffers has also been used, e.g., triethanolamine, acetate, but there is probably little difference in their efficacy. The recent adoption of hypotonic borate buffers (0.02 M boric acid, 0,2 mM EDTA adjusted to pH 10.2) (Thom et al. 1977) helps to strengthen the plasma membrane. The use of membrane 'stabilising' agents by Warren et al. (1966) involves pretreating the cells with fluorescein mercuric acetate (FMA), or zinc chloride (§ 2.3.3) and is unsatisfactory when membrane enzymes are to be analysed (see § 2.4.5). It has been shown that the effects of zinc pretreatment can be subsequently reversed by addition of chelating agents, e.g., treatment with 1 mM EDTA or EGTA restores both Mg^{2+} and (Na^+-K^+)-ATPase activity (Colombini and Johnstone 1973). However, this has to be done with care, since an excess of chelating agents may in turn inhibit other plasma membrane enzymes, e.g., 5′-nucleotidase.

Cell disruption A range of methods has been devised to disrupt

TABLE 5.3.

Summary of selected examples of various methods used to prepare plasma membranes from cultured cells.

Cell type	Pretreatment	Method of cell breakage	Centrifugation details[a]	Characterisation				Form and yield	Density[b] (g/cm³)	References
				Morph.	Enz.	Chem.	Other			
HeLa	washed in 10 mM Tris, 1 mM $MgCl_2$	swollen cells homog. in tight Dounce to give 'ghosts'	discontinuous gradient, sonication, continuous sucrose gradient	e.m.	1 + 1 −	no	antibody binding	vesicles n.s.	1.14	1
HeLa	washed in saline buffer	swollen cells homog. in tight Dounce to give 'ghosts'	continuous sucrose gradient	e.m.	2 +	yes	–	vesicles, 21% (Na^+-K^+)-ATPase 6% 5′-nucleotidase	1.16–1.20	2
BHK	washed in sucrose Tris–$MgCl_2$	N_2-cavitation	dextran-$MgSO_4$ gradient	–	2 + 3 −	no	–	vesicles, n.s.	multiple	3
MF2	washed in saline buffer	swollen cells homog. in loose Potter to give ghosts	discontinuous sucrose gradient	e.m.	3 + 5 −	no	–	vesicles, 30% protein	1.12–1.14 1.14–1.16	4
Chicken embryo fibroblasts	washed in saline Tris–$MgCl_2$	monolayers homog. in Dounce	discontinuous and continuous sucrose gradients	–	1 + 1 −	yes	fucose incorporation	n.s., 60% 5′-nucleotidase	1.12 1.09	5

Chicken embryo fibroblasts	collagenase, washed in saline buffer	Potter homog.	discontinuous sucrose gradient	e.m.	3 +	yes	–	vesicles, 1% protein; 30% CTPase	1.08 1.13	6
Chicken embryo muscle fibroblasts	monolayers scraped and cells washed in sucrose-triethanolamine buffer	washed cells Dounce homog.	discontinuous sucrose gradient	–	5 + 2 –	yes	–	n.s. 16% 5′-nucleotidase, 30% bung-arotoxin	1.06–1.17	7
Cultured neuroblastoma	spinner or monolayer cells treated with $ZnCl_2$	swollen cells Dounce homog.	discontinuous sucrose-Ficoll gradient	e.m.	–	yes	iodination	vesicles and strips, 4% protein,	1.14	8
KB	washed in saline buffer	swollen cells Dounce homog.	discontinuous sucrose gradient	e.m.	2 + 2 –	yes	adenovirus binding	vesicles 2–4% protein, 10–15% 5′-nucleotidase	1.14	9
Murine sarcoma	washed in saline buffer	swollen cells Dounce homog.	2-phase separation[c]	e.m.	1 + 1 –	yes	Con A[d] binding	7% protein; 62% ATPase	–	10

In noting enzymic characterisation, + or – denotes the number of positive or negative markers used, respectively.

n.s. = not stated.

[a] In most instances, nuclei and mitochondria are removed first by a low-speed centrifugation step.

[b] Density applies to medium used to isolate membranes. Approx. isopycnic in most cases.

[c] Two-phase polyethyleneglycol–dextran gradient (see § 3.2).

[d] Concanavalin A.

e.m.: Electron micrographs, usually of stained sections.

1. Boone et al. (1969); 2. Johnsen et al. (1975); 3. Graham (1972); 4. Lelievre (1973); 5. Hay (1974); 6. Perdue et al. (1971); 7. Schimmel et al. (1973); 8. Truding et al. (1974); 9. Butters and Hughes (1974); 10. Hourani et al. (1973).

cultured cells (see Table 5.3). A widely used method involves homogenising in a tight-fitting Dounce homogeniser cells previously swollen by standing for 10 min in hypotonic buffers. Gas cavitation is more difficult to control and produces small membrane vesicles that require a long time to equilibrate when centrifuged in sucrose gradients (see § 2.1.4). The conditions necessary to obtain efficient cell disruption can vary according to cell type, the growth conditions and the cell to medium ratio. Hence, conditions described in Table 5.3 serve only as a guide, and the efficacy of the disruption procedure adopted requires close monitoring by using phase contrast microscopy. When approx. 80% of the cells have been ruptured, an equal volume of 20% sucrose (w/w) buffered with 10 mM Tris–HCl, pH 7.4, is added.

Subcellular fractionation Nuclei are removed by centrifugation at 1000 *g* for approx. 3 min. The efficiency of removal of nuclei can be monitored by determination of the DNA content of the supernatant. A flow chart for the preparation of plasma membrane 'ghosts' is shown in Fig. 5.8. The use of a continuous sucrose gradient is always advised in preliminary experiments, although it can be replaced later by introducing a density discontinuity at the appropriate sucrose concentration, thus speeding up the collection of membranes (see § 3.1.3).

The plasma membrane 'ghosts' collected by the above procedure (Fig. 5.8) can then be resuspended in a tight-fitting Dounce homogeniser and the isopycnic centrifugation step repeated. The above procedure follows that described by Atkinson and Summers (1971), Butters and Hughes (1974) and Hay (1974). Johnsen et al. (1974) showed that the plasma membranes produced by a similar procedure are of high purity as judged by monitoring the following markers: 5′-nucleotidase, ouabain-sensitive (Na^+-K^+)-ATPase, and the negative markers cytochrome *c* oxidase, NADPH cytochrome *c* reductase, DT diaphorase (a mitochondrial enzyme catalysing the oxidation of NADH and NADPH, Ernster, 1967) and *β*-*N*-acetylglucosaminidase. Morphological and chemical analyses were also indicative

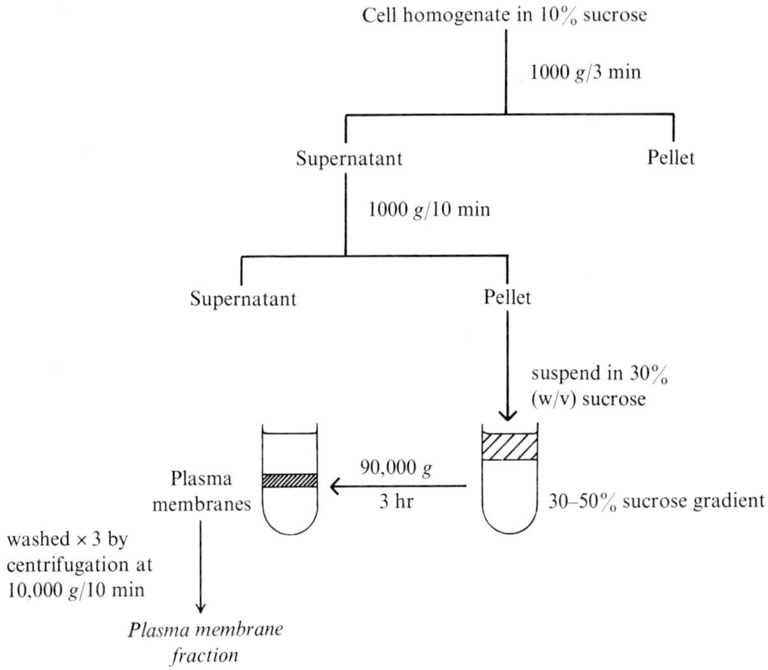

Fig. 5.8. Preparation of plasma membranes from HeLa cells. See text for further details.

of a high purity. Reliable use of 5′-nucleotidase as a plasma membrane marker can only be made when EDTA is absent from the incubation mixtures. The above method, with modifications, was used successfully by Hubbard and Cohn (1975) to prepare plasma membranes from L-cells (§ 2.4.5).

5.6.2. Maintaining the intactness of plasma membrane 'ghosts'

The efficient disruption of the cultured cells to produce membrane ghosts (i.e., large plasma membrane fragments incorporating a greater part of the cell surface) is a major factor influencing the feasibility of the methods. The degree of breakage of the surface membrane can be followed by measuring the distribution of plasma

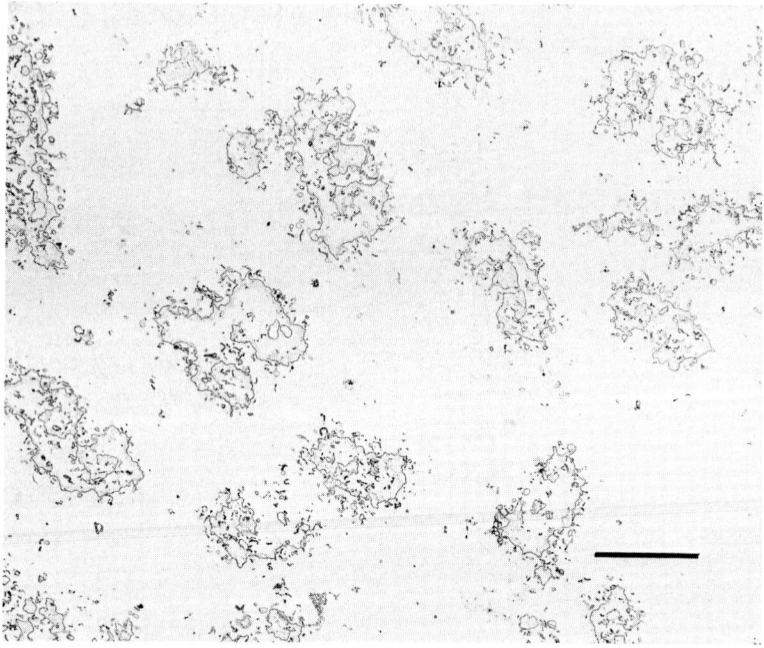

Fig. 5.9. Electron-micrograph of HeLa plasma membrane ghosts. These ghosts may also be observed by phase-contrast microscopy (see Atkinson and Summers 1971). For further details see text. Bar = 5 μm. By courtesy of Dr. P.H. Atkinson and reproduced by permission of Academic Press.

membrane markers between the 1000 *g*/10 min supernatant and the pellet, thus complementing the information obtained by monitoring the disruption procedure by phase contrast microscopy. On the same note, it is also important to monitor the distribution of plasma membrane marker components in all the fractions obtained en route to the plasma membrane fraction(s). Atkinson (1973) reported that methods that depend on the production of 'ghosts' work best with cells grown in 'spinner' culture, e.g., HeLa, KB, ascites cells, and that 'ghosts' were difficult to obtain from mouse cell lines grown as monolayers. However, in applying this method to fibroblasts, following their release from their attachment to glass, the cells should be allowed to swell in hypotonic media before homogenising.

5.6.3. *Preparation of plasma membrane 'micro-vesicles'*

(a) *Gas cavitation* The production of a range of sizes of surface membrane fragments constitutes the greatest hazard in obtaining acceptable yields of plasma membranes (§ 2.1.4). Other approaches have aimed at breaking up the surface membrane completely into small vesicles which are then separated from vesicles derived from intracellular membranes. This intensive fragmentation, usually achieved by using gas cavitation has been applied to fibroblasts because they produce ghosts with difficulty on hypotonic lysis and Dounce homogenisation (Kamat and Wallach 1965). In this method the vesicles produced by extensive fragmentation and resealing of the plasma and intracellular membranes are separated on Mg^{2+}-containing dextran or Ficoll gradients. Some parameters governing the separation were investigated in 4 tissue culture cell types (baby hamster normal kidney cells, polyoma virus transformed kidney cells, rat embryo normal fibroblasts and spontaneously transformed fibroblasts) by Graham (1972) who showed that this method of cell disruption is fraught with difficulties. Hence a generalised scheme for preparation of fibroblastic plasma membranes will not be presented. The use of iodinated cells to monitor the co-purification of plasma membrane radioactivity and 5′-nucleotidase is described in § 4.6.1.2.

(b) *Exposure to formaldehyde and disulphide-blocking reagents* Monolayer cell cultures of density-inhibited and transformed fibroblasts, macrophages and monocytes release plasma membrane vesicles following exposure to formaldehyde and disulphide-blocking reagents (Scott 1976). The vesicles are formed and released within 15 min to 2 hr of exposure of cells to 250 mM formaldehyde or 10 mM *N*-ethylmaleimide dissolved in phosphate-buffered saline, pH 7.4, containing 0.5 mM Ca^{2+}. The vesicles can be removed from cells by decanting and are then sedimented by centrifugation at 30,000 *g* for 30 min. Analysis of enzymes (e.g., 5′-nucleotidase), lipids (cholesterol and sphingomyelin) and proteins (by polyacrylamide gel electrophoresis) indicates that released vesicles have a similar composition to plasma membranes prepared by alternative methods. However, the question of how representative of the total plasma membrane these vesicular membranes are remains to be established. This method appears a promising and simple adjunct to the subcellular fractionation approaches described in other sections.

5.6.4. *Preparation of plasma membranes by use of latex-bead ingestion*

A novel approach using latex-bead ingestion for preparing plasma membranes from KB cells and L-cells is described by Charalampous et al. (1973) and Hunt and Brown (1975; see § 3.6.2). 'Plasma' membranes prepared by this method compare favourably with the membranes prepared from 'ghosts' by using the $ZnCl_2$ method. However, there are important reservations to be borne in mind regarding the representiveness of the membranes prepared by using latex beads. There is evidence accumulating from studies on the phagocytosis of inert particles by macrophages and polymorphonuclear leukocytes (Tsan and Berlin 1971) that only a specific part of the plasma membrane is internalised. If this is a general phenomenon, i.e. the surface membrane of these cells is a mosaic of topographically-separated transport and phagocytotic sites, then plasma membranes prepared by the latex-bead ingestion method are only representative of a specific part of unknown unit area of the cell surface. A further reservation concerns the question of the speed of modification of the surface membrane following internalisation. It is essential to rigorously control the time allowed for the internalisation of latex-beads, since internalised plasma membranes may fuse ultimately ($t_{\frac{1}{2}} = 2$ hr) with the lysosomes (Werb and Cohn 1972). The vesicles are of the inside-out configuration (§ 3.6.2).

5.6.5. *Enzymic and chemical heterogeneity of plasma membranes*

When a microsomal fraction of cultured baby hamster kidney cells prepared from homogenates produced by gas cavitation was fractionated on dextran gradients, separate peaks of the plasma membrane markers 5′-nucleotidase and $(Na^+$-$K^+)$-ATPase activities were obtained (Graham 1972). Other examples of two or more plasma membrane fractions being identified on sucrose gradients include chicken embryo fibroblasts (Perdue et al. 1971) and chicken embryo muscle cells (Schimmel et al. 1973), and are indicative of either cellular heterogeneity, or an uneven distribution of markers on the fibroblast surface, e.g., the plasma membrane attached to a substructure may differ from the unattached area.

Comments There are relatively few good markers that can be used in the preparation of the plasma membranes from cultured cells. This probably accounts for the shortcomings of many of the various methods used in attempts to prepare pure and representative membranes, especially from fibroblastic cell lines. To overcome the relative lack of indigenous markers and the problem of plasma membrane heterogeneity, increasing use is being made of radioactive markers attached to the outside of the plasma membrane of cells. Examples of the use of ^{125}I as a covalently-attached marker for the plasma membrane of fibroblasts were described in § 4.6.2.5.

5.7. Endocrine tissue plasma membranes

Despite the cellular complexity of mammalian endocrine tissues, plasma membrane fractions have been prepared that have proved useful in investigating the interaction and consequences of the binding of hormones to cell surface receptors. Owing to the low mass of some endocrine tissues and organs, many investigators have used a bovine or porcine source. Preparative methods are summarised in Table 5.4.

Thyroid The method of Neville (1960) as applied to liver (§ 2.4.3 and 5.1) also yields a plasma membrane fraction from thyroid tissue homogenates. In general, a slightly higher *g* force is required to sediment thyroid plasma membrane fragments than with liver homogenates, prior to further separation in density gradients (see Hilderson et al. 1975 for an analysis of bovine thyroid membranes by density gradient centrifugation in zonal rotors).

TABLE 5.4
Preparations of plasma membranes from endocrine and exocrine tissue.

Tissue	Animal	Comments	References
Thyroid	Bovine	Membrane isolated at density 1.16 and 1.18[a] adenylate cyclase stimulated by TSH[b] and F^-. (Na^+-K^+)-ATPase	Yamashita and Field (1970) Wolff and Jones (1971) Suzuki et al. (1977)
	Porcine		Nagai and Hosoya (1974)
Corpus luteum	Bovine	Membranes isolated at densities 1.16 and 1.18[a] Bind luteinising hormone and Prostaglandin $F_{2\alpha}$	Gospodarowicz (1973) Powell et al. (1976)
	Bovine	Hypo- and iso-osmotic media used, density of membranes 1.13[a] Bind luteinising hormone and prostaglandin E_2	Sidhu et al. (1975)
Pancreas	Rat	Membranes isolated at density 1.12–1.16[a] Adenylate cyclase stimulated by gastro-intestinal hormones	Svoboda et al. (1976)
	Rat	Islets isolated; membrane density 1.10[a]	Lernmark et al. (1976)
Adrenal glands	Bovine	Membranes isolated at density 1.11[a] from medulla	Nijjar and Hawthorne (1974)
	Bovine	Vesicles, highly enriched in corticotropin-sensitive adenylate cyclase and angiotensin binding sites	Schlegel and Schwyzer (1977)
Placenta	Human	Membranes isolated at density 1.16–1.17[a]	Carlson et al. (1976)
Parotid	Mouse	Membranes prepared from a post-mitochondrial fraction. Adenylate cyclase studied.	Durham et al. (1975)

[a] Density in sucrose gradients (g/cm^3). [b] TSH, thyroid stimulating hormone.

Corpus luteum A similar methodology to that of thyroid is used. The cellular and/or plasma membrane topographical heterogeneity is illustrated by the isolation of subfractions of different densities by Gospodarowitz (1973) and Sidhu et al. (1975). Co-purification of 5′-nucleotidase and prostaglandin $F_{2\alpha}$-binding activities on plasma membranes by Powell et al. (1976) is given as an example of expressing subcellular fractionation results in § 4.1.6. Sidhu et al. (1975) compare methods using hypotonic and iso-osmotic media for the preparation of corpus luteum plasma membranes with adenylate cyclase activity, and show that homogenates prepared in hypotonic media yielded the better product.

Pancreas Mammalian pancreas is a notoriously difficult tissue from which to prepare undegraded subcellular fractions, owing to the presence of trypsin and chymotrypsin, etc. in the exocrine cells. Thus, continual proteolysis occurs during fractionation. However, pancreatic subcellular fractions, including a plasma membrane fraction, were prepared by Meldolesi et al. (1971). The inclusion of Trasylol® (aprotonin proteinase inhibitor) at 500 Kallikrein inactivator units (Kl.U.)/ml (available in ampoules of 10 ml containing 200,000 Kl.U., Bayer) helps to minimise degradation. Phospholipase A_2 (Ca^{2+}-dependent) activity is high in pancreatic homogenates, and the action of this enzyme on membranes may be diverted by inclusion of a phospholipid mixture at 2 mg/ml in media. Inclusion of such additives helps to maintain the sensitivity of plasma membrane adenylate cyclase activity, especially towards the gastro-intestinal hormones cholecystokinin and pancreozymin. The method of Lernmark et al. (1976) commences with rat islets of Langerhans prepared by digesting pancreas with crude collagenase and the islet homogenates are then resolved in sucrose gradients to yield a partially purified plasma membrane fraction.

Adrenal medulla Most subcellular fractionations of this organ secreting and storing catecholamines have been directed at the isolation of chromaffin granules (e.g., Smith and Winkler 1972; Trifaro et al. 1976). Nijjar and Hawthorne (1974) have described the isolation of a plasma membrane fraction from bovine adrenal medulla homogenates that is enriched in three types of ATPase activities. The properties of adrenal medulla Golgi apparatus membranes are described by Trifaro and Duerr (1976).

Placenta A plasma membrane fraction, presumably derived mainly from syncytial trophoblasts was prepared from human placenta by Carlson et al. (1976) by a minor variation of methods applicable to liver (Neville 1960) and kidney homogenates (Wilfong and Neville 1970). As an alternative to the use of Dounce homogenisers, the tissue press (§ 2.1.6) can also be used to disrupt placenta, and a vesicular plasma membrane fraction is obtained by a scheme similar to that shown in Fig. 5.7. A partially purified plasma membrane fraction of pregnant guinea pig uterus was used by Shami and Radde (1971) and Messer et al. (1975) to study membrane-bound phosphatases.

5.8. *Muscle tissue plasma membranes*

These are difficult to prepare owing to the resistance of muscular tissue to cell disruption methods (§ 2.3.1), an extensive connective tissue network and the multiple layers of sarcolemma. Although earlier methods for the preparation of skeletal muscle plasma membranes (e.g., Kono and Colowick 1961; Boegman et al. 1970) have been modified and improved, the purity and representiveness of the membranes still leaves much to be desired. Two general methods may be distinguished, i.e., those that first remove myofibrils (mainly actin and myosin) by salt extraction of homogenates (e.g., Schapira et al. 1974; Festoff and Engel 1974) and those that involve immediate fractionation of the muscle tissue homogenate (Kidwai et al. 1971a). The 12 hr extraction of Polytron-dispersed homogenates (§ 2.1.8) with 0.5 M LiBr, 0.05 mM EDTA, 16 mM Tris–HCl, pH 8.5, to remove myofibrils can result in proteolysis of membrane proteins. Alternative methods (e.g., Tada et al. 1972) use 0.1 M to 0.6 M KCl alone or in the presence of 0.5 M LiBr or 0.05 M pyrophosphate to extract the subcellular fractions before sucrose gradient centrifugation (Engelhardt et al. 1976; Smith and Appel 1977).

The membrane markers used to locate plasma membrane are the (Na^+-K^+)-ATPase (Ca^{2+}-ATPase concentration gives the extent of contamination by sarcoplasmic reticulum), adenylate cyclase (Severson et al. 1972) and binding of ^{125}I-bungarotoxin (Andrew et al. 1974). Discrepancy, possibly related to the use of chaotropic agents in the extraction of myofibrils, exists in the usefulness of 5′-nucleotidase as a marker. Barchi et al. (1977) and Heller and Harary (1977) have used cell surface labelling techniques (§ 4.6) to identify skeletal muscle plasma membrane fractions. Similar methods for the preparation of plasma membrane fraction from cardiac tissue (Stam et al. 1969; Kidwai et al. 1971b; Tada et al. 1972) and smooth muscle (Oliviera and Holzhacker 1974) have been described. Intercalated discs are useful markers in cardiac plasma membranes.

5.9. *Preparation of plasma membrane junctional specialisations*

Small areas (3% or less) of the plasma membrane at the point of interaction or close opposition with the plasma membrane of adjacent cells show typical morphological features. The major junctional specialisations of vertebrate tissues are classified as gap junctions, tight junctions and desmosomes. In addition, the plasma membrane of the cells of excitable tissue show characteristic features at the synaptic cleft region. The morphological features of these specialised membrane regions have already been described in view of their usefulness as markers in identifying plasma membrane fragments (see § 4.4.5). Methods that allow the preparation of fractions enriched in gap junctions, desmosomes and synaptic thickenings are now described. Essentially, methods for preparing cell junctions start with isolated plasma membranes, and use solvents to selectively dissolve away the non-junctional membranes. These solvents are usually detergents, urea or in the case of desmosomes, acidic aqueous media.

5.9.1. Gap communicating junctions

Gap junctional enriched fractions were prepared from rat and mouse liver homogenates by Benedetti and Emmelot (1968), Goodenough and Stoeckenius (1972) and Evans and Gurd (1972). The procedures for preparation of gap junctions commence with isolated plasma membrane fractions prepared in bulk by using zonal rotors (see § 3.1.8) to fractionate large amounts of tissue homogenates (Evans 1970; Goodenough 1974). Deoxycholate and sarcosyl (*N*-dodecyl sarcosinate) are the detergents of choice for selectively solubilising non-junctional membranes. The procedure shown in Fig. 5.10 utilises sarcosyl to solubilise non-junctional membranes.

To obtain morphologically and biochemically satisfactory isolated gap junctions, a number of problems have to be overcome. First, it is necessary to prepare from tissues a plasma membrane fraction in large amounts containing the correct area of the surface membrane, i.e., the lateral or contiguous area. Plasma membrane fractions prepared by variants of the methods of Neville (1960, 1968) are derived mainly from the lateral surface membrane of hepatocytes. Second, the persistence of collagen fibres in gap junctional enriched fractions has led to the introduction of a collagenase treatment step in the procedure for their elimination. However, the use of crude enzyme preparations (e.g., Goodenough 1974) results not only in the limitation of collagen contamination, but also in proteolysis of the constituent polypeptides of the junctional complex to yield peptides of low and variable molecular weight. The scheme shown in Fig. 5.9 uses low amounts of a highly purified collagenase (e.g., Sigma Type III). Assessment of the purity of the isolated gap junction is made by observing in the electron microscope negatively stained samples (Fig. 4.9) or fixed and stained sections (Fig. 5.11). Further, polyacrylamide gel analysis can indicate the presence of undigested collagen (MW 110,000–130,000) and bands of apparent molecular weight 36,000–40,000. The presence of major bands below the 30,000 molecular weight limit described above is indicative of proteolysis (Culvenor and Evans 1977). Isolated gap junctions, probably due to their being lipid-depleted structures, adsorb to glass,

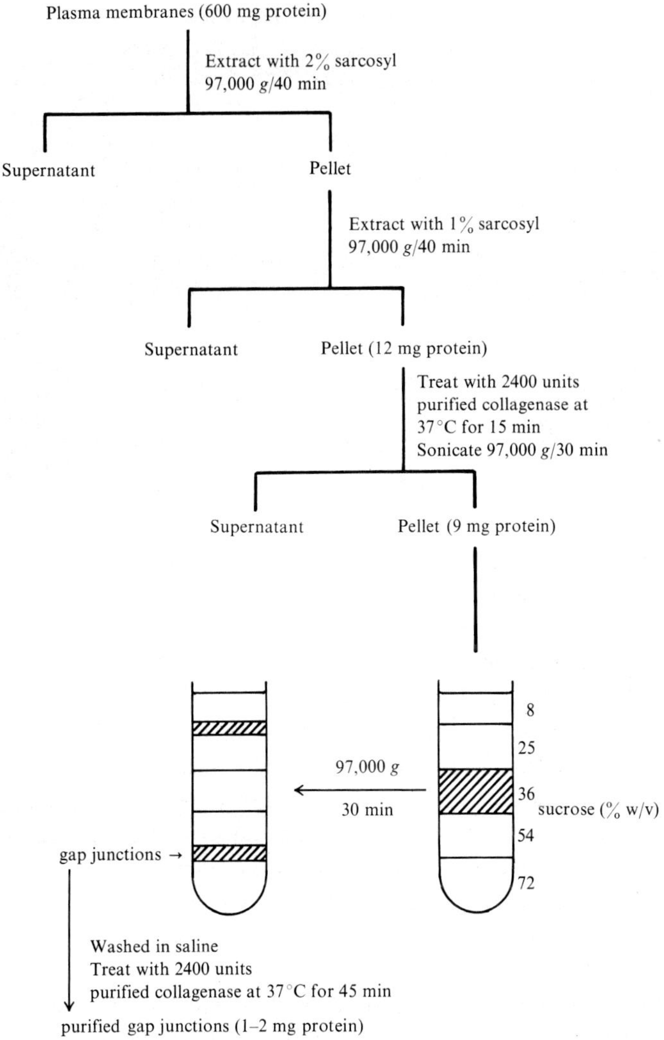

Fig. 5.10. Preparation of hepatic gap junctions. The gap junctions prepared by this method are relatively free of collagen and the constituent polypeptides are undegraded. Purified collagenase, e.g., Sigma Type 111; Sarcosyl, *N*-dodecyl sarcosinate (Geigy).

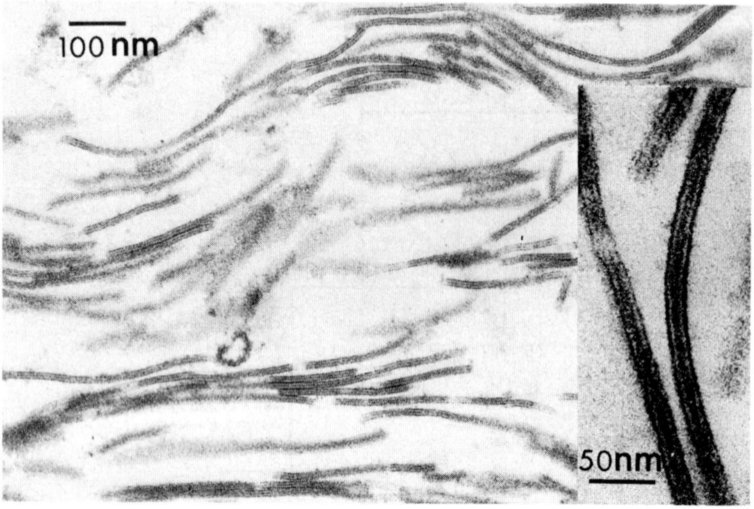

Fig. 5.11. Electron-micrograph of isolated gap junctions. Non-junctional membranes have been removed, and all that remains are the gap junctions. At higher magnification (inset) a 'gap' may be observed between the two membranes.

and thus plastic tubes or siliconised glass should be used for washing them by low speed centrifugation.

The plasma membranes of eye lens fibres are a further rich source of gap junctions. The isolation procedures use detergents (Dunia et al. 1974) or urea (Alcala et al. 1975) to remove non-junctional components. These methods do not require the use of collagenases and are less prone to proteolysis artifacts. A polypeptide of 34,000–38,000 molecular weight is a major component.

5.9.2. *Desmosomes*

These are prominent surface specialisations, particularly in epithelium and myocardium. Skerrow and Matoltsy (1974) prepared desmosomes from cow nose epidermal tissue. Cow nasal tissue is dispersed in a blender in 0.1 M citric acid/sodium citrate buffer, pH 2.6, which dissolves non-junctional membranes, and the filtered residue is then resuspended in pH 2.6 buffer by using a loose-fitting Dounce homogeniser and washed 4 times by low speed centrifugation. The residue is then resolved on sucrose gradients and the desmosome fraction collects at the 55–60% sucrose interface. Analysis by polyacrylamide gel electrophoresis of isolated desmosomal plaque indicated the presence of several polypeptides of molecular weight 60,000–230,000.

5.9.3. Synaptic junctions

The synaptic junctions of mammalian brain have a characteristic ultrastructure of closely apposed pre- and post-synaptic membranes (Fig. 5.5). They are prepared from the 'synaptosomes' or plasma membrane fractions of cerebral cortex (see § 5.2.1) by treatment with detergents that dissolve away the synaptosomal membrane and associated structures leaving an insoluble residue consisting mainly of post-synaptic densities (Cotman et al. 1974; De Robertis 1971; Davis and Bloom 1973; Matus and Walters 1975). A crude synaptic density fraction is prepared rapidly by dispersing rodent brain synaptosomal plasma membranes in 10 mM Tris–HCl, pH 7.4, containing 1–2% (w/v) sodium deoxycholate. The insoluble residue is enriched in synaptic junctions and is pelleted by centrifugation at 4°C for 1 hr at 100,000 *g*. In the procedure of Cotman et al. (1974), synaptosomal plasma membranes are dispersed in 0.5% sarcosyl (*N*-dodecyl sarcosinate) dissolved in 10 mM Bicine, pH 7.5 (10 mg membrane protein/ml detergent) and applied to a discontinuous sucrose gradient containing equal volumes of 1.0, 1.4 and 2.2 M sucrose dissolved in 0.05 mM $CaCl_2$. After centrifugation at 60,000 *g* for 75 min (e.g., in a Beckman SW27 rotor), the synaptic-density enriched fraction collects mainly at the 1.4–2.2 M sucrose interface. The synaptic densities are then washed by low speed centrifugation. Approx. 0.5 mg of post-synaptic densities are obtained from 20 g of rat forebrain. Analysis by polyacrylamide gel electrophoresis of synaptic densities indicates polypeptides present corresponding in electrophoretic mobility to myosin, tubulin, actin and α- and β-actinin and neurofilament protein.

In a further modification, Therien and Mushynski (1976) prepared rat brain synaptic junctional complexes by fractionating synaptosomal plasma membranes in a biphasic system consisting of Freon-113 and an aqueous phase containing 0.2% Triton X-100. Morphologically well preserved synaptic junctions were recovered in the aqueous phase and purified further by centrifugation in discontinuous sucrose gradients. Finally, Bartfai et al. (1976) used highly purified phospholipase A_2 to remove selectively non-junctional synaptic membranes to yield a fraction enriched in the post-synaptic marker proteins, muscarinic and nicotinic acetylcholine receptors and 3′,5′-cyclic nucleotide phosphodiesterase.

Appendices

APPENDIX I

Some properties of aqueous sucrose solutions

Sucrose concentration		Refractive Index	Molarity	Density	
% w/w	% w/v	20 °C		5 °C	20 °C
0	0	1.3330	0	1.000	0.998
1	1.00	1.3344	0.03	–	1.002
2	2.01	1.3359	0.06	1.008	1.006
3	3.03	1.3374	0.09	–	1.010
4	4.06	1.3388	0.12	1.016	1.014
5	5.09	1.3404	0.15	–	1.018
6	6.13	1.3418	0.18	1.024	1.022
7	7.18	1.3433	0.21	–	1.206
8	8.24	1.3448	0.24	1.032	1.030
9	9.31	1.3464	0.27	–	1.034
10	10.38	1.3479	0.030	1.040	1.038
11	11.47	1.3494	0.34	–	1.042
12	12.56	1.3510	0.37	1.048	1.046
13	13.66	1.3526	0.40	–	1.051
14	14.77	1.3541	0.43	1.057	1.055
15	15.89	1.3557	0.47	–	1.059
16	17.02	1.3573	0.50	1.065	1.064
17	18.15	1.3590	0.53	–	1.068
18	19.30	1.3606	0.56	1.074	1.072
19	20.45	1.3622	0.60	–	1.076
20	21.62	1.3639	0.63	1.083	1.081
21	22.79	1.3655	0.67	–	1.085
22	23.98	1.3672	0.70	1.092	1.090
23	25.17	1.3689	0.74	–	1.094

Subject index p. 261

Some properties of aqueous sucrose solutions (*continued*)

Sucrose concentration		Refractive Index	Molarity	Density	
% w/w	% w/v	20 °C		5 °C	20 °C
24	26.38	1.3706	0.77	1.010	1.099
25	27.59	1.3723	0.80	–	1.104
26	28.81	1.3740	0.84	1.110	1.108
27	30.05	1.3758	0.88	–	1.113
28	31.29	1.3775	0.92	1.120	1.118
29	32.54	1.3793	0.94	–	1.122
30	33.81	1.3811	0.99	1.129	1.127
31	35.08	1.3829	1.03	–	1.132
32	36.37	1.3847	1.06	1.139	1.137
33	37.67	1.3865	1.10	–	1.142
34	38.98	1.3883	1.14	1.148	1.147
35	40.29	1.3902	1.18	–	1.151
36	41.62	1.3920	1.22	1.158	1.156
37	42.97	1.3939	1.26	–	1.161
38	44.32	1.3958	1.29	1.168	1.166
39	45.68	1.3978	1.34	–	1.171
40	47.06	1.3997	1.38	1.179	1.176
41	48.45	1.4016	1.42	–	1.182
42	49.84	1.4036	1.46	1.189	1.187
43	51.26	1.4056	1.50	–	1.192
44	52.68	1.4076	1.54	1.199	1.197
45	54.11	1.4096	1.58	–	1.202
46	55.56	1.4117	1.63	1.210	1.208
47	57.02	1.4137	1.67	–	1.213
48	58.49	1.4159	1.71	1.221	1.219
49	59.98	1.4179	1.75	–	1.224
50	61.48	1.4205	1.80	1.232	1.230
51	62.99	1.4221	1.84	–	1.235
52	64.51	1.4242	1.88	1.243	1.241
53	66.05	1.4264	1.93	–	1.246
54	67.60	1.4285	1.98	1.254	1.252
55	69.19	1.4307	2.02	–	1.258
56	70.74	1.4329	2.07	1.266	1.263
57	72.33	1.4351	2.11	–	1.269
58	73.94	1.4373	2.16	1.277	1.275
59	75.57	1.4396	2.21	–	1.281
60	77.19	1.4418	2.26	1.289	1.286

Some properties of aqueous sucrose solutions (*continued*)

Sucrose concentration		Refractive Index	Molarity	Density	
% w/w	% w/v	20 °C		5 °C	20 °C
61	78.83	1.4441	2.31	–	1.292
62	80.49	1.4464	2.36	1.300	1.298
63	82.17	1.4486	2.41	–	1.304
64	83.86	1.4509	2.45	1.312	1.310
65	85.56	1.4532	2.50	–	1.316
66	87.28	1.4558	2.56	1.324	1.322

Density and viscosity of aqueous Ficoll solutions

Concentration (g/ml)	Density (g/cm³)	Viscosity (centipoises)	l_n viscosity
(A) at 4 °C			
3.032	1.0106	2.788	1.025
4.058	1.0145	3.397	1.223
5.090	1.0180	4.102	1.411
6.129	1.0215	5.014	1.612
7.177	1.0253	6.017	1.795
8.232	1.0290	7.372	1.998
9.295	1.0328	8.579	2.149
10.37	1.0365	10.35	2.337
12.53	1.0441	14.27	2.658
14.73	1.0518	20.21	3.006
16.96	1.0597	27.42	3.311
19.21	1.0673	38.33	3.646
21.50	1.0752	52.31	3.957
23.84	1.0837	69.28	4.238
26.21	1.0922	95.03	4.554
28.61	1.1004	125.7	4.834
31.06	1.1093	172.0	5.147
33.53	1.1176	225.6	5.419
36.04	1.1263	308.8	5.733
38.57	1.1345	407.2	6.009
41.19	1.1442	565.4	6.338
43.83	1.1534	762.2	6.636
46.52	1.1629	1020.0	6.928

Subject index p. 261

Density and viscosity of aqueous Ficoll solutions (*continued*)

Concentration (g/ml)	Density (g/cm^3)	Viscosity (centipoises)	l_n viscosity
(B) at 25 °C			
3.021	1.0071	1.54	0.432
6.107	1.0178	2.63	0.967
9.258	1.0287	4.393	1.480
12.47	1.0391	7.145	1.966
15.76	1.0505	11.27	2.422
19.12	1.0621	17.78	2.878
22.54	1.0733	26.92	3.293
26.05	1.0855	41.22	3.719
29.64	1.0977	61.09	4.112
33.31	1.1103	91.63	4.518
37.05	1.1228	137.1	4.921
40.90	1.1362	209.2	5.343
44.83	1.1495	313.8	5.749

Data from Pretlow et al. (1969) Anal. Biochem. *29*, 230.

(i) Conventional rotors useful in membrane preparation

Manufacturers	Centrifuge type	Rotor type	Volume	Maximum *g* force	Comments
(A) Large volume general purpose rotors. Useful for sedimenting nuclei, mitochondria, etc. from large volumes of tissue homogenates					
Beckman	J-21B	JA14	6 × 260 ml	30,500	
Beckman	L-2 series	Type 19	6 × 250 ml	53,700	
M.S.E.	Hi-spin 21	43114-112	6 × 300 ml	31,700	
Sorvall	RC2B, RC5	G.S.A.	6 × 250 ml	27,000	
(B) Intermediate speed rotors. Useful for sedimenting large volumes of membranes in intermediate range centrifuges					
Beckman	J-21B	JA20	8 × 50 ml	48,300	
M.S.E.	Hi-spin 21	43114-143	8 × 50 ml	49,640	
Sorvall	RC2B, RC5	SS-34	8 × 50 ml	48,240	
(C) High *g* force rotors for sedimenting membranes (angle) or density gradient centrifugation					
Beckman	L-2, L5	Type 35	6 × 94 ml	142,000	Very large capacity high-performance angle rotor
Beckman	L-2, L5	Type 50	8 × 13.5 ml	>200,000	Smaller capacity angle rotors
		Type 65	12 × 13.5 ml		
		Type 42-1	8 × 38.5 ml		
M.S.E.	'50', '65'	43114-118	8 × 14 ml	>200,000	Smaller capacity angle rotors
		43114-123	8 × 35 ml		
Beckman	L-2, L5	SW27	6 × 38,5 ml	131,000	Most popular for gradients (1 × 3½″)
		SW27	6 × 17 ml	135,000	Extra long slender tubes (5/8′ × 4″)
M.S.E.	'50', '65', '75'	43127-310	6 × 38 ml	115,300	'SW27' equivalent

Similar capacity and performance rotors are available from I.E.C., Dupont-Sorvall, etc.
Under approved warranty conditions, Beckman rotors may be used in Sorvall OTD-5 ultracentrifuges.

(D) Small volume rotors. < 5 ml $> 20,000$ *g*

(a) Adaptors are available that fit in high-speed angle-rotors that allow small volume tubes to be centrifuged.
(b) High *g* force refrigerated bench-top centrifuges are available, e.g., Beckman 'Airfuge' miniature fixed angle rotor 6 × 175 μl tubes. Up to 160,000 *g* (accelerates in 30 s); operated by compressed air can be used in a cold room. Burkard 'Koolspin', 8 × 1.5–10 ml 20,000 *g* 'refrigerated'.

APPENDIX 2 (*continued*)

(ii) Batch-type zonal rotors used in membrane preparation

	Type	Maximum speed (revs/min)	Centrifuge	Core configuration	Gradient vol. (ml)	Radius at edge (cm)	g_{max}	Comments on use
(A) 'Low' speed	A-XII	4,500	M.S.E. 6L	Centre loading	1300	18	4,200	Transparent. Used for liver plasma membrane separation
	H.S.	10,000	M.S.E. 18, 21, 25	Centre loading	695	10.3	11,400	Transparent. Used for lysosome separation
	SZ-14	20,000 14,500	Sorvall RC2B RC5	Reorientating gradient	1373	9.5	40,500 22,400	Static and dynamic loading possible. No rotating seals
(B) 'High' speed	B-XIV Al Ti	35,000 47,000	M.S.E. Beckman I.E.C.	Centre loading	650	6.7	91,000 165,000	Large-scale isopycnic runs. Analytical centrifugation
	B-XV Al Ti	25,000 32,000	M.S.E. Beckman I.E.C.	Centre[a] loading	1670	8.9	62,000 102,000	Used for plasma membrane isolation from liver
	B-XXIX	32–39,000	M.S.E. Beckman I.E.C.	Edge unloading	570–1350	6.7–8.4	95,000–116,000	Sample may be unloaded from centre or edge of rotor

[a] These rotors are now adapted for centre or edge unloading and hence overlap with B-XXIX rotors.

Properties of some detergents commonly used in membrane characterisation

Detergent	Critical micellar concentration (mM)	Medium	Formula
Sodium dodecyl sulphate (SDS)	8.2 0.52	H_2O 0.5 M NaCl	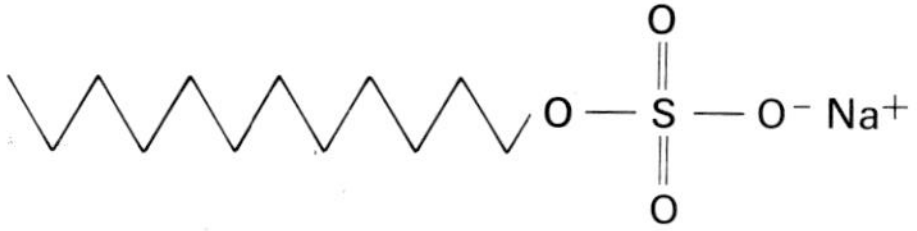
Sarcosyl; sodium dodecyl-*N*-sarcosinate			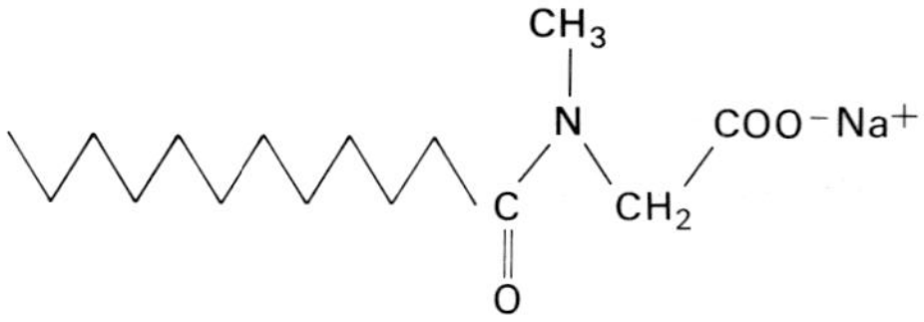
Sodium cholate	13–15	H_2O	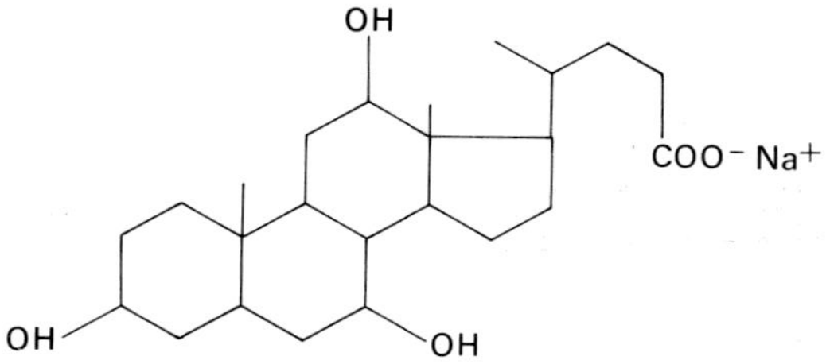
Sodium taurocholate	10–15	H_2O	conjugated with taurine NH_2-CH_2-CH_2-SO_4

APPENDIX 3 (*continued*)

Properties of some detergents commonly used in membrane characterisation

Detergent	Critical micellar concentration (mM)	Medium	Formula
Sodium deoxycholate	4–6	H_2O	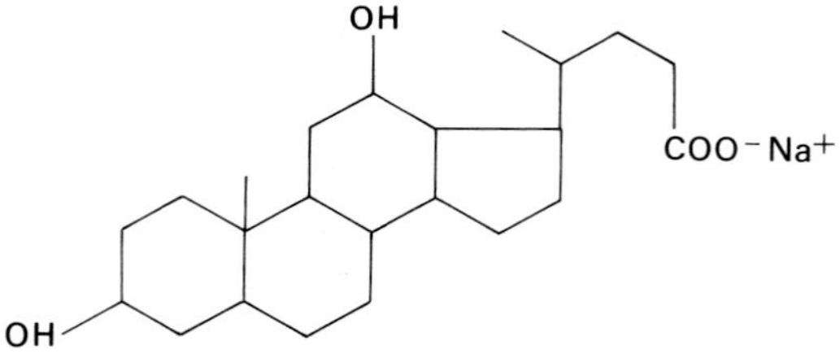
Sodium taurodeoxycholate	2–6	H_2O	conjugated with taurine NH_2-CH_2-CH_2-SO_4
Triton X-100	0.24	H_2O	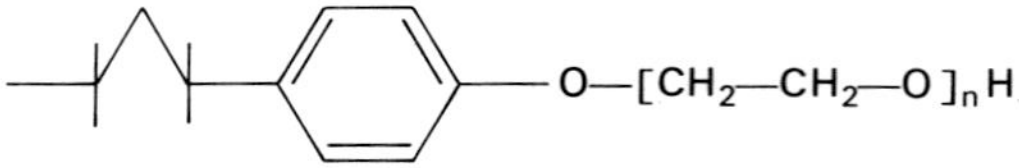
Nonidet P-40	0.29	H_2O	polyethyleneglycol-(9)*p*-t-octylprenol
Brij 56	0.002		polyethyleneglycol cetyl alcohol
76	0.03		polyethyleneglycol stearyl alcohol
96	<0.04		polyethyleneglycol oleyl alcohol
Tween 20	60[a]	H_2O	polyethyleneglycol sorbitan monolaurate
40	29[a]	H_2O	polyethyleneglycol sorbitan monopalmitate
Lubrol WX	0.02–0.06	H_2O	polyethyleneglycol cetylstearyl alcohol

Data from Helenius and Simons (1975)

Note that most commercial detergents are probably chemically impure, containing varying amounts of H_2O and additives. Batches may differ and after prolonged storage the composition at the bottom of the container may differ from that on the top

[a] mg/l

Appendix 4

List of Suppliers

Source	*Specific item mentioned in text*
American Instrument Co. 8030 Georgia Avenue Silver Springs Maryland 20910 U.S.A.	French press
Artisan Industries Inc. 73 Pond Street Waltham Massachusetts 02154 U.S.A.	Gas cavitation bomb
Baskerville and Lindsay Ltd. Chorlton-cum-Hardy Manchester U.K.	Gas cavitation bomb
Bayer U.K. Ltd. Pharmaceutical Division Haywards Heath West Sussex U.K.	Trasylol®-inhibitor of proteolysis
Beckman Instruments Spinco Division 1117 California Avenue Palo Alto California 94304 U.S.A. Subsidiaries and agents world-wide	Centrifuges
Becton-Dickinson & Co. York House Empire Way Wembley U.K.	Syringes
Bender und Hobein Lindwurmstrasse 71–73 D-8000 München G.F.R.	Free-flow electrophoresis apparatus FF5

Subject index p. 261

Blaessig Glass Co. Rochester New York U.S.A.	Dounce homogenisers
Boehringer Mannheim GmbH. Biochemica P.O. Box 51 D-68 Mannheim 31 G.F.R. *U.K.:* **Boehringer Corp. (London) Ltd.** Bell Lane Lewes East Sussex U.K. Agencies in most countries	cAMP, cGMP measuring kits, biochemicals
Branson Sonic Power Co. Eagle Road Danbury Connecticut 06810 U.S.A. *U.K. agents:* **Dawe Instruments** Concord Road Western Avenue London U.K.	Sonicators
B. Braun Postfach 110–120 D-3508 Melsungen G.F.R. *U.K. agents:* **F.T. Scientific Instruments Ltd.** Rosebank Parade Yateley Camberley Surrey U.K. *U.S.A. agents:* **Fischer Scientific** Springfield New Jersey U.S.A.	Tissue homogenisers, gas cavitation bomb

Brinkmann Instruments Inc. Westbury New York U.S.A.	Free-flow electrophoresis apparatus, 'Polytron' homogenisers
British Drug Houses Poole Dorset U.K. *U.S.A. agents:* see Gallard–Schlesinger	Biochemicals, molecular weight markers for gel electrophoresis
Buck and Hickman Ltd. Bank House 100 Queen Street Sheffield U.K.	Jones and Shipman 'Mandrel press' used in tissue disruption
Burkard Scientific Ltd. Rickmansworth Hertfordshire U.K.	Refrigerated small bench centrifuge
Collaborative Research Inc. 1365 Main Street Waltham Massachusetts 02154 U.S.A.	Measuring kits for cAMP, cGMP
Cyclo Chemical Corp. Los Angeles California U.S.A.	Fluorescently labelled chemicals
Desaga Rohrgraben 7 Freiburg im Br. G.F.R. *U.S.A. agents:* **Molecular Instruments Co.** P.O. Box 1652 Evanson Illinois 60201 U.S.A.	Gradient maker, free-flow electrophoresis apparatus

Subject index p. 261

Dow Chemical Co. Midland Michigan U.S.A.	Polystyrene beads
Du Pont de Nemours Wilmington Delaware 19898 U.S.A.	'Ludox HS' silica sols for density gradients
Eastman Kodak Co. Rochester New York 14650 U.S.A. Agencies in most countries	Cetyl trimethylammonium bromide
Electro-Nucleonic Inc. Clifton New Jersey U.S.A.	Glass beads for chromatography
Flow Laboratories Ltd. Victoria Park Heatherhouse Road Irvine Ayrshire Scotland	Media for culturing cells
Gilson Medical Electronics Inc. P.O. Box 27 3000 West Beltline Highway Middleton Wisconsin 53562 U.S.A.	'Minipuls' peristaltic pump, gel sheet fractionators
Gilson (France) 72 rue Gambetta P.O. Box 5 F-95400 Villiers-le-Bel France *U.K. agent:* **Anachem Ltd.** 20 A North St. Luton Bedfordshire U.K.	

G.A.F. Corp.
New York
U.S.A.
U.K.: 268 North Street
Romford
U.K.

Alipal CO-433 detergent

Gallard–Schlesinger Mfg. Corp.
584 Mineola Avenue
Carle Place
Long Island
New York 11514
U.S.A.

Biochemicals, molecular weight markers for gel electrophoresis

Gradient Pty. Ltd.
Sydney
Australia
U.K. agent:
see Universal Scientific Ltd.

'Gradipore' polyacrylamide gel electrophoresis apparatus

Hoffman–La Roche
Grenzacherstrasse
CH-4058 Basel
Switzerland
U.K.: **Roche Products Ltd.**
Broadwater Road
Welwyn Garden City
Hertfordshire
U.K.

Fluorescamine

A.H. Horwell Ltd.
2 Grangeway
Kilburn High Road
London NW 6
U.K.

'Thomas' tissue homogenisers, syringes

Houston Glass Fabricating Co.
Houston
Texas
U.S.A.

Tissue homogenisers

I.E.C. Damon
300, 2nd Avenue
Needham Heights
Massachusetts 02194
U.S.A.

Centrifuges, density gradient makers

Subject index p. 261

ISCO Instrument Specialities Co. 4700 Superior Lincoln Nebraska 68504 U.S.A. *U.K. agent:* see M.S.E.	Apparatus for construction and fractionation of density gradients
Jankel & Kunkel KG D-7813 Staufen im Br. G.F.R. *U.K. agent:* **Sartorius Instruments** 18 Avenue Road Belmont Surrey U.K.	'Ultraturrax' tissue homogenisers
Jenkons (Scientific) Ltd. Mark Road Hemel Hempstead Hertfordshire U.K.	Homogenisers
Kanthal AB Halstammer Sweden	'Nitrothal' wire for microhomogenisation
Kinematica GmbH. Steinhofhalde 272 CH-6005 Luzern Switzerland *U.S.A. agent:* see Brinkmann Instruments	'Polytron' tissue homogenisers, sonicators
Kontes Vineland New Jersey 08360 U.S.A.	Tissue homogenisers
LKB Produkter AB LKB Biotec S-161 Bromma Sweden Sales companies in most countries	Gradient makers, X-press cell disruptor

E. Merck D-61 Darmstadt G.F.R. *U.K. agent:* see B.D.H.	Biochemicals
Mickle Laboratory Engineering Co. Mill Works Gomshall Surrey U.K.	Tissue chopper
Millipore Corp. Bedford Massachusetts 01730 U.S.A. Subsidiaries in most countries	Swinny filter holders
Minnesota Mining and Manufacturing Co. St. Paul Minnesota 55101 U.S.A. *U.K.:* 3 M House, Wigmore House London W.1 Subsidiaries in most countries	'Superbrite' glass beads
M.S.E. Scientific Instruments Ltd. Manor Royal Crawley Sussex U.K.	Centrifuges, density gradient fractionators, sonicators
New England Nuclear 575 Albany Street Boston Massachusetts 02118 U.S.A.	Radiochemicals
Nyanza Inc. Ashland Massachusetts U.S.A.	'Nyacol' aqueous silica sols

Subject index p. 261

Nyegaard & Co. Nycoveien 2 Oslo 4 Norway *U.K. agent:* **Vestric** Stonefield Way Ruislip Middlesex U.K.	Isopaque, Metrizamide for cell separation by centrifugation
Packard Instruments Co. Ltd. 2200 Warrenville Road Downers Grove Chicago Illinois 60515 U.S.A. *U.K.:* Caversham Bridge House 15–17 Church Road Caversham Reading Berkshire U.K. Agents world-wide	Radiochromatogram scanners
Parr Instruments Co. 211, 53rd Street Molline Illinois 61265 U.S.A.	Gas cavitation bomb
Pierce Chemical Co. Rockford Illinois 61105 U.S.A.	Biochemicals, fluorescently labelled chemicals
Pharmacia Fine Chemicals AB P.O. Box 175 S-751 64 Upsala 1 Sweden Agencies in most countries	Gel filtration apparatus for gel electrophoresis, peristaltic pumps, Ficoll
F.W. Potter and Soar Ltd. Beaumont Road Banbury Oxfordshire U.K.	Stainless steel gauze

Radiochemical Centre
Amersham
Buckinghamshire
U.K.
U.S.A.: **Amersham Corp.**
2636 S Clearbrook Drive
Arlington Heights
Illinois 60005
U.S.A.
Agencies in most countries

Kits for measuring cAMP and cGMP, radiochemicals

Research Engineers Ltd.
Orsman Road
Shoreditch
London N1
U.K.

Cell rupturing pump

Rohm and Haas
Philadelphia
Pennsylvania
U.S.A.
U.K.: Lennig House
2 Masons Avenue
Croydon
Surrey
U.K.

Detergents

Sandoz Ltd.
CH-4002 Basel
Switzerland
Subsidiaries in most countries

Scillaren

Schering Chemicals Ltd.
Burgess Hill
Sussex
U.K.

Urografin

Sigma Chemical Co.
St. Louis
Missouri
U.S.A.
Agents world-wide

Biochemicals

Subject index p. 261

Small Parts Inc. 6901 N.E. 3rd Avenue Miami Florida U.S.A.	Stainless steel gauze
Smith and Nephew-Southalls Ltd. Alum Rock Road Birmingham B8 3DY U.K.	Hypodermic tubing (stainless steel) for unloading gradients
Sorvall Operations Dupont Instruments Norwalk Connecticut U.S.A. *U.K.*: 64 Wilbury Way Hitchin Herfordshire U.K.	'Omni-mixer' tissue homogeniser, centrifuges
Stansted Fluid Power Ltd. 70 Bentfield Road Stansted Essex U.K.	Cell rupturing pump
Swiss Silk Bolting Cloth Manufacturing Co. Ltd. P.O. Box CH-8027 Zürich 2 Switzerland *U.K. agent:* **John Stanier & Co. Ltd.** Sherbourne Street Manchester 3 U.K.	Nylon bolting cloth
A.H. Thomas Vine Street at 3rd P.O. Box 779 Philadelphia Pennsylvania U.S.A.	Homogenisers

Tri-R Instruments Inc. 144–13 Jamaica Avenue Jamaica New York 11435 U.S.A.	Homogenisers
Union Carbide Corp. New York U.S.A.	Polyethyleneglycol for coating glass beads and two-phase separation systems
Universal Scientific Ltd. 231 Planchet Road London E 13 U.K.	Felt sheets for rapid destaining of gels 'Gradipore' gel sheets
Whatman Lab Sales Ltd. Springfield Mill Maidstone Kent U.K. Agencies in most countries	Discs for filtration
Winthrop Laboratories Winthrop House Surbiton Surrey U.K.	Sodium diatrizoate
Yeda Research and Development Co. Ltd. Yeda Scientific Instruments P.O. Box 95 Kiryat Rehovot Israel	Tissue chopper, gas cavitation bomb

References

ABNEY, E., W.H. EVANS and R.M.E. PARKHOUSE (1976) Biochem. J. *159*, 293.

ABOOD, L.G. (1969) *In:* Handbook of Neurochem, Vol. 2 (Plenum Press, New York) p. 303.

AGRAWAL, H.C., R.M. BURTON, M.A. FISHMAN, R.F. MITCHELL and A.L. PRENSKY (1972) J. Neurochem. *19*, 2083.

ALBERTSSON, P.A. (1971) *In:* Partition of Cell Particles and Macromolecules, 2nd ed. (Wiley-Interscience, New York).

ALBANO, J.D.M., G.D. BARNES, D.V. MAUDSLEY, B.L. BROWN and R.P. EKINS (1974) Anal. Biochem. *60*, 130.

ALCALA, J., N. LIESKA and H. MAISEL (1975) Exp. Eye Res. *21*, 581.

ALEXANDER, E.L. and B. WETZEL (1975) Science *188*, 732.

ALLAN, D. and M.J. CRUMPTON (1970) Biochem. J. *120*, 133.

ALLAN, D., M.M. BILLAH, J.B. FINEAN and R.H. MICHEL (1976) Nature (London) *261*, 58.

AMAR-COSTESEC, A., H. BEAUFAY, M. WIBO, D. THINES-SEMPOUX, E. FEYTMANS, M. ROBBI and J. BERTHET (1974) J. Cell Biol. *64*, 201.

AMAR-COSTESEC, A., M. WIBO, D. THINES-SEMPOUX, H. BEAUFAY and J. BERTHET (1974) J. Cell Biol. *62*, 717.

AMINOFF, D. (1961) Biochem. J. *81*, 384.

ANDERSON, N.G. (ed.) (1966) Zonal Centrifugation; (Natl. Cancer Inst. Monograph) *21*.

ANDERSON, N.G., A.I. LANSING, I. LIEBERMAN, C.T. RANKIN and H. ELROD (1968) *In:* L. Manson (ed.) Biological Properties of the Mammalian Surface Membrane (Wistar, Philadelphia) p. 23.

ANDREW, C.G., R.R. ALMON and S.H. APPEL (1974) J. Biol. Chem. *249*, 6163.

ANSELSTETTER, V. and H. HORSTMANN (1975) Eur. J. Biochem. *56*, 259.

AOYAGI, T., H. SUDA, M. NAGAI, K. OGAWA, J. SUZUKI, T. TAKEUCHI and H. UMEZAWA (1976) Biochim. Biophys. Acta *452*, 131.

ARBORGH, B.A.M., H. GLAUMANN and J.L.E. ERICSSON (1974) Lab. Invest. *30*, 674.

ASSMANN, G., R.M. KRAUSS, D.S. FREDERICKSON and R.I. LEVY (1973) J. Biol. Chem. *248*, 1992.

ATKINSON, P.H. (1973) *In:* D.M. Prescott (ed.) Methods in Cell Biology, Vol. 7, (Academic Press, New York) p. 158.

Atkinson, P.H. and D.F. Summers (1971) J. Biol. Chem. *246*, 5162.

Avruch, J. and D.F.H. Wallach (1971) Biochim. Biophys. Acta *233*, 334.

Axelsen, N.H. (1975) (ed.) Quantitative Immunoelectrophoresis; New Developments and Applications, Scand. J. Immunol. 2, Suppl.

Babitch, J.A., T.B. Breithaupt, T.C. Chiu, R. Garadi and D.L. Helseth (1976) Biochim. Biophys. Acta *433*, 75.

Bachorik, P.S. and L.S. Dietrich (1972) J. Biol. Chem. *247*, 5071.

Baenziger, N.L., G.N. Brodie and P.W. Majerus (1972) J. Biol. Chem. *247*, 2723.

Barber, A.J. and G.A. Jamieson (1970) J. Biol. Chem. *245*, 6357.

Barber, B.H. and M.J. Crumpton (1976) FEBS Lett. *66*, 215.

Barchi, R.L., E. Bonilla and M. Wong (1977) Proc. Natl. Acad. Sci. USA *74*, 34.

Barondes, S.H. (1968) J. Neurochem. *15*, 343.

Barrett, E.J. and D.R. Headon (1975) Biochem. Soc. Trans. *3*, 1248.

Bartfai, T., P. Berg, M. Schultzberg and E. Heilbronn (1976) Biochim. Biophys. Acta *426*, 186.

Bartlett, G.R. (1959) J. Biol. Chem. *234*, 466.

Baudhuin, P., H. Beaufay, Y. Rahman-Li, O.Z. Sellinger, R. Wattiaux, P. Jacques and C. De Duve (1964) Biochem. J. *92*, 179.

Baudhuin, P., P. Evrard and J. Berthet (1967) J. Cell Biol. *32*, 181.

Beaufay, H., A. Amar-Costesec, D. Thines-Sempoux, M. Wibo, M. Robbi and J. Berthet (1974) J. Cell Biol. *61*, 213.

Beechey, R.B., A.M. Robertson, C.T. Holloway and J.G. Knight (1967) Biochemistry *6*, 3867.

Bell, M.L., H.M. Lazarus, A.H. Herman, R.H. Egdahl and A.M. Rutenberg (1971) Proc. Soc. Exp. Biol. Med. *136*, 298.

Bell, R.D. and E.A. Doisy (1920) J. Biol. Chem. *44*, 55.

Benedetti, E.L. and P. Emmelot (1968) J. Cell Biol. *38*, 15.

Bennett, G., C.P. Leblond and A. Haddad (1974) J. Cell Biol. *60*, 258.

Bennett, V. and P. Cuatrecasas (1973) Biochim. Biophys. Acta *311*, 362.

Berger, S.J. and B. Sacktor (1970) J. Cell Biol. *47*, 637.

Bergeron, J.J.M., J.H. Ehrenreich, P. Siekevitz and G.E. Palade (1973a) J. Cell Biol. *59*, 73.

Bergeron, J.J.M., W.H. Evans and I.I. Geshwind (1973b) J. Cell Biol. *59*, 771.

Bernath, P. and T.P. Singer (1962) Methods Enzymol. *5*, 597.

Bhakdi, S., H. Knufermann and D.F.H. Wallach (1974) Biochim. Biophys. Acta *345*, 448.

Bischoff, E., T. Tran-Thi and K. Decker (1975) Eur. J. Biochem. *51*, 353.

Bischoff, E., J. Wilkening, T. Tran-Thi and K. Decker (1976) Eur. J. Biochem. *62*, 279.

Blanchet, J.P. (1974) Exp. Cell Res. *84*, 159.

Bligh, E.G. and W. Dyer (1959) Can. J. Biochem. Physiol. *37*, 911.

Blomberg, F. and K. Berzius (1975) Eur. J. Biochem. *56*, 319.

BLOMBERG, F. and P. PERLMANN (1971) Biochim. Biophys. Acta *233*, 53.

BLOMBERG, F. and M. RAFTELL (1974) Eur. J. Biochem. *49*, 21.

BLOOMSTRAND, C. and A. HAMBERGER (1970) J. Neurochem. *17*, 1187.

BLOUIN, A., R.P. BALANDER and E.R. WEIBEL (1977) J. Cell Biol. *72*, 441.

BODE, F., K. BAUMANN and R. KINNE (1976) Biochim. Biophys. Acta *433*, 294.

BODMER, W.F., M. TRIPP and J. BODMER (1967) *In:* Histocompatibility testing, Munksgraad, Copenhagen.

BOEGMAN, R.J., J.F. MANERY and L. PINTERIC (1970) Biochim. Biophys. Acta *203*, 506.

BOLTON, A.E. and W.M. HUNTER (1973) Biochem. J. *133*, 529.

BONNER, W.M. and R.A. LASKEY (1974) Eur. J. Biochem. *46*, 83.

BOONE, C.W., L.E. FORD, H.E. BOND, D.C. STUART and D. LORENZ (1969) J. Cell Biol. *41*, 378.

BOOTH, A.G. (1977) Biochem. J. *163*, 165.

BOOTH, A.G. and A.J. KENNY (1974) Biochem. J. *142*, 575.

BOOTH, A.G. and A.J. KENNY (1976a) Biochem. Soc. Trans *4*, 270.

BOOTH, A.G. and A.J. KENNY (1976b) J. Cell Sci. *21*, 449.

BOSMANN, H.B., A. HAGOPIAN and E.H. EYLER (1969) Arch. Biochem. Biophys. *130*, 573.

BOXER, D.H., R.E. JENKINS and M.J.A. TANNER (1974) Biochem. J. *137*, 531.

BOYER, J.L. and D. RENO (1975) Biochim. Biophys. Acta *401*, 59.

BRADFORD, H.F. (1974) Biochem. Soc. Trans. *2*, 680.

BRADFORD, H.F., G.W. BENNETT and A.J. THOMAS (1973) J. Neurochem. *21*, 495.

BRAMLEY, T.A., R. COLEMAN and J.B. FINEAN (1971) Biochim. Biophys. Acta *241*, 752.

BRAY, D. and C. THOMAS (1975) Biochem. J. *147*, 221.

BRETSCHER, M.S. (1971) J. Mol. Biol. *58*, 775.

BRETZ, U., M. BAGGIOLINI, R. HAUSER and C. HODEL (1974) J. Cell Biol. *61*, 466.

BRIDGEN, J., D. SNARY, M.J. CRUMPTON, C. BARNSTABLE, P. GOODFELLOW and W.F. BODMER (1976) Nature (London) *261*, 200.

BRIVIO-HAUGLAND, R.P., S.L. LOUIS, K. MUSCH, N. WALDECK and M.A. WILLIAMS (1976) Biochim. Biophys. Acta *433*, 150.

BROEKMAN, M.J., N.P. WESTMORELAND and P. COHEN (1974) J. Cell Biol. *60*, 507.

BROMER, W.W., M.E. BOUCHER and J.M. PATTERSON (1973) Biochem. Biophys. Res. Commun. *53*, 134.

BROWN, B.L., J.D.M. ALBANO, R.P. EKINS and A.M. SGERZI (1971) Biochem. J. *121*, 561.

BRUNETTE, D.M. and J.E. TILL (1971) J. Membr. Biol. *5*, 215.

BURGER, R.M. and J.M. LOWENSTEIN (1970) J. Biol. Chem. *245*, 6274.

BUTTERS, T.D. and R.C. HUGHES (1974) Biochem. J. *140*, 469.

CAREY, F. and W.H. EVANS (1977) Biochem. Soc. Trans. *5*, 103.

CARLSON, D.M., G.W. JOURDIAN and S. ROSEMAN (1973) J. Biol. Chem. *248*, 5742.

CARLSON, R.W., H.G. WADA and H.H. SUSSMAN (1976) J. Biol. Chem. *251*, 4139.

Carraway, K. (1975) Biochem. Biophys. Acta *415*, 379.

Carson, S.D. (1976) Anal. Biochem. *75*, 472.

Castle, J.D., J.D. Jamieson and G.E. Palade (1975) J. Cell Biol. *64*, 182.

Chacko, G.K., D.E. Goldman, H.C. Malhotra and M.M. Dewey (1974) J. Cell Biol. *62*, 831.

Chandrabose, K.A., J.M. Graham and I.A. MacPherson (1976) Biochim. Biophys. Acta *429*, 112.

Chang, K.J., V. Bennett and P. Cuatrecasas (1975) J. Biol. Chem. *250*, 488.

Chao, S. and M. Rumsby (1977) Brain Res. *124*, 347.

Charalampous, F.C., N.K. Gonatas and A.D. Melbourne (1973) J. Cell Biol. *59*, 421.

Cheetham, R.D., D.J. Morré, C. Pannek and D.S. Friend (1971) J. Cell Biol. *49*, 899.

Chen, P.S., T.Y. Toribara and H. Warner (1956) Anal. Chem. *28*, 1756.

Cheng, H. and M.G. Farquhar (1976) J. Cell Biol. *70*, 660, 671.

Chrambach, A., R.A. Reisfeld, M. Wyckoff and J. Zaccari (1967) Anal. Biochem. *20*, 150.

Chung, L.C. and J. Hochstadt (1976) J. Biol. Chem. *251*, 1175.

Clausen, J. (1969) Immunochemical Techniques for the Identification and Estimation of Macromolecules. *In:* Lab. Techniques in Biochemistry and Molecular Biology, T.S. Work and E. Work (eds.) Vol. 1, Part 2 (North-Holland, Amsterdam).

Clarke, S. (1976) J. Biol. Chem. *251*, 950.

Claude, A. (1946) J. Exp. Med. *84*, 61.

Cleland, K.W. and E.C. Slater (1953) Biochem. J. *53*, 547.

Coackley, T. (1974) Brain Res. *70*, 281.

Cohen, J., G.R. Dutton, G.P. Wilkin, J.E. Wilson and R. Balazs (1974) J. Neurochem. *23*, 899.

Colbeau, A., J. Nachbaur and P. Vignais (1971) Biochim. Biophys. Acta *249*, 462.

Colombini, M. and R.M. Johnstone (1973) Biochim. Biophys. Acta *323*, 69.

Conrad, M.J. and J.T. Penniston (1976) J. Biol. Chem. *251*, 253.

Cooper, P.H. and D.R. Stanworth (1976) Biochem. J. *156*, 691.

Cotman, C.W. (1972) *In:* Research Methods in Neurochemistry, N. Marks and R. Rodnight (eds.) Vol. 1, p. 45 (Plenum Press, New York).

Cotman, C.W., H.R. Mahler and N.G. Anderson (1968) Arch. Biochem. Biophys. *163*, 172.

Cotman, C.W. and D.A. Matthews (1971) Biochim. Biophys. Acta *249*, 380.

Cotman, C.W., G. Banker, L. Churchill and D. Taylor (1974) J. Cell Biol. *63*, 441.

Cotman, C.W., R.E. McCaman and S.A. Dewhurst (1971) Biochim. Biophys. Acta *249*, 395.

Cotman, C.W. and D. Taylor (1972) J. Cell Biol. *55*, 696.

Cotman, C.W., H. Herchman and D. Taylor (1971) J. Neurobiol. *2*, 169.

CRUMPTON, M.J. and R.M.E. PARKHOUSE (1972) FEBS Lett. *22*, 210.

CRUMPTON, M.J. and D. SNARY (1974) *In:* R.R. Porter and G.L. Ada (eds.) Contemp. Top. Mol. Immunol. *3*, 27.

CUATRECASAS, P. (1971) Proc. Natl. Acad. Sci. USA *68*, 1264.

CULP, L.A. (1976) Biochemistry *15*, 4094.

CULVENOR, J.G. and W.H. EVANS (1977) Biochem. J., *168*, 475.

DALLNER, G.P., P. SIEKEVITZ and G.E. PALADE (1966) J. Cell Biol. *30*, 97.

DANIEL, A. and G. GUTH (1975) Exp. Neurol. *47*, 181.

DAVID, G.S. (1972) Biochim. Biophys. Res. Commun. *48*, 464.

DAVIDSON, W.F. and C.R. PARISH (1975) J. Immunol. Methods *7*, 291.

DAVIS, G.A. and F.E. BLOOM (1970) J. Cell Biol. *47*, 46a.

DAVIS, G.A. and F.E. BLOOM (1973) Brain Res. *62*, 135.

DE DUVE, C. (1959) *In:* Subcellular Particles (Ronald Press, New York).

DE DUVE, C. (1965) Harvey Lect. *59*, 49.

DE DUVE, C. (1967) *In:* D.B. Roodyn (ed.) Enzyme Cytology (Academic Press, London), p. 1.

DE DUVE, C. (1975) Science *189*, 186.

DEHEER, D.H., M.S. OLSON and R.N. PINCKARD (1974) J. Cell Biol. *60*, 460.

DEMUS, H. (1973) Biochim. Biophys. Acta *291*, 93.

DEN, H., B. KAUFMANN, E.J. MCGUIRE and S. ROSEMAN (1975) J. Biol. Chem. *250*, 739.

DEPIERRE, J.W. and M.L. KARNOVSKY (1974) J. Biol. Chem. *249*, 7111.

DEPIERRE, J.W. and G. DALLNER (1975) Biochim. Biophys. Acta *415*, 411.

DEPPERT, W., H. WERCHAU and G. WALTER (1974) Proc. Natl. Acad. Sci. USA *71*, 3068.

DE ROBERTIS, E. (1971) Science *171*, 963.

DE ROBERTIS, E., G.R. DELORES-ARNAIZ, L. SALGONICOFF, A.P. DE IRALDI and L.M. ZIEHER (1963) J. Neurochem. *10*, 225.

DODGE, J.T., C. MITCHELL and D.J. HANAHAN (1963) Arch. Biochem. Biophys. *100*, 119.

DODS, R.F., E. ESSNER and M. BARCLAY (1972) Biochim. Biophys. Res. Commun. *46*, 1074.

DOUNCE, A.L., R.F. WITTER, K.J. MONTY, S. PATE and M.A. COTTONE (1955) J. Biophys. Biochem. Cytol. *1*, 139.

DREYER, W.J., D.S. PAPERMASTER and H. KUHN (1972) Ann. N. Y. Acad. Sci. *195*, 61.

DULANEY, J.T. and O. TOUSTER (1970) Biochim. Biophys. Acta *196*, 29.

DUNIA, I., C.S. GOSH, E.L. BENEDETTI, A. ZWEERS and H. BLOEMENDAL (1974) FEBS Lett. *45*, 139.

DURHAM, J.P., N. GALANTI and N.W. REVIS (1975) Biochim. Biophys. Acta *394*, 388.

EARL, D.C.N. and A. KORNER (1965) Biochem. J. *94*, 721.

EBEL, H., E. AULBERT and H.J. MERKER (1976) Biochim. Biophys. Acta *433*, 531.

ECKHARDT, A.E., C.E. HAYES and I.J. GOLDSTEIN (1976) Anal. Biochem. *73*, 192.

EDELSON, P.J. and Z.A. COHN (1976) J. Exp. Med. *144*, 1596.

EDWARDS, M.J. and M.H. MCGUIRE (1970) Mol. Pharmacol. *6*, 641.

EHRENREICH, J.H., J.J.M. BERGERON, P. SIEKEVITZ and G.E. PALADE (1973) J. Cell Biol. *59*, 45.

EIBL, H. and W.E.M. LANDS (1969) Anal. Biochem. *30*, 51.

EICHHOLZ, A. and R.K. CRANE (1965) J. Cell Biol. *26*, 678.

EICHNER, D. (1966) Experientia *22*, 620.

EL-AASER, A.A., J.T.R. FITZSIMMONS, R.H. HINTON, E. REID, E. KLUCIS and P. ALEXANDER (1966) Biochim. Biophys. Acta *127*, 553.

ELDER, J.H. and D.J. MORRÉ (1976) J. Biol. Chem. *251*, 5054.

ELLMAN, G.C., K.D. COURTNEY, V. ANDRES and R.M. FEATHERSTONE (1961) Biochem. Pharmacol. *7*, 88.

EMANUEL, C.F. and I.L. CHAIKOFF (1957) Biochim. Biophys. Acta *24*, 254.

EMMELOT, P., C.J. BOS, E.L. BENEDETTI and P. RUMKE (1964) Biochim. Biophys. Acta *90*, 126.

EMMELOT, P. and C.J. BOS (1966) Biochim. Biophys. Acta *121*, 434.

EMMELOT, P. and C.J. BOS (1974) Biochim. Biophys. Acta *249*, 285.

ENGELHARD, V.H., D.A. PLUT and D.R. STORM (1976) Biochim. Biophys. Acta. *451*, 48.

ERNSTER, L. (1967) Methods Enzymol. *10*, 309.

ESSNER, E., A.B. NOVIKOFF and B. MASEK (1958) J. Biophys. Biochem. Cytol. *4*, 711.

EVANS, E.M., J.M. WRIGGLESWORTH, K. BURDETT and W.F.R. POVER (1971) J. Cell Biol. *51*, 452.

EVANS, W.H. (1970) Biochem. J. *116*, 833.

EVANS, W.H. (1974) Nature *250*, 391.

EVANS, W.H. and J.W. BRUNING (1970) Immunology *19*, 735.

EVANS, W.H. and J.W. GURD (1971) Biochem. J. *125*, 615.

EVANS, W.H. and J.W. GURD (1972) Biochem. J. *128*, 691.

EVANS, W.H. and J.W. GURD (1973) Biochem. J. *133*, 189.

EVANS, W.H., D.O. HOOD and J.W. GURD (1973) Biochem. J. *135*, 819.

EVANS, W.H., T. KREMMER and J.G. CULVENOR (1976) Biochem. J. *154*, 589.

EY, P.L. and E. FERBER (1977) Biochim. Biophys. Acta *480*, 403.

FAIRBANKS, D. and J. AVRUCH (1972) J. Supramol. Structure *1*, 66.

FARQUHAR, M.G. and G.E. PALADE (1963) J. Cell Biol. *17*, 375.

FARQUHAR, M.G., J.J.M. BERGERON and G.E. PALADE (1974) J. Cell Biol. *60*, 8.

FERBER, E., K. RESCH, D.F.H. WALLACH and W. IMM (1972) Biochim. Biophys. Acta *266*, 494.

FERGUSON, K.A. (1964) Metabolism *13*, 985.

FESTOFF, B.W. and W.K. ENGEL (1974) Proc. Natl. Acad. Sci. USA *71*, 2435.

FISH, W.W. (1975) *In:* E.D. Korn (ed.) Methods in Membrane Biology, Vol. 4 (Plenum Press, New York), p. 189.

FISK, C.H. and Y. SUBBAROW (1925) J. Biol. Chem. *66*, 375.

FITZPATRICK, D.F., G.R. DAVENPORT, L. FORTE and E.J. LANDON (1967) J. Biol. Chem. *244*, 3561.

FLANAGAN, S.D. and S.H. BARONDES (1975) J. Biol. Chem. *250*, 1484.

FLANAGAN, S.D., P. TAYLOR and S.H. BARONDES (1975) Nature (London) *254*, 441.

FLECK, A. and H.N. MUNRO (1962) Biochim. Biophys. Acta *55*, 571.

FLEISCHER, B. and F. ZAMBRANO (1974) J. Biol. Chem. *249*, 5995.

FLEISCHER, B., S. FLEISCHER and H. Ozawa (1969) J. Cell Biol. *43*, 59.

FLEISCHER, S. and B. FLEISCHER (1967) Methods Enzymol. *10*, 427.

FOLCH, J., M. LEES and G.H. SLOAN-STANLEY (1957) J. Biol. Chem. *226*, 497.

FORSTNER, G.G. (1971) Biochem. J. *121*, 781.

FORSTNER, G.G., S.M. SABESIN and K.J. ISSELBACHER (1968) Biochem. J. *106*, 381.

FRANK, R.N. and D. RODBARD (1975) Arch. Biochem. Biophys. *171*, 1.

FRASER, D. (1951) Nature (London) *167*, 33.

FREYTAG, J.W., M. OHNO and B.G. HUDSON (1976) Biochem. Biophys. Res. Commun. *72*, 796.

FRITZSON, P. and I. SMITH (1971) Biochim. Biophys. Acta *235*, 128.

FUJITA, M., H. OHTA, K. KAWAI, H. MATSUI and M. NAKAO (1972) Biochim. Biophys. Acta *274*, 336.

GAHMBERG, C.G. and S.I. HAKOMORI (1973) J. Biol. Chem. *248*, 4311.

GAHMBERG, C.G. and K. SIMONS (1970) Acta Path. Microbiol. Scand., Sect. B, *78*, 176.

GEMMELL, R.T. and T. HEATH (1972) Anat. Reç. *172*, 57.

GEORGE, S.G. and A.J. KENNY (1973) Biochem. J. *134*, 43.

GIANETTO, R. and C. DE DUVE (1955) Biochem. J. *59*, 433.

GIBBONS, R.A., S.N. DIXON and R. SELLWOOD (1976) Eur. J. Biochem. *66*, 243.

GILES, K.W. and A. MYERS (1965) Nature (London) *206*, 93.

GILMAN, A.G. (1970) Proc. Natl. Acad. Sci. USA *67*, 305.

GLOSSMANN, H. and D.M. NEVILLE (1971) J. Biol. Chem. *246*, 6339.

GLOSSMANN, H. and H. GIPS (1974) Naunyn-Schmiederberg's Arch. Pharmacol. *282*, 439.

GLOSMANN, H. and H. GIPS (1976) Naunyn-Schmiedeberg's Arch. Pharmacol. *292*, 199.

GOLDBERG, J.A. and A.M. RUTENBERG (1958) Cancer *11*, 283.

GONATAS, N.K., J.O. GONATAS, A. STIEBER, J.C. ANTOINE and S. AVRAMEAS (1976) J. Cell Biol. *70*, 477.

GOODENOUGH, D.A. (1974) J. Cell Biol., *61*, 557.

GOODENOUGH, D.A. and W. STOECKENIUS (1972) J. Cell Biol. *54*, 646.

GORDON, D.J., J.L. BOYER and E.D. KORN (1977) J. Biol. Chem. *252*, 8300.

GORDON, A.H. (1975) *In:* Electrophoresis of Proteins in Polyacrylamide and Starch Gels, Lab. Techniques in Biochem. and Mol. Biol., T.S. Work and E. Work (eds.) Vol. 1, Part 1 (North-Holland, Amsterdam)

GOROVSKY, M.A., K. CARLSON and J.L. ROSENBAUM (1970) Anal. Biochem. *35*, 359.

GOSPODAROWICZ, D. (1973) J. Biol. Chem. *248*, 5050.

GOTTLIEB, D., R. MERRELL and L. GLASER (1974) Proc. Natl. Acad. Sci. USA *71*, 1800.

GRAHAM, J.M. (1972) Biochem. J. *130*, 1113.

GRAHAM, J.M., R.O. HYNES, E.A. DAVISON and D.F. BAINTON (1975) Cell *4*, 353.

GRAY, E.G. and V.P. WHITTAKER (1962) J. Anat. *96*, 79.

GREEN, D.E., S. MII and P.M. KOHOUT (1955) J. Biol. Chem. *217*, 551.

GREEN, J.R., M.J. DUNN and A.H. MADDY (1975) Biochim. Biophys. Acta *382*, 457.

GREEN, M.R., J.V. PESTEWKA and A.C. PEACOCK (1973) Anal. Biochem. *56*, 43.

GREENFIELD, S., W.T. NORTON and P. MORREL (1971) J. Neurochem. *18*, 2119.

GREENGARD, O., M. FEDERMAN and W.E. KNOX (1972) J. Cell Biol. *52*, 261.

GREENWOOD, F.C., N.M. HUNTER and J.S. GLOVER (1963) Biochem. J. *89*, 114.

GREENWOOD, M.R.C. and J. HIRSCH (1974) J. Lipid Res. *15*, 474.

GREFRATH, S.P. and J.A. REYNOLDS (1974) Proc. Natl. Acad. Sci. USA *71*, 3913.

GRUENSTEIN, E., A. RICH and R.R. WEIHING (1975) J. Cell Biol. *64*, 223.

GURD, J.W. and W.H. EVANS (1974) Arch. Biochem. Biophys. *164*, 305.

GURD, J.W. and W.H. EVANS (1976) Can. J. Biochem. *54*, 477.

GURD, J.W., W.H. EVANS and H.R. PERKINS (1972) Biochem. J. *130*, 271.

GURD, J.W., W.H. EVANS and H.R. PERKINS (1973) Biochem. J. *135*, 827.

GURD, J.W. and H.R. MAHLER (1974) Biochemistry *13*, 5193.

GURD, J.W., L.R. JONES, H.R. MAHLER and W.J. MOORE (1974) J. Neurochem. *22*, 281.

HAJOS, F., R. TAPIA, G. WILKIN, A.L. JOHNSON and R. BALAZS (1974) Brain Res. *70*, 261.

HANAHAN, D.J., J.E. EKHOLM and M.G. LUTHRA (1974) Biochim. Biophys. Acta *363*, 283.

HANNIG, K. (1972) *In:* Techniques of Biochem. Biophys. Morphology, D. Glick and R.M. Rosenbaum (eds.) Vol. 1 (Wiley, New York) p. 191.

HAWKES, S.P., T.D. MEEHAN and M.J. BISSELL (1976) Biochim. Biophys. Res. Commun. *68*, 1226.

HAY, A.J. (1974) Virology *60*, 398.

HAYASHI, O. (1962) Methods Enzymol. *5*, 807.

HEIDRICH, H. and G. LEUTNER (1974) Eur. J. Biochem. *41*, 37.

HEIDRICH, H., R. KINNE, E. KINNE-SAFFRAN and K. HANNIG (1972) J. Cell Biol. *54*, 232.

HEINE, J.W. and C.A. SCHNAITMAN (1971) J. Cell Biol. *48*, 703.

HELENIUS, A. and K. SIMONS (1975) Biochim. Biophys. Acta *415*, 29.

HELLER, M. and I. HARARY (1977) Biochim. Biophys. Acta *467*, 29.

HELWIG, J.J., P. MANDEL and C. BOLLACK (1974) Arch. Int. Physiol. Biochim. *82*, 907.

HELWIG, J.J., C. BOLLACK, P. MANDEL and C. GORIDIS (1975) Biochim. Biophys. Acta *377*, 463.

HEMMINKI, K. and O. SUOVANIEMI (1973) Biochim. Biophys. Acta *298*, 75.
HENN, F.A. and A. HAMBERGER (1976) Neurochem. Res. *1*, 261.
HENN, F.A., D.J. ANDERSON and D.G. RUSTAD (1976) Brain Res. *101*, 341.
HENN, F.A., H. HALJAMAE and A. HAMBERGER (1972) Brain Res. *43*, 437.
HENRY, R.J., C. SOBEL and J. KIM (1957) Am. J. Clin. Path. *28*, 572.
HERBERMAN, R. and C.A. STETSON (1965) J. Exp. Med. *121*, 533.
HEYWOOD, S.M., R.M. DOWBEN and A. RICH (1968) Biochemistry *7*, 3289.
HILDERSON, H.J.J., M.J.S. DE WOLFE, A.R. LAGROU and W.S.H. DIERICK (1975) Biochem. J. *152*, 601.
HINTON, R.H., M.L.E. BURGE and G.L. HARTMAN (1969) Anal. Biochem. *29*, 248.
HINTON, R. and M. DOBROTA (1976) Centrifugation in density gradients. *In:* Lab. Techniques in Biochemistry and Molecular Biology; T.S. Work and E. Work (eds.) Vol. 6 (North-Holland, Amsterdam).
HIRAOKA, T. and D. GLICK (1963) Anal. Biochem. *5*, 497.
HOGEBOOM, G.A., W.C. SCHNEIDER and G.E. PALADE (1948) J. Biol. Chem. *172*, 619.
HORVAT, A. and O. TOUSTER (1967) Biochim. Biophys. Acta *148*, 725.
HOURANI, B.T., N.M. CHANCE and J.H. PINCUS (1973) Biochim. Biophys. Acta *328*, 520.
HUBBARD, A.L. and Z.A. COHN (1972) J. Cell Biol. *55*, 390.
HUBBARD, A.L. and Z.A. COHN (1975) J. Cell Biol. *64*, 438.
HUGGINS, J.W. and K.L. CARRAWAY (1976) J. Supramol. Biol. *5*, 59.
HUGHES, D.E. (1951) Brit. J. Exp. Pathol. *32*, 97.
HUGHES, R.C. and R.W. JEANLOZ (1964) Biochemistry *3*, 1535.
HUNT, R.C. and J.C. BROWN (1975) J. Mol. Biol. *97*, 413.
HUNT, R.C., E. GOLD and J.C. BROWN (1975) Biochim. Biophys. Acta *413*, 453.
HUNTER, M.J. and S.L. COMMERFORD (1961) Biochim. Biophys. Acta *47*, 580.
HUNTER, W.M. and F.C. GREENWOOD (1962) Nature (London) *194*, 495.
HYNES, R.O. (1973) Proc. Natl. Acad. Sci. (USA) *70*, 3170.
HYNIE, S. (1974) Collect. Czech. Chem. Commun. *35*, 2325.
INOYE, M., S. HORIUCHI and Y. MORINO (1977) Eur. J. Biochem. *73*, 335.
IPATA, P.I. (1967) Anal. Biochem. *20*, 30.
IPATA, P.I. (1968) Biochemistry *7*, 507.
ITOH, A., K. TSUSHIMA and R. ITOH (1968) J. Biochem. (Tokyo) *63*, 165.
JACKSON, R.C. (1976) Biochemistry *15*, 5641.
JETT, M., T.M. SEED and G.A. JAMIESON (1977) J. Biol. Chem. *252*, 2134.
JOHANNESEN, K., J. DEPIERRE, A. BERGSTRAND, G. DALLNER and L. ERNSTER (1977) Biochim. Biophys. Acta *496*, 115.
JOHNSEN, S., T. STOKKE and H. PRYDZ (1974) J. Cell Biol. *63*, 357.
JOHNSEN, S., T. STOKKE and H. PRYDZ (1975) Exp. Cell Res. *93*, 245.
JOHNSON, R.A., J.G. HARDMAN, A.E. BROADUS and E.W. SUTHERLAND (1970) Anal. Biochem. *35*, 91.
JONES, D.H. and A.I. MATUS (1974) Biochim. Biophys. Acta *356*, 276.

JONES, L.C. and L. ERNSTER (1962) J. Cell Biol. *15*, 563.
JOURDIAN, G.W., L. DEAN and S. ROSEMAN (1971) J. Biol. Chem. *246*, 430.
JULIANO, R.L. and M. BEHAR-BANNELIER (1975a) Biochemistry *14*, 3816.
JULIANO, R.L. and M. BEHAR-BANNELIER (1975b) Biochim. Biophys. Acta *375*, 249.
KAMAT, V.B. and D.F.H. WALLACH (1965) Science *148*, 1343.
KAMATH, S.A. and E. RUBIN (1972) Biochem. Biophys. Res. Commun. *49*, 52.
KANIG, K. and H. KUNKEL (1957) Hoppe-Seyler's Z. Physiol. Chem. *309*, 166.
KASHNIG, D.M. and C. B. KASPER (1969) J. Biol. Chem. *244*, 3786.
KATES, M. (1972) *In:* Techniques in Lipidology (North-Holland, Amsterdam).
KAUFMAN, R.L. and V. GINSBURG (1968) Exp. Cell Res. *50*, 127.
KAULEN, H.D., R. HENNING and W. STOFFEL (1970) Hoppe-Seyler's Z. Physiol. Chem. *351*, 1555.
KAWAI, K., M. FUJITA and M. NAKAO (1974) Biochim. Biophys. Acta *369*, 222.
KAY, R.R., D. FRASER and I.R. JOHNSTON (1972) Eur. J. Biochem. *30*, 145.
KEAN, E.L. and K.J. BIGHOUSE (1974) J. Biol. Chem. *249*, 7813.
KEENAN, T.W. and MORRÉ, D.J. (1970) Biochemistry *9*, 19.
KEENAN, T.W. and D.J. MORRÉ (1975) FEBS Lett. *55*, 8.
KEMPSON, S.A., R.G. PRICE and J.L. STIRLING (1974) *In:* E. Reid (ed.) Methodological Developments in Biochemistry, Vol. 4 (Longmans, London) p. 297.
KIDWAI, A.M., M.A. RADCLIFFE and E.E. DANIEL (1971a) Biochim. Biophys. Acta *233*, 538.
KIDWAY, A.M., M.A. RADCLIFFE, G. DUCHON and E.E. DANIEL (1971b) Biochem. Biophys. Res. Commun. *45*, 901.
KING, E.J. (1932) Biochem. J. *26*, 292.
KING, L.E. and M. MORRISON (1976) Anal. Biochem. *71*, 223.
KINNE-SAFFRAN, E. and R. KINNE (1974) J. Membr. Biol. *17*, 261.
KLENIG, H. (1970) J. Cell Biol. *46*, 396.
KLENK, H. and CHOPIN, P.W. (1969) Virology *38*, 255.
KONO, T. and S.P. COLOWICK (1961) Arch. Biochem. Biophys. *93*, 520.
KOPACZYK, K., J. PERDUE and D.E. GREEN (1966) Arch. Biochem. Biophys. *115*, 215.
KORNFIELD, R. and C. SIEMENS (1974) J. Biol. Chem. *249*, 1295.
KREMMER, T., M.H. WISHER and W.H. EVANS (1976) Biochim. Biophys. Acta *455*, 655.
KRISHNA, G., B. WEISS and B.B. BRODIE (1968) J. Pharm. Exp. Therap. *163*, 379.
KRISHNAN, N. and G. KRISHNA (1976) Anal. Biochem. *70*, 18.
LAEMMLI, U.K. (1970) Nature (London) *227*, 680.
LAICO, M.T., E.I. RUOSLAHTI, D.S. PAPERMASTER and W.J. DREYER (1970) Proc. Nat. Acad. Sci. USA *67*, 120.
LASKEY, R.A. and A.D. MILLS (1975) Eur. J. Biochem. *56*, 335.
LAUTER, C.J., A. SOLYOM and E.G. TRAMS (1972) Biochim. Biophys. Acta *266*, 511.
LEHTINEN, P., E. VUORIO and E. KULONEN (1975) Biochem. J. *146*, 565.
LEIGHTON, F., B. POOLE, H. BEAUFAY, P. BAUDHUIN, J.W. COFFEY, S. FOWLES and C. DE DUVE (1968) J. Cell Biol. *37*, 482.

LELIEVRE, L. (1973) Biochim. Biophys. Acta *291*, 662.

LERNMARK, A., A. NATHANS and D.F. STEINER (1976) J. Cell Biol. *71*, 606.

LESCO, L., M. DONLON, G.V. MARINETTI and J.D. HARE (1973) Biochim. Biophys. Acta *311*, 173.

LEWIS, B.A., A. ELKIN, R.H. MICHELL and R. COLEMAN (1975) Biochem. J. *152*, 71.

LEVITAN, I.B., W.E. MUSHYNSKI and G. RAMIREZ (1972) J. Biol. Chem. *247*, 5376.

LIANG, C.T. and B. SACKTOR (1977) Biochim. Biophys. Acta *466*, 474.

LIAO, S., A.H. LIN and J.L. TYMOCZKO (1971) Biochim. Biophys. Acta *230*, 535.

LIEBERMAN, I., A.I. LANSING and W.E. LYNCH (1967) J. Biol. Chem. *242*, 736.

LIM, R.W., R.S. MOLDAY, H.V. HUANG and S.-P.S. YEN (1975) Biochim. Biophys. Acta *394*, 377.

LIN, C.W., M. SASAKI, M.L. ORCUTT, H. MIYAYAMA and R.M. SINGER (1976) J. Histochem. Cytochem. *24*, 659.

LINDQUIST, I., O. NILSSON and R. RONQUIST (1974) Uppsala J. Med. Sci. *79*, 1.

LIS, H. and N. SHARON (1973) Ann. Rev. Biochem. *42*, 541.

LONG, C. and D.A. STAPLES (1959) Biochem. J. *73*, 385.

LOPES, J., M. NACHBAR, D. ZUCKER-FRANKLIN and R. SILBER (1973) Blood *41*, 131.

LOPEZ, R.M.Y. and P. SIEKEVITZ (1973) Anal. Biochem. *53*, 594.

LOTAN, R., E. SKUTELSKY, D. DANON and N. SHARON (1975) J. Biol. Chem. *250*, 8518.

LOUD, A.V. (1962) J. Cell Biol. *15*, 481.

LOWRY, O.H. and J.A. LOPEZ (1946) J. Biol. Chem. *162*, 421.

LOWRY, O.H., N.J. ROSEBROUGH, A.L. FARR and R.J. RANDALL (1951) J. Biol. Chem. *193*, 265.

LULY, P. and P. EMMELOT (1975) Chem.-Biol. Interac. *11*, 377.

LUZIO, P., A.C. NEWBY and C.N. HALES (1976) Biochem. J. *154*, 11.

MADDY, A.H. (1964) Biochim. Biophys. Acta *88*, 390.

MAGUIRE, M.C. and A.G. GILMAN (1974) Biochim. Biophys. Acta *358*, 154.

MAIZEL, J.V. (1971) *In:* K. Maramorosch and H. Koprowski (eds.) Methods in Virol, Vol. 5 (Academic Press, New York), p. 180.

MARCUS, A.J., D. ZUCKER-FRANKLIN, L.B. SAIFER and H.L. ULLMAN (1966) J. Clin. Invest. *45*, 14.

MARCHALONIS, J.J. (1971) Biochem. J. *124*, 921.

MARGOLIASH, E. and N. FROHWIRT (1959) Biochem. J. *71*, 570.

MARJANEN, L.A. and I.J. RYRIE (1974) Biochim. Biophys. Acta *371*, 442.

MARIQUE, D. and J. HILDEBRAND (1973) Cancer Res. *33*, 2761.

MARINETTI, G.V. and G.M. GRAY (1967) Biochim. Biophys. Acta *135*, 580.

MARTIN, J.B. and D.M. DOTY (1949) Anal. Chem. *21*, 965.

MARX, S.J., S.A. FEDAK and G.D. AURBACH (1972) J. Biol. Chem. *247*, 6931.

MASTRO, A.M., C.T. BEER and G.C. MUELLER (1974) Biochim. Biophys. Acta *352*, 38.

MATHER, I.H. and T.W. KEENAN (1974) FEBS Lett. *44*, 79.

MATUS, A.I. and B.B. WALTERS (1975) J. Neurocytol. *4*, 369.

MELDOLESI, J., J.D. JAMIESON and G.E. PALADE (1971) J. Cell Biol. *49*, 109, 130, 150.

MELLORS, A. (1969) Can. J. Biochem. *47*, 173.
MESSER, H.H., Y. SHAMI and D.H. COPP (1975) Biochim. Biophys. Acta *391*, 61.
MESSER. M. and H. DAHLQVIST (1966) Anal. Biochem. *14*, 376.
MCCOLLESTER, D.L. (1970) Cancer Res. *30*, 2832.
MCILWAIN, H. and H.L. BUDDLE (1953) Biochem. J. *53*, 412.
MILES, L.E.M. and C.N. HALES (1968) Biochem. J. *154*, 11.
MILLER, D. and R.K. CRANE (1961) Biochim. Biophys. Acta *52*, 293.
MILLER, L.S., S.B. EVANS, J.L. ROSSIO and M.C. DODD (1974) Prep. Biochem. *4*, 489.
MILNER, H.W., N.S. LAWRENCE and C.S. FRENCH (1950) Science *111*, 633.
MILSON, J.P. and C.H. WYNN (1973) Biochem. J. *132*, 493.
MISRA, D.N., T.J. GILL and L.W. ESTES (1974) Biochim. Biophys. Acta *352*, 455.
MOLDAY, R.S., W.J. DREYER, A. REMBAUM and S.P.S. YEN (1975) J. Cell Biol. *64*, 75.
MOLNAR, J., G. MARKOVIC, H. CHAO and Z. MOLNAR (1969) Arch. Biochem. Biophys. *134*, 524.
MONNERON, A. (1974) J. Histochem. Cytochem. *22*, 1128.
MONNERON, A., G. BLOBEL and G.E. PALADE (1972) J. Cell Biol. *55*, 104.
MOOKERJEA, S. and J.W.M. YUNG (1975) Arch. Biochem. Biophys. *166*, 223.
MOOSEKER, M.S. (1976) J. Cell Biol. *71*, 417.
MONTMORY, E. and M.M. BOURGES (1975) C.R. Acad. Sci., Paris, Series D, *281*, 1333.
MORGAN, I.G. (1976) Neurosci. *1*, 159.
MORGAN, I.G., L.S. WOLFE, P. MANDEL and G. GOMBOS (1971) Biochim. Biophys. Acta *241*, 737.
MORRISON, M. and D.E. HULTQUIST (1963) J. Biol. Chem. *238*, 2847.
MUNRO, J.R., S. NARASIMHAN, S. WETMORE, J.R. RIORDAN and H. SCHACHTER (1975) Arch. Biochem. Biophys. *169*, 269.
MUNTHE-KAAS, A.C. and P.O. SEGLEN (1974) FEBS Lett. *43*, 252.
MURER, H., E. AMMANN, J. BIBER and U. HOPFER (1976) Biochim. Biophys. Acta *433*, 509.
NAGAI, Y. and T. HOSOYA (1974) J. Biochem. (Tokyo) *75*, 1091.
NAGY, A., R.R. BARBER, S.J. MORRIS and V.P. WHITTAKER (1976) Brain Res. *109*, 285.
NAKAZAWA, K. and M. SANO (1974) J. Biol. Chem. *249*, 4207.
NEER, E.J. (1973) J. Biol. Chem. *248*, 4775.
NEILANDS, J. (1955) Methods Enzymol. *1*, 449.
NELSON, C.A. (1971) J. Biol. Chem. *246*, 3895.
NEUHOFF, V. (1973) *In:* Micromethods in Molecular Biology (Springer Verlag, Berlin, Chapman and Hall, London) p. 399.
NEVILLE, D.M. (1960) J. Biophys. Biochem. Cytol. *8*, 413.
NEVILLE, D.M. (1968) Biochim. Biophys. Acta *154*, 540.
NEVILLE, D.M. (1971) J. Biol. Chem. *246*, 6328.
NEVILLE, D.M. and H. GLOSSMANN (1971) J. Biol. Chem. *246*, 6335.
NEWBY, A.C., J.P. LUZIO and C.N. HALES (1975) Biochem. J. *146*, 625.

NEWKIRK, J.D. and M. WAITE (1971) Biochim. Biophys. Acta *225*, 224.
NICOLSON, G.L. and J. BLAUSTEIN (1972) Biochim. Biophys. Acta *266*, 543.
NICOLSON, G.L. and S.J. SINGER (1974) J. Cell Biol. *60*, 236.
NIELSON, C.S. and O.J. BJERRUM (1977) Biochim. Biophys. Acta *466*, 496.
NIJJAR, M.S. and J.N. HAWTHORNE (1974) Biochim. Biophys. Acta *367*, 190.
NILSSON, R., E. PETERSON and G. DALLNER (1973) J. Cell Biol. *56*, 762.
NORMAN, R.L. and B.D. HOWARD (1977) Exp. Brain Res. *28*, 279.
NORTON, W.T. and S.E. PODULSO (1972) Res. Meth. Neurochem. *1*, 19.
NORTON, W.T. and S.E. PODULSO (1973) J. Neurochem. *21*, 759.
NOVOGRODSKY, A., S.S. TATE and A. MEISTER (1976) Proc. Natl. Acad. Sci. USA *73*, 2414.
NOWAKOWSKI, M., P.H. ATKINSON and D.F. SUMMERS (1972) Biochim. Biophys. Acta *266*, 154.
NYQUIST, S.N. and D.J. MORRÉ (1971) J. Cell Physiol. *78*, 9.
O'BRYAN, D. and L.M. LOWENSTEIN (1974) Biochim. Biophys. Acta *339*, 1.
ODA, M., V.M. PRICE, M.M. FISHER and M.J. PHILLIPS (1974) Lab. Invest. *31*, 314.
O'FARRELL, P.H. (1975) J. Biol. Chem. *250*, 4007.
OGOWA, H., M. SAWADA and M. KAWADA (1966) J. Biochem. (Tokyo) *59*, 126.
OLDEN, K. and K.M. YAMADA (1977) Anal. Biochem. *78*, 483.
OLIVIERA, M.M. and S. HOLZHACKER (1974) Biochim. Biophys. Acta *332*, 221.
PARHAM, P., R.E. HUMPHREYS, M.J. TURNER and J.L. STROMINGER (1975) Proc. Nat. Acad. Sci. USA *71*, 3998.
PATT, L.M. and W.J. GRIMES (1974) J. Biol. Chem. *249*, 4157.
PATTON, S. and T.W. KEENAN (1975) Biochim. Biophys. Acta *415*, 273.
PEKARTHY, J.M., J. SHORT, A.I. LANSING and I. LIEBERMAN (1972) J. Biol. Chem. *247*, 1767.
PENNINGTON, R.J. (1961) Biochem. J. *80*, 649.
PERDUE, J.F., R. KLETZIEN and K. MILLER (1971) Biochim. Biophys. Acta *249*, 419.
PETERS, T.J. (1976) Clin. Sci. Mol. Med. *51*, 557.
PETERS, T.J., M. MULLER and C. DE DUVE (1972) J. Exp. Med. *136*, 117.
PFLEGER, R.C., N.G. ANDERSON and F. SNYDER (1968) Biochemistry *7*, 2826.
PHILIPSON, L., K. LONBERG-HOLM and U. PETTERSSON (1968) J. Virol. *2*, 1064.
PHILLIPS, A.H. and R.G. LANGDON (1962) J. Biol. Chem. *237*, 2652.
PHILLIPS, D.R. and M. MORRISON (1971) Biochemistry *10*, 1766.
PHILLIPS, J.H. and A. SLATER (1975) FEBS Lett. *56*, 327.
PINCUS, J.H., S.I. CHUNG, N.M. CHACE and M. GROSS (1975) Arch. Biochem. Biophys. *169*, 724.
PITT-RIVERS, R. and F.S.A. IMPIOMBATO (1968) Biochem. J. *109*, 825.
PODULSO, J.F., C.S. GREENBERG and M.C. GLICK (1972) Biochemistry *11*, 2616.
PODULSO, S.E. (1975) J. Neurochem. *24*, 647.
POHL, S.L., L. BIRNBAUMER and M. RODBELL (1971) J. Biol. Chem. *246*, 1849.

POIRIER, G., A. DE LEAN, G. PELLETIER, A. LEMAY and F. LABRIE (1974) J. Biol. Chem. *249*, 316.

POLLARD, T.D. and R.R. WEIHING (1974) Critical Rev. Biochem. *2*, 1.

PORTER, K., D. PRESCOTT and J. FREY (1973) J. Cell Biol. *57*, 815.

POTTER, V.L. (1955) Methods Enzymol. *1*, 10.

POTTER, V.L. and C.A. ELVEHJEM (1936) J. Biol. Chem. *114*, 495.

POWELL, W.S., S. HAMMERSTROM and B. SAMUELSSON (1976) Eur. J. Biochem. *61*, 605.

PRICE, C.A. (1973) *In:* Centrifugation in Density Gradients (Academic Press, New York).

PRICE, C.A. (1974) *In:* Subcellular Particles, Structures and Organelles (Dekker, New York), p. 156.

PRICE, R.G., D.G. TAYLOR and D. ROBINSON (1972) Biochem. J. *129*, 919.

QUIGLEY, J.P. and S.G. GOTTERER (1972) Biochim. Biophys. Acta *255*, 107.

QUIRK, S.J. and G.B. ROBINSON (1972) Biochem. J. 1319.

RAY, T.K. (1970) Biochim. Biophys. Acta *196*, 1.

RAZZEL, W.E. (1963) Methods Enzymol. *6*, 236.

REID, E. (ed.) (1971) Separations with Zonal Rotors (University of Surrey, U.K.).

RENKONEN, O., C.G. GAHMBERG, K. SIMONS and L. KAARIANEN (1970) Acta Chem. Scand. *24*, 733.

REVEL, J.P. and M.J. KARNOVSKY (1967) J. Cell Biol. *33*, C7.

REYNOLDS, J. and C. TANFORD (1970) Proc. Nat. Acad. Sci. USA *66*, 1002.

RICKWOOD, D. and G.D. BIRNIE (1975) FEBS Lett. *50*, 102.

RIEMER, B.L. and C.C. WIDNELL (1975) Arch. Biochem. Biophys. *171*, 343.

ROBINSON, N.C. and C. TANFORD (1975) Biochemistry *14*, 369.

RODBELL, M. (1964) J. Biol. Chem. *239*, 375.

RODRIGUES DE LORES ARNAIZ, G., M. ALBERICI and E. DE ROBERTIS (1967) J. Neurochem. *14*, 215.

ROSALKI, S.B. and D. TARLOW (1974) Clin. Chem. *20*, 1121.

ROSANO, T.G., S.K. LEE and D.H. JONES (1976) J. Cell Biol. *69*, 581.

ROTH, J. (1967) Methods Cancer Res. *3*, 203.

RYAN, K.J., H. KALAUT and E.L. THOMAS (1971) J. Cell Biol. *49*, 235.

SACCOMANI, G., H.B. STEWART, D. SHAW, M. LEWIN and G. SACHS (1977) Biochim. Biophys. Acta *465*, 311.

SAKAI, T., W.J. THOMPSON, V.R. LAVIS and R.H. WILLIAMS (1974) Arch. Biochem. Biophys. *162*, 331.

SALOMON, Y., C. LONDOS and M. RODBELL (1974) Anal. Biochem. *58*, 541.

SANDERSON, A.R. (1964) Br. J. Exp. Path. *45*, 398.

SCHAPIRA, G., I. DOBOCZ, J.P. PIAU and E. DELAIN (1974) Biochim. Biophys. Acta *345*, 348.

SCHENGRUND, C.L., A. ROSENBERG and M.A. REPMAN (1976) J. Cell Biol. *70*, 555.

SCHIMMEL, S.D., C. KENT, R. BISCHOFF and P.R. VAGELOS (1973) Proc. Nat. Acad. Sci. USA *70*, 3195.

SCHIZUTA, Y., P.J.A. DAVIES, K. OLDEN and I. PASTAN (1976) Nature (London) *261*, 414.

SCHLEGEL, W. and R. SCHWYZER (1977) Eur. J. Biochem. *72*, 415.

SCHNAITMAN, C. (1969) Proc. Nat. Acad. Sci. USA *63*, 412.

SCHNAITMAN, C., V.G. ERWIN and J.W. GREENAWALT (1967) J. Cell Biol. *32*, 719.

SCHNEIDER, W.C. and G.H. HOGEBOOM (1950) J. Biol. Chem. *183*, 123.

SCHMIDT-ULLRICH, R.H., E. FERBER, H. KNUFERMAN, H. FISCHER and D.F.H. WALLACH (1974) Biochim. Biophys. Acta *332*, 175.

SCHMIDT-ULLRICH, R.H., D.F.H. WALLACH and J. HENDRICKS (1975) Biochim. Biophys. Acta *382*, 295.

SCHMITZ, J., H. PREISER, D. MAESTRACCI, B.K. GHOSH, J.J. CERDA and R.K. CRANE (1973) Biochim. Biophys. Acta *323*, 98.

SCHWARTZ, I.L., L.J. SCHLATZ, E. KINNE-SAFFRAN and R. KINNE (1974) Proc. Nat. Acad. Sci. USA *71*, 2595.

SCOTT, R.E. (1976) Science *194*, 743.

SEGREST, J.P. and R.L. JACKSON (1972) Methods Enzymol. *28B*, 54.

SELA, B., H. LIS and L. SACHS (1972) J. Biol. Chem. *247*, 7575.

SELLINGER, O.Z. and J.M. AZCURRA (1974) Res. Methods Neurochem. *2*, 3.

SEVERSON, D.L., G.I. DRUMMOND and P.V. SULAKHE (1972) J. Biol. Chem. *247*, 2949.

SHAMI, Y. and I.C. RADDE (1971) Biochim. Biophys. Acta *249*, 345.

SHARON, N. and H. LIS (1972) Science *177*, 949.

SIDHU, K.S., T.Y. LIU, C.E. CAMP and J.B. MARSH (1975) Biochim. Biophys. Acta *397*, 254.

SIKSTROM, R., J. LANOIX and J.J.M. BERGERON (1976) Biochim. Biophys. Acta *448*, 88.

SILVERMAN, M. (1974) Biochim. Biophys. Acta, 339, 92.

SKERROW, C.J. and A.G. MATOLSTY (1974) J. Cell Biol. *63*, 515, 524.

SKOU, J.C., K.W. BUTLER and O. HANSEN (1971) Biochim. Biophys. Acta *241*, 443.

SMITH, A.D. and H. WINKLER (1972) *In:* Catecholamines (Springer, Berlin) p. 538.

SMITH, R.P. and S.H. APPEL (1977) Biochim. Biophys. Acta *466*, 109.

SNARY, D., P. GOODFELLOW, M.J. HAYMAN, W.F. BODMER and M.J. CRUMPTON (1974) Nature *247*, 457.

SNARY, D., F.R. WOODS and M.J. CRUMPTON (1976) Anal. Biochem. *74*, 457.

SNOKE, R.E. and R.C. NORDLIE (1956) Biochem. Biophys. Acta *139*, 190.

SNYDER, S.H., A.B. YOUNG, J.P. BENNETT and A.H. MULDER (1973) Fed. Proc. Fed. Am. Soc. Exp. Biol. *32*, 2039.

SOHN, R. and G.V. MARINETTI (1974) Chem. Phys. Lipids *12*, 17.

SOIFER, D. and O. HECHTER (1971) Biochim. Biophys. Acta *230*, 539.

SOLYOM, A. and E.G. TRAMS (1973) Enzyme *13*, 329.

SONG, C.S. and O. BODANSKY (1967) J. Biol. Chem. *242*, 694.

SONG, C.S., W. RUBIN, A. RIFKIND and A. KAPPAS (1969) J. Cell Biol. *41*, 124.

SOTTOCASA, G.L., B. KUYLENSTIERNA, L. ERNSTER and A. BERGSTRAND (1967) J. Cell Biol. *32*, 415.

STAHN, R., K.-P. MAIER and K. HANNIG (1970) J. Cell Biol. *46*, 576.

STAM, A.C., J.W. SHELBURNE, D. FELDMAN and E.H. SONNENBLICK (1969) Biochim. Biophys. Acta *189*, 304.

STECK, T.L. and G. DAWSON (1974) J. Biol. Chem. *249*, 2135.

STECK, T.L. and J.A. KANT (1974) Methods Enzymol. *31*, 172.

STECK, T.L. (1974) J. Cell Biol. *62*, 1.

STEENSGAARD, J. (1970) Eur. J. Biochem. *16*, 66.

STEFANOVIC, V., P. MANDEL and A. ROSENBERG (1975) J. Biol. Chem. *250*, 7081.

STEINER, A.L., D.M. KIPNIS, R. UTIGER and C. PARKER (1969) Proc. Nat. Acad. Sci. USA *64*, 367.

STEINHARDT, J., J.R. SCOTT and K.S. BIRDI (1977) Biochemistry *16*, 718.

STEINMAN, R.M., S.E. BRODIE and Z.A. COHN (1976) J. Cell Biol. *68*, 665.

STRAUSS, R.G. and S.E. BURROWS (1975) Blood *46*, 655.

SUN, F.F. and F.L. CRANE (1969) Biochim. Biophys. Acta *172*, 417.

SUZUKI, S., C.C. WIDNELL and J.B. FIELD (1977) J. Biol. Chem. *252*, 3074.

SVOBODA, M., P. ROBBERECHT, J. CAMUS, M. DESCHODT–LANCKMAN and J. CHRISTOPHE (1976) Eur. J. Biochem. *69*, 185.

SWANSON, M.A. (1955) Methods Enzymol. *2*, 541.

SWICK, R.W., S.L. TOLLAKSEN, S.L. NANCE and J.F. THOMSON (1975) Arch. Biochem. Biophys. *168*, 281.

TADA, M., J.O. FINNEY, M.H. SWARTZ and A.M. KATZ (1972) J. Mol. Cell Cardiol. *4*, 417.

TAMIR, H., M.M. RAPPORT and L. ROIZIN (1974) J. Neurochem. *23*, 943.

TAMIR, H., S.P. MAHADIK and M.M. R. PAPPORT (1976) Anal. Biochem. *76*, 634.

TANNER, M.J.A. and D.J. ANSTEE (1976) Biochem. J. *153*, 265.

TAPIA, R., F. HAJOS, G. WILKIN, A.L. JOHNSON and R. BALAZS (1974) Brain Res. *70*, 280.

TAYLOR, D.G., R.G. PRICE and D. ROBINSON (1971) Biochem. J. *122*, 641.

THERIEN, H.M. and W.E. MUSHYNSKI (1976) J. Cell Biol. *71*, 807.

THINES–SEMPOUX, D., A. AMAR–COSTESEC, H. BEAUFAY and J. BERTHET (1969) J. Cell Biol. *43*, 189.

THOM, D., A.J. POWELL, C.W. LLOYD and D.A. REES (1977), Biochem. J. *168*, 187.

THORELL, J. (1972) *In:* Protein and Polypeptide Hormones, M. Margoulies and F.C. Greenwood (eds.) (Excerpta Medica, Amsterdam) part 1.

THUNEBERG, L. and J. ROSTGAARD (1968) Exp. Cell Res. *51*, 123.

TOOZE, J. (1973) Molecular Biology of Tumour Viruses (Cold Spring Harbor Monograph) Chapter 2.

TOUSTER, O. (1974) *In:* E. Reid (ed.) Subcellular Studies (Longmans, London), p. 247.

TOUSTER, O., N.M. ARONSON, J.T. DULANEY and H. HENDRICKSON (1970) J. Cell Biol. *47*, 604.

TRIFARO, J.M. and A.C. DUERR (1976) Biochim. Biophys. Acta *421*, 153.

TRIFARO, J.M., A.C. DUERR and J.E.B. PINTO (1976) Mol. Pharmacol. *12*, 536.

TRUDING, R., M.L. SHELANSKI, M.P. DANIELS and P. MORELL (1974) J. Biol. Chem. *249*, 3973.

TSAN, M.-F. and R.D. BERLIN (1971) J. Exp. Med. *134*, 1016.

TSOU, K.C., H.P. MORRIS, K.W. LO and J.J. MUSCATO (1974) Cancer Res. *34*, 1295.

TRAMS, E.G. and C.J. LAUTER (1974) Biochim. Biophys. Acta *345*, 180.

TSUZUKI, J. and R.W. NEWBURGH (1975) J. Neurochem. *25*, 895.

TULKENS, P., H. BEAUFAY and A. TROUET (1974) J. Cell. Biol. *63*, 383.

TUNG, I.S. and C.A. KNIGHT (1972) Anal. Biochem. *48*, 153.

UUSITALO, R.J. and M.J. KARNOVSKY (1977) J. Histochem. Cytochem. *25*, 87.

VAN BLITTERSWIJK, W.J., P. EMMELOT and C.A. FELTKAMP (1973) Biochim. Biophys. Acta *298*, 577.

VAN DEN BERGHE, G., C. VAN POTTELSBERGHE and H.G. HERS (1977) Biochem. J. *162*, 611.

VANDENBURGH, H.H. (1977) Biochim. Biophys. Acta *466*, 302.

VAN DIJK, W., H. MAIER and D.H. VAN DEN EIJNDEN (1976) Biochim. Biophys. Acta *444*, 816.

VAN LEEUWEN, C., H. STAM and A. OSTREICHER (1976) Biochim. Biophys. Acta *436*, 53.

VICTORIA, E.G., L.M.G. VAN GOLDE, K.Y. HOSTETLER, G.L. SCHERPHOF and L.L.M. VAN DEENAN (1971) Biochim. Biophys. Acta *239*, 443.

VILHARDT, H., R.V. BAKER and D.P. HOPE (1975) Biochem. J. *148*, 57.

VORBRODT, A., M. KONWINSKI, D. SOLTER and H. KOPROWSKI (1977) Dev. Biol. *55*, 117.

WACKER, H. (1974) Biochim. Biophys. Acta *334*, 417.

WACKER, H., P. LEHKY, F. VANDERHAEGHE and E.A. STEIN (1976) Biochim. Biophys. Acta *429*, 546.

WALLACH, D.F.H. (1967) *In:* B.D. Davis and L. Warren (eds.) Specificity of Cell Surfaces (Prentice Hall, Engelwood Cliffs, N.J.), p. 129.

WALLACH, D.F.H. and M. CONLEY (1977) J. Mol. Med. *2*, 119.

WALLACH, D.F.H. and V.B. KAMAT (1966) Methods Enzymol. *8*, 164.

WALLACH, D.F.H., B. KRANZ, E. FERBER and H. FISCHER (1972) FEBS Lett. *21*, 29.

WALSH, F.S. and M.J. CRUMPTON (1977) Nature (London), *269*, 307.

WALSH, F.S., B.H. BARBER and M.J. CRUMPTON (1976) Biochemistry *15*, 3557.

WALTER, H., E.J. KROB and D.E. BROOKS (1976) Biochemistry *15*, 2959.

WARLEY, A. and G.M.W. COOK (1973) Biochim. Biophys. Acta *323*, 55.

WARLEY, A. and G.M.W. COOK (1976) Biochem. J. *156*, 245.

WARREN, L. (1959) J. Biol. Chem. *234*, 1971.

WARREN, L. (1963) Methods Enzymol. *6*, 463.

WARREN, L., M.C. GLICK and M.K. NASS (1966) J. Cell Physiol. *68*, 269.

WATSON, M.L. (1965) J. Biophys. Biochem. Cytol. *1*, 257.

WATTIAUX, R., M. WIBO and P. BAUDHUIN (1963) *In:* A.V.S. de Reuck and M.P. Cameron (eds.) CIBA Foundation Series on Lysosomes (Churchill, London) p. 176.

WATTIAUX, R. and C. DE DUVE (1956) Biochem. J. *63*, 606.

WATTIAUX, R., S. WATTIAUX-DE CONINK and M.F. RONVEAUX–DUPAL (1971) Eur. J. Biochem. *22*, 31.

WEAVER, R.A. and W. BOYLE (1969) Biochim. Biophys. Acta *173*, 377.

WEBER, K. and M. OSBORN (1969) J. Biol. Chem. *244*, 4406.

WEIBEL, E.R., W. STAUBLI, H.R. GNAGI and F.A. HESS (1969) J. Cell Biol. *42*, 68.

WEINSTEIN, D.B., J.B. MARSH, M.C. GLICK and L. WARREN (1969) J. Biol. Chem. *244*, 4103.

WEISER, M.M. (1973) J. Biol. Chem. *248*, 2536.

WERB, Z. and Z.A. COHN (1972) J. Biol. Chem. *247*, 2439.

WETZEL, M.G. and E.D. KORNE (1969) J. Cell Biol. *43*, 90.

WHITTAKER, V.P. (1969) *In:* A. Lajthe (ed.) Handbook of Neurochemistry, Vol. 2 (Plenum, New York), p. 327.

WHITTAKER, V.P. (1970) *In:* The Neurosciences, 2nd Study Program. (Rockefeller Press, New York) p. 761.

WHITTAKER, V.P. and G.H.C. DOWE (1965) Biochem. Pharmacol. *14*, 194.

WHITTAKER, V.P., I.A. MICHAELSON and R.J.A. KIRKLAND (1964) Biochem. J. *90*, 293.

WIDNELL, C.C. (1972) J. Cell Biol. *52*, 542.

WIDNELL, C.C., T.H. HAMILTON and J.R. TATA (1967) J. Cell Biol. *32*, 766.

WIGZELL, H. (1965) Transpl. 3, 423.

WILFONG, R.F. and D.M. NEVILLE (1970) J. Biol. Chem. *245*, 6106.

WILLIAMS, J.G. and W.B. GRATZER (1971) J. Chromatogr. *57*, 121.

WISHER, M.H. and W.H. EVANS (1975) Biochem. J. *146*, 375.

WISHER, M.H. and W.H. EVANS (1977) Biochem. J. *164*, 415.

WOLFF, J. and A.B. JONES (1971) J. Biol. Chem. *246*, 3939.

WRIGHT, B.M., A.J. EDWARDS and V.E. JONES (1974) J. Immunol. Methods *4*, 281.

WU, H.C., E. MEEZAN, P.H. BLACK and P.W. ROBBINS (1969) Biochemistry *8*, 2509.

YAHARA, I. and G.M. EDELMAN (1975) Exp. Cell Res. *91*, 125.

YAMASHITA, K. and J.B. FIELD (1970) Biochem. Biophys. Res. Commun. *40*, 171.

YUNGHANS, W.N. and D.J. MORRÉ (1973) Prep. Biochem. *3*, 301.

ZACHARIAS, R.M., R.E. ZEU, J.H. MORRISON and J.J. WOODLOCK (1969) Anal. Biochem. *30*, 148.

ZACKOWSKI, A. and A. PARAF (1974) Biochem. Biophys. Res. Commun. *57*, 787.

ZAK, R., J. ETLINGER and D.A. FISCHMAN (1973) *In:* Research in Muscle Development and the Muscle Spindle; Excerpta Medica Congress Series *240*, 163.

ZALIK, S.E. and G.M.W. COOK (1976) Biochim. Biophys. Acta *419*, 119.

ZANETTA, J.O., P. BENDA, G. GOMBOS and I.G. MORGAN (1972) J. Neurochem. *19*, 881.

ZATZ, M. and S.H. BARONDES (1970) J. Neurochem. *17*, 157.

ZINDER, O., O. NIKODIJEVIC, P.G. HOFFMANN and H.B. POLLARD (1976) J. Biol. Chem. *251*, 2179.

Subject index

AN INTRODUCTION TO AFFINITY CHROMATOGRAPHY

C.R. Lowe

Department of Biochemistry,
School of Biochemical and Physiological Sciences,
University of Southampton,
Bassett Crescent East,
Southampton SO9, 3TU, U.K.

Contents

Preface

Within recent years, affinity chromatography has become established as one of the most potent separatory techniques available to the biochemist. The technique has been exploited in almost every area of biochemistry and its applications are rapidly infiltrating cellular biology, immunology, medicine and technology. Because of this explosive development, this book is not intended as a comprehensive account of all the available systems but rather as a guide to current trends. However, whilst the detailed methodology of these systems vary, they are based on common principles. The aim of this book is to delineate these common principles and show how they are applied in practice. Thus, following a brief introduction to chromatographic terminology, Chapter 2 deals with the basic concepts in the design of affinity adsorbents. Appreciation of these basic principles should permit the reader to select a suitable ligand for immobilisation for his particular system. Chapter 3 details most of the widely used immobilisation techniques and should thus carry the concept into the reality of a prospective adsorbent. The chromatographic techniques involved in testing the adsorbent and its subsequent application to the purification in question are given in Chapter 4. These three chapters contain all the information necessary to design, construct and run an affinity adsorbent. The remainder of the book deals with a variety of potential applications of the technique, both preparative and analytical, and a number of related techniques where the principle of affinity chromatography is exploited in slightly different ways. The examples have been chosen with a view to illustrating some of the parameters involved in affinity separations and no attempt has been made to present a comprehensive review of each subject. Where applicable, references are given to more complete reviews of each specific area.

March 1978 C.R. Lowe

An introduction to chromatographic logic

1.1. The principles of chromatography

The history of biochemistry to a large extent parallels the history of the resolution and isolation of substances from cellular sources. However carefully the cellular organisation is disrupted, a complex mixture of biochemical substances will always result. In many cases a research problem is only brought to fruition by judicious choice of separation methods. The success of future investigations will thus depend on refined techniques for the separation of sensitive biochemical substances.

The optimal resolution of a complex mixture of biochemical substances will largely reflect a combination of techniques which will sort the substances according to different principles. Thus a number of techniques such as electrophoresis, isoelectric precipitation and ion exchange chromatography exploit the overall charge of the molecules. Other techniques such as preparative ultracentrifugation fractionate the molecules according to their size or diffusion coefficient. A successful purification regimen will incorporate a combination of these techniques such that the complex mixture is successively fractionated according to several different molecular principles. Almost without exception, a preparative isolation scheme will involve some form of chromatography.

Chromatography involves the separation of the components of a mixture using a medium, the *stationary phase*, through which a flow of liquid, the *mobile phase*, is passed to achieve a differential migration of the components. In practice the separation is effected on a chromatographic bed or column. The bed comprises minute particles of the chromatographic medium usually packed into a tube. The space between the particles is occupied by a liquid which is made to flow through the bed by gravity, pressure or some other mechanical means. The substances to be resolved are carried through the chromatographic bed by the flow of the mobile phase. The stationary phase retards the percolation of substances through the column, with different substances being retarded to differing degrees and thus migrating through the bed at different velocities. The chromatographic bed will thus resolve a multicomponent mixture by differential partioning of the components between the liquid mobile phase and the stationary phase. This type of chromatography is generally referred to as *partition chromatography* since the separation is achieved by partitioning of the substances between solvent or liquid immobilised on a solid such as cellulose, Sephadex, agarose or polyacrylamide and the liquid mobile phase which flows round the solid particles. The separation thus depends on solubility differences between the stationary and mobile phases and specific interactions between the components to be separated on the chromatographic medium or solvent are minimal. In contrast, the chromatographic medium in *adsorption chromatography* is designed or selected to interact more or less specifically with some or all of the components of the mixture to be resolved, and the liquid mobile phase is chosen to increase or decrease these specific interactions.

In principle, partition chromatography is based on the ideal thermodynamic behaviour of all the components involved, whilst in adsorption chromatography the opposite is the case. In practice, however, this distinction is rarely obtained since most natural substances interact with each other to some extent. Consequently, irrespective of the selected chromatographic technique, both partition and adsorption processes can mutually assist or interfere and thus assume importance in effecting a particular separation. This feature of chromatography makes the selection of chromatographic materials for a particular separation of great importance.

1.2. Partition chromatography

For successful separations by partition chromatography, the stationary phase should be inert and non-adsorbing and be of an adequate mesh size to achieve a good balance between flow rate and resolution. The chromatographic support, often silica gel, alumina, cellulose powder or kieselguhr, may be used for the separation of both apolar and polar substances. For example, if the support is coated with a hydrophobic liquid such as benzene and the mobile phase is a hydrophilic solvent such as methanol or formamide, the chromatographic bed may be used to resolve apolar materials. In contrast, if the bed is coated with a more hydrophilic solvent such as *n*-butanol and water is used as the mobile phase the chromatographic bed will resolve more polar materials.

In this type of liquid–liquid partition chromatography the separation depends on differences in solubility between the stationary and mobile phases. The *partition coefficient, K*, is the ratio at equilibrium of the amounts of a substance dissolved in two immiscible solvents which are in contact. Thus, if an ideal substance A is dissolved in two ideal immiscible solvents, 1 and 2, then at equilibrium the partition coefficient, K, is a constant:

$$K = \frac{(A)_1}{(A)_2} = \frac{(C_A)_1}{(C_A)_2}$$

Needless to say, for the partition coefficient to be constant over a range of solute concentrations, adsorption effects must be minimal. If adsorption is not involved and the distribution of solute between the two phases is ideal then symmetrical peaks of the substances to be separated will be eluted from the bottom of the chromatographic bed. However, since equilibrium conditions must be relatively rapidly attained between the two phases and this involves diffusion in liquids, the flow rate of the mobile phase through the packed bed is important. It must be sufficiently low to allow equilibrium to be attained and yet not too low as to permit diffusion to broaden the peaks of the eluted materials.

Partition chromatography is employed largely for the separation of small molecular weight substances. Its application in the resolution of sensitive biochemical substances such as proteins has been largely superceded by more refined chromatographic techniques based on molecular principles other than solubility.

1.3. Gel filtration

Gel filtration (Flodin, 1962) is a technique of partition chromatography in which the partitioning is based on the molecular size of the substances to be separated rather than their solubility. The technique employs a mobile liquid phase and a stationary phase comprising the same liquid entrapped within an uncharged gel lattice. The chromatographic gels used in gel filtration comprise macromolecules with a high affinity for the solvent. These gels usually have a covalent cross-linked structure which forms a three-dimensional insoluble network. The gels are allowed to swell in the same liquid that percolates through the bed and in so doing imbibe large amounts of the liquid.

Gel filtration separates substances according to their molecular size; large molecules emerge from the bed first followed by the smaller molecules. For most practical purposes it suffices to say that the elution volume is determined almost entirely by the molecular weight. Flodin (1962) proposed a simple model for gel filtration to account for these observations. The partition coefficient of the solute between the gel phase and the mobile phase was deemed to be governed exclusively by steric effects. Large molecules cannot penetrate into regions close to the cross-links in the gel lattice because of steric obstruction. In contrast, small molecules can approach these regions more closely and thus have access to most of the space between the chains of the gel matrix. As a result small molecules are distributed fairly evenly between the free solvent and the solvent entrapped within the gel matrix, whilst large molecules are more restricted within the gel. The partition coefficient of large molecules is thus shifted in favour of the liquid outside the gel

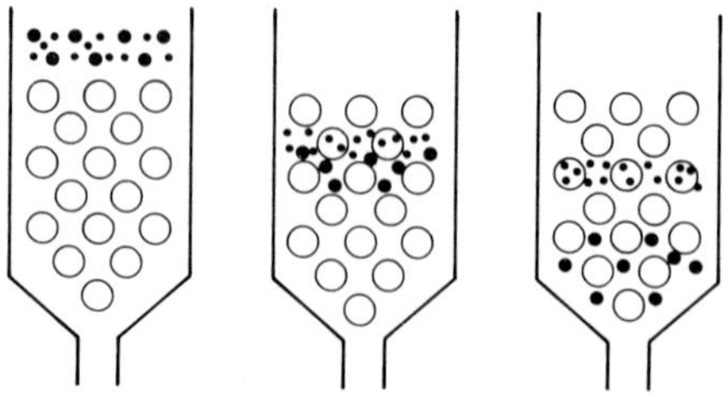

Fig. 1.1. The principle of gel filtration. A sample containing a mixture of large and small molecules is applied to the top of a gel column. The large dots represent large molecules, the small dots small molecules and the open circles the gel beads. As the elution proceeds, the small molecules penetrate the gel beads and lag behind the large molecules that pass round the beads. Reproduced with permission from *Sephadex®-Gel Filtration in Theory and Practice* by Pharmacia Fine Chemicals.

particles with the consequence that large molecules emerge from the gel bed earlier than small ones. Figure 1.1 illustrates the principle of gel filtration.

The solvent in a chromatographic column packed with swollen gel beads may be regarded as being in two phases (Flodin, 1961); in the spaces between the gel beads, the *void volume*, V_0, and entrapped within the gel matrix, the *internal volume*, V_i. A solute introduced into the column will equilibrate between the solvent contained in these two phases, although only a fraction of the internal volume, represented by the partition coefficient, K, is available to the solute. The total volume accessible to solute within the gel matrix is thus $K \times V_i$ and the solute will thus emerge in the mobile phase after a volume V_e, the *elution volume*, has been displaced

$$V_e = V_0 + K\,V_i$$

The partition coefficient,

$$K = \frac{V_e - V_0}{V_i}$$

is characteristic for chromatography of a given solute on a given gel under specified operating conditions and is independent of bed geometry. For solutes completely excluded from the internal volume

$K = 0$ and hence $V_e = V_0$, i.e., the solute emerges in the minimum possible volume, corresponding to the void volume. For very small molecules with unlimited access to the internal volume of the gel beads, K approaches 1 and V_e approaches a maximum value equivalent to $V_0 + V_i$.

In practice the volume of the solvent entrapped within the gel particles (V_i) is difficult to evaluate realistically. Consequently, an alternative means of expressing solute behaviour in terms of an *available partition coefficient*, K_{av}, is evoked. The internal volume, V_i, is replaced by the total volume of the gel ($V_i + V_g$), where V_g is the volume occupied by the gel matrix itself. Thus,

$$K_{av} = \left(\frac{V_e - V_0}{V_i + V_g}\right) = \left(\frac{V_e - V_0}{V_t - V_g}\right)$$

since the total packed volume of the gel bed, V_t, is given by:

$$V_t = V_0 + V_i + V_g$$

The available partition coefficient, K_{av}, is readily evaluated in terms of the bed parameters V_t and V_0, and the solute elution volume, V_e (Laurent and Killander, 1964). The total bed volume, V_t, may be deduced by calculation from the bed diameter and the bed height or by direct calibration with water. The column is filled with water and subsequently withdrawn from the column in portions and weighed. A plot of the weight of water against the level of water in the column will allow interpolation of the volume of any given bed height. The void volume, V_0, may be determined by chromatography of a substance that is completely excluded from the gel beads and measurement of its elution volume. In practice, a polysaccharide with a weight average molecular weight of 2×10^6, Blue Dextran 2000, is commonly employed for this purpose and is commercially available. The most important variable to be measured is the elution volume of the solute of interest, V_e, which should, within limits, be independent of the flow rate through the bed. If, as under conditions of ideal chromatography, the elution profile is symmetrical, the elution volume, V_e, is the volume of liquid that has passed through the column between the application of the sample and the elution of the maximum concentration of the substance.

The chromatographic behaviour of a solute in gel filtration can be related to its molecular size or a molecular parameter closely associated with molecular size such as molecular weight or Stoke's radius. It can be empirically shown that a plot of elution volume, V_e, or a suitable function of it such as K_{av}, of a number of globular proteins was approximately a linear function of the logarithm of their molecular weights (Ackers, 1964; Andrews, 1964). This relationship is depicted in Fig. 1.2. Below a certain molecular weight the curve is almost horizontal and in this region all solutes are eluted close together in a volume which is maximal for the given bed geometry and approximately $V_0 + V_i$. The central part of the curve is inclined downwards in such a way that a variation in molecular weight corresponds to a significant alteration in the elution volume. This part of the curve represents the working or fractionation range of the gel. Clearly, a gel that has a steep curve within the working range will efficiently fractionate molecules within that range. For practical purposes, however, an acceptable compromise between steepness and fractionation range is required. At molecular weights above the fractionation range, the curve becomes horizontal again with all solutes in this region moving with the void volume. The molecules are so large that $K_{av} = 0$ and $V_e = V_0$. This point is termed the *exclusion limit* of the gel.

Gel filtration may be used to determine molecular weights of globular proteins by interpolating the elution volume or K_{av} of the unknown protein on a plot of K_{av} versus log molecular weight constructed with standard proteins of known molecular size. Siegel and Monty (1966) showed, however, that the elution volume of proteins are better correlated with Stoke's radii than with molecular weights. Consequently, the method assumes that the asymmetry and extent of hydration of the proteins being analysed and of the standard calibration proteins are approximately the same. This assumption appears to hold true for most globular proteins but not for proteins

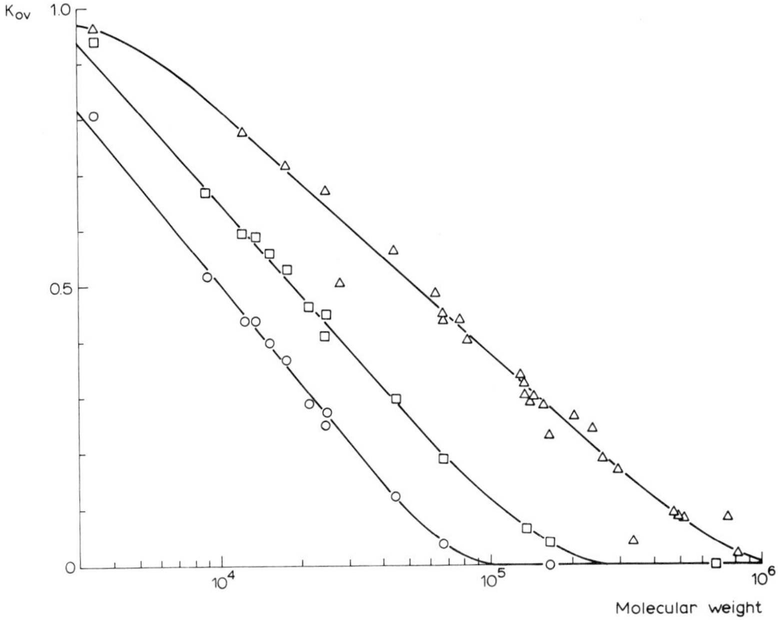

Fig. 1.2. The relationship between elution behaviour and molecular properties: K_{av} versus molecular weight for globular proteins on different types of Sephadex®. △ Sephadex G-200, □ Sephadex G-100, ○ Sephadex G-75. Reproduced with permission from *An Introduction to Gel Chromatography* by L. Fischer, North-Holland Publishing Co., Amsterdam, 1969.

containing large amounts of carbohydrate or for proteins that can interact with the gel matrix itself. Thus certain dextranases (Porath, 1968), polyglucanases (Pettersson, 1968) and plant haemagglutinins (So and Goldstein, 1968) may interact specifically and in some cases very strongly, with gel filtration matrices comprising polysaccharide backbones. In these cases the available partition coefficient, K_{av}, may exceed unity and become infinite in extreme cases. In such extreme cases the chromatography ceases to be partition chromatography and becomes adsorption chromatography. Full details of the theory and practice of gel filtration are given in the companion volume by L. Fischer, *An introduction to gel chromatography*.

1.4. Adsorption chromatography

Adsorption chromatography is not only one of the oldest and most popular of the various chromatographic techniques but is also one of the least well understood. This situation arises in part from the greater complexity of the adsorption process but also from its amazing versatility.

In adsorption chromatography, the chromatographic medium is chosen or designed to permit a more or less specific interaction with some or all of the components of the mixture to be resolved. Adsorption is merely an expression of the intermolecular forces between the surface atoms of a solid and an external substance. The intermolecular forces which are thought to be primarily responsible for chromatographic adsorption include:

(a) Van der Waal's forces between all surfaces and adsorbed molecules

(b) Electrostatic forces

(c) Hydrogen bonds

(d) Hydrophobic forces.

Furthermore, a chromatographic adsorbent will possess many types of adsorption site available for interaction with any given substance. This heterogeneity of adsorption sites means that as the concentration of solute in the liquid phase (c) is increased, the amount adsorbed (q) for each increment in c, decreases. A plot of amount adsorbed (q) versus the concentration (c) will thus be convex (Fig. 1.3) since the binding centres with greater affinities tend to be populated first so that additional increments of solute are less firmly bound. This behaviour may be expressed mathematically in terms of the *Langmuir Adsorption Isotherm*:

$$q = \frac{k_1 c}{1 + k_2 c}$$

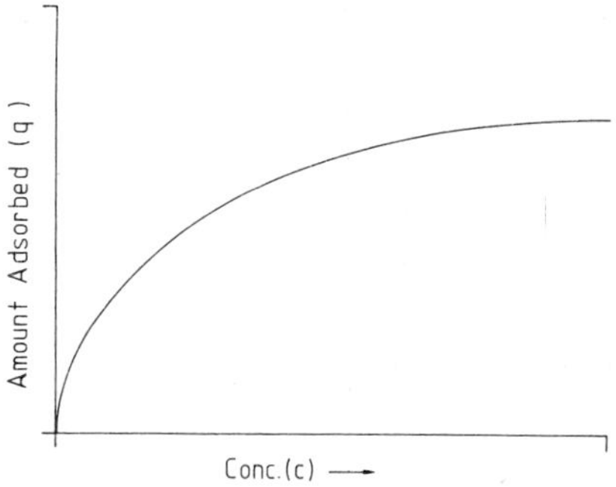

Fig. 1.3. The Langmuir adsorption isotherm.

where q represents the amount of solute adsorbed (x) per unit mass of adsorbent (m), c is the equilibrium concentration of the solute and k_1 and k_2 are constants. At sufficiently low solute concentrations such that $k_2c \ll 1$, the equation assumes the form of a linear adsorption isotherm:

$$q = k_1 c$$

that is, the amount adsorbed is directly proportional to the solute concentration. At very high concentrations of solute, such that $k_2c \gg 1$,

$$q = \frac{k_1}{k_2}$$

and the amount adsorbed, q, has a finite upper limit, k_1/k_2, which characterises the maximum amount of solute adsorbed per unit mass of adsorbent. The latter parameter is generally referred to as the *capacity* of the adsorbent.

It should be noted that where the adsorption is weak the isotherm will tend to have less curvature than when it is strong. This is a consequence of the fact that the curvature is a function of the ratio of spaces available to the concentration of molecules available to fill them. Adsorption generally involves weak forces which are essential for good chromatographic separation since they are readily reversible. The various types of adsorption chromatography are classified on the basis of these interactions.

1.4.1. Classical adsorption chromatography

Classical adsorption chromatography employs a mobile liquid phase and a stationary solid phase such as alumina, silica, charcoal and a number of other inorganic materials. The technique, however, finds few applications in the purification of proteins.

1.4.2. Ion-exchange chromatography

Ion-exchange chromatography is a type of adsorption chromatography in which the interactions between the chromatographic medium and the solute are based primarily on ionic charge. Ion exchangers are generally partly or entirely synthetic resins often based on cross-linked polystyrene, cellulose or other polysaccharides that can exchange ions with aqueous solutions. When placed in water they imbibe water such that the resin-bound ionic groups become hydrated and the osmotic balance between the internal and external ions is maintained. When the resin beads are immersed in solutions containing other ions, the ions inside the beads exchange with those outside in the bulk medium:

$$\text{Resin}^- \, \text{Na}^+ + \text{K}^+ \rightleftharpoons \text{Resin}^- \, \text{K}^+ + \text{Na}^+$$

The polystyrene resin based ion exchangers, however, tend to exclude large molecules such as proteins and thus exhibit a low effective exchange capacity. Furthermore, some proteins are denatured by them. These disadvantages of the resin exchangers were largely circumvented by the development of ion exchangers based on cellulose (Peterson and Sober, 1956) or other polysaccharides which were found to be far more suitable for protein separations. The most common types available are:

(a) DEAE (diethylaminoethyl)–cellulose, Sephadex or Sepharose; an anion exchange resin used primarily for neutral and acidic proteins.

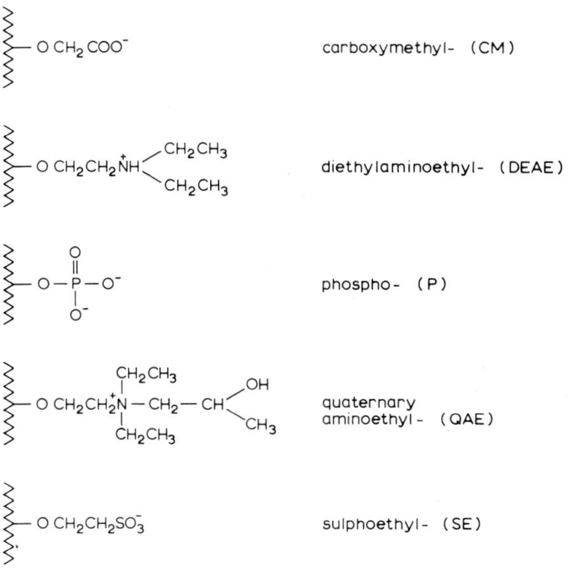

Fig. 1.4. The structures of the active groups of some common ion exchangers.

(b) CM (carboxylmethyl)–cellulose, Sephadex or Sepharose; a cation exchanger used primarily for the separation of neutral and basic proteins.

Figure 1.4 illustrates the structures of these and other types of ion-exchange adsorbent. The Sephadex ion exchangers, available commercially from Pharmacia, Sweden, have a considerably higher degree of substitution with ion exchange groups and thus exhibit a 3.5–5 times enchanced adsorption capacity.

The complex amphoteric nature of most proteins does not allow a precise representation of their interaction with the polysaccharide ion exchangers. In general, however, the major attractive force between proteins and these adsorbents is believed to be electrostatic (Wolf, 1969). The adsorbed protein could in theory be eluted by altering the ambient pH such that either the protein or the adsorbent loses its net charge. Elution by altering the pH should, however, be avoided if it approaches the pK of the exchange group.

This is because all the adsorbed proteins would be released simultaneously when the charge on the exchanger was neutralized and thus produce poor selectivity. As a general rule, proteins adsorbed to cation exchangers may be eluted by gradually increasing the pH, whilst proteins adsorbed to anion exchangers may be eluted by decreasing the pH, in each case away from the pK of the exchange group. Increasing ionic strength also reduces the electrostatic interactions between proteins and ion-exchange media by exerting a 'damping' effect. Increasing the ionic strength appears to produce sharper elution patterns than alteration of the pH and thus is the preferred procedure. The practical methodology of cellulosic ion-exchange media is adequately covered in the companion volume by E.A. Peterson, *Cellulosic Ion Exchangers*.

1.4.3. Affinity chromatography

Affinity chromatography represents the ultimate extension of adsorption chromatography since it embodies the complex set of Van der Waal's, hydrophobic, steric and electrostatic forces involved in the specific binding of substrates and other ligands to proteins. The technique of affinity chromatography exploits the unique biological specificity inherent in a ligand–macromolecule interaction. In this context, *ligand* refers to a substrate, product, inhibitor, coenzyme, allosteric effector or any other molecule that interacts specifically and reversibly with the protein or other macromolecule to be purified. The biological functions of proteins are based on their ability to adsorb these ligands specifically and reversibly. Thus, for example, enzymes are well known for their ability to bind substrates, products, inhibitors, coenzymes and allosteric modulators. Hormones form complexes with specific binding proteins and cellular receptors in a highly specific fashion and with high affinity. Genes interact with nucleic acids and repressor proteins. Antibodies combine with their complementary antigens and plant lectins bind to specific cell surface antigens on erythocytes and lymphocytes and to certain

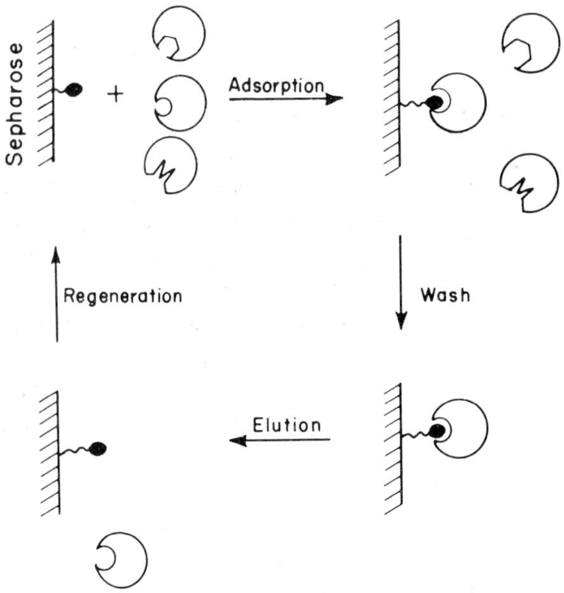

Fig. 1.5. The principle of affinity chromatography. Reproduced with permission from M. Wilchek and D. Givol, *Peptides, 1971*, p. 204, North-Holland Publishing Co., Amsterdam, 1973.

carbohydrates and polysaccharides. These and other interactions may be exploited in the purification of a wide variety of biological substances.

The concept of affinity chromatography is realised by covalently attaching the ligand to an insoluble support and packing the support into a chromatographic bed. In principle, as depicted in Fig. 1.5, if a mixture comprising several proteins is applied to the column, only that protein that displays appreciable affinity for the ligand will be retained or retarded; others which show no recognition of the insolubilised ligand will pass through the bed unretarded. The specifically adsorbed protein can subsequently be eluted by altering the composition of the solvent to permit dissociation from the insoluble ligand. In principle, affinity chromatography can be applied wherever a specific interaction occurs between any two biological molecules. Needless to say, the potential applications of this type of process for the purification and isolation of biological macromolecules are almost unlimited. Thus, for example, specific adsorbents can be designed for the purification of enzymes, antibodies, nucleic acids, cofactors, and proteins involved in the recognition, storage or transport of compounds of biochemical or pharmacological interest. Furthermore, the technique may be exploited for the purification or resolution of supramolecular structures such as cells, organelles and viruses. The technique also has a number of other applications besides those exploited in protein purification. Thus, affinity chromatography may be employed for concentrating dilute protein solutions in much the same way as ion exchange chromatography, and for a number of analytical applications. These and other applications of affinity chromatography are listed in Table 1.1.

Clearly, affinity chromatography has a number of inherent advantages over the classical methods of protein isolation. Firstly, the approach introduces a degree of rationale hitherto lacking in

TABLE 1.1
Some applications of affinity chromatography.

1. Protein purification
    - Enzymes
    - Antibodies
    - Binding, transport and receptor proteins
    - Repressor proteins
2. Separative procedures
    - Cells and viruses
    - Nucleic acids and complementary nucleotides
    - Isoenzymes
    - Denatured, chemically modified and synthetic proteins from native proteins
    - Mutant proteins
    - Affinity labelled peptides
3. Concentration of dilute protein solutions
4. Investigation of kinetic sequences and binding mechanisms
5. Determination of dissociation and equilibrium constants

protein purification in that an adsorbent is designed and constructed specifically for the protein to be purified. Secondly, the specific adsorbent permits a rapid and facile separation of the desired protein from inhibitors and destructive contaminants such as proteolytic enzymes. Thirdly, the operation of the technique, in many cases, as a 'single step' procedure, leads to a high yield of purified protein. This is partly because of the reduced time involved and partly because of protection of the protein from denaturation by stabilisation of the tertiary structure. Fourthly, the dependence of the technique on biological specificity rather than on physico-chemical properties means that the technique is ideally suited to the isolation of proteins present in very low concentrations such as the serum binding or transport proteins. Finally, the principle of affinity chromatography may be exploited in a number of other related applications such as affinity electrophoresis, affinity histochemistry and affinity density perturbation.

1.5. Nomenclature of affinity chromatography

A variety of terms have been evolved to emphasise the dependence of the technique on natural biological interactions and to distinguish it from non-specific or hydrophobic interactions. For example, the technique has been termed 'biospecific adsorption chromatography' (Porath, 1973), 'bioselective adsorption chromatography' (Scouten, 1974), 'bio-affinity chromatography' (O'Carra, 1974), 'ligand specific chromatography' (May and Zaborsky, 1974) and 'biospecific affinity chromatography' (Porath and Kristiansen, 1975). The present author prefers and adopts the two worded phrase, *affinity chromatography*, as recommended by the ad hoc committee for the standardization of nomenclature in affinity chromatography (Sundaram et al., 1976). Furthermore, the individual components of the chromatographic system suffer an equal confusion in terminology. The polymer to which one of the interacting species is covalently attached has been termed the 'solid support', 'carrier', 'matrix', 'gel' or 'insoluble support' (Cuatrecasas and Anfinsen, 1971; May and Zaborsky, 1974; Porath and Kristiansen, 1975). The interacting species attached to the matrix has been termed the 'ligand' (Cuatrecasas and Anfinsen, 1971; Lowe and Dean, 1974), 'effector' (Brümmer, 1974) or 'affinant' (Turkova et al., 1973) whereas the substance to be isolated is occasionally called the 'affiner partner' (Brümmer, 1974) or 'ligate' (Nishikawa, 1975). The present author prefers the terms 'complementary enzyme or macromolecule' and generally adopts the recommendations of Sundaram et al. (1976) throughout this monograph.

Basic concepts in the design of adsorbents for affinity chromatography

2.1. *The matrix*

2.1.1. *Qualities of the ideal matrix*

The solid support or matrix to which the affinity ligand is attached is generally a gel. In everyday scientific parlance a gel comprises a three dimensional structure, lattice or network of a material such as a cross-linked polymer. The space within the gel not occupied by the lattice back-bone comprises entrapped liquid. In most cases the liquid entrapped within the lattice forms the major proportion of the weight of the gel. The nature of the structural material and the proportion of imbibed water determine the mechanical properties of the gels. Some matrices are soft and readily deform under hydrostatic pressure whilst others tend to be rigid or even brittle.

Almost any macromolecule, synthetic or natural, may form a gel in a suitable liquid when suitably cross-linked with a bifunctional reagent. Most chromatographic matrices used for affinity chromatography are xerogels. Such gels shrink on drying to a compact solid comprising only the gel matrix. When the dried xerogel is resuspended in the liquid, the gel matrix imbibes liquid, swells and returns to the gel state. For these types of gels, the water regain, W_r, is an indication of the swelling capacity of the gel. The water regain is the volume of water (in millilitres) taken up by 1 g dry xerogel on swelling but does not include the interstitial liquid between the gel particles. For some gels, particularly dextran and polyacrylamide, there is a good correlation between the water regain and the fractionation range and exclusion limit of the gel.

Subject index p. 519

Some matrices used for affinity chromatography are *aerogels*. These gels, of which porous glass is an example, do not shrink on drying but merely allow penetration of the surrounding air. When the dry gel is exposed to liquid, the latter displaces the air in the gel.

Of the many types of gel matrix potentially available, two distinct gel types may be distinguished; some matrices exhibit a strongly heterogeneous microstructure with regions of highly aggregated matrix material and regions where the matrix backbone is almost entirely absent. Such gels, generally referred to as *macroreticular* gels, contain large open spaces and thus permit the unimpeded entry and exit of large macromolecules. The regions of aggregated matrix form a skeleton which imparts some mechanical stability to the gel. *Microreticular gels*, on the other hand, are characterised by a relatively homogeneous distribution of the matrix backbone throughout the gel. The microreticular gels are usually xerogels and often exhibit a lower exclusion limit than the macroreticular gels.

There are many polymers, both organic and inorganic, capable of forming gels and that could potentially be used as matrices for affinity chromatography. However, very few are suitable for affinity chromatography since the technique imposes a number of chromatographic and practical restrictions on their use. The success of the technique depends largely on mimicking the interaction between the two components that occurs when both components are in free solution. Careful consideration must therefore be given to the nature of the solid matrix which must have a number of desirable characteristics (Cuatrecasas et al., 1968).

(a) *The insoluble support should form a loose, porous network* which permits the uniform and unimpaired entry and exit of large macromolecules. A high degree of porosity is essential for good flow properties and for the unhindered penetration of macromolecules into the matrix. The penetration of the gel matrix by the macromolecule will also determine the concentration of ligand freely available to the macromolecule. The latter, in turn, determines the behaviour of the system under chromatographic conditions. This consideration is important for ligand–protein systems displaying relatively weak affinity, i.e., those with dissociation constants ⩾ 10^{-4} M, since a high gel-bound ligand concentration is required to permit interactions strong enough to effect retardation of the downward migration of the protein through the chromatographic bed.

(b) *The gel particles should be uniform, spherical and rigid.* The diffusion equilibria of substances of low diffusion rates, such as proteins, are considerably assisted by fractionation in a gel of fine particles. Ideally, the gel particles should be uniform and spherical. A bed consisting of small spherical particles will generally give good resolving power although at the expense of an acceptable flow rate. In contrast, a bed packed with large particles retards the attainment of diffusion equilibrium and the applied sample zone is thus broadened. Furthermore, the resistance to flow in a bed containing large particles is lower and the maximum attainable flow rate is thus higher. Consequently, a compromise between particle size, flow rate and optimal resolution is required.

The gel beads should also display a degree of mechanical rigidity such that they will not be deformed by the hydrostatic pressure required to generate a flow of liquid through the bed. Weak, flexible beads tend to compact, thus increasing the resistance to flow and hence limiting the maximal flow rate.

(c) *The gel matrix must be chemically inert.* The matrix backbone must interact very weakly, if at all, with proteins or other substances of biochemical interest, to minimize non-specific adsorption. Even weak interactions between the matrix backbone and sensitive proteins or enzymes may lead to denaturation and hence sub-quantitative yields. In this context, a low content of ionic groups is essential to to avoid ion-exchange effects, particularly when the chromatography is performed at low ionic strength. The content of charged groups in most commercially available matrices is comparatively low and should produce no undesirable effects providing that the ionic strength is routinely kept above about 0.02.

(d) *The gel must be physically and chemically stable.* The solid support must be mechanically, physically and chemically stable to the conditions employed for covalent coupling of the selected ligand

Subject index p. 519

and for the adsorption and subsequent elution of the complementary macromolecule. The gels should be stable for extended periods of time such that they can be reused many times for months or even years. The support should resist bacterial degradation and should be stable to a reasonable range of pH and temperature that may be encountered during the preparation and operation of the affinity adsorbent. This is important to permit a relatively unrestricted choice of experimental conditions. Solubilisation or leaching of soluble materials from the matrix should be minimal under the conditions employed in the chromatography.

(e) *The matrix must be capable of functionalisation.* One of the most important features of the prospective matrix for affinity chromatography is that it must possess functional groups which can be activated or modified, under conditions which are not detrimental to its structure, to allow the covalent attachment of a variety of ligands. These functional groups on the matrix backbone should be sufficiently abundant to allow a high concentration of coupled ligand and thus effect a satisfactory retardation of proteins that display low affinity for the immobilised ligand. The rigidity and hence the porosity of the gel beads should not be altered under the conditions used for functionalisation.

Desirable features of the insoluble support material have been reviewed by a number of authors (Cuatrecasas and Anfinsen, 1971a,b; Lowe and Dean, 1974).

2.1.2. Properties of the available matrices

In recent years many water-insoluble carriers have become available for potential use in affinity chromatography. Many have properties which deviate considerably from those of the ideal matrix and are thus suitable for special applications only.

2.1.2.1. Cellulose

The usefulness of cellulose derivatives as adsorbents for affinity chromatography is limited by their fibrous and non-uniform character which impedes penetration by large macromolecules. Cellulose fibres comprise an aggregation of glucosidic chains bound together by their high capacity for hydrogen bonding. This high degree of molecular structure leads to 'crystalline' regions and, where the glucosidic chains leave these areas, to 'amorphous' or disordered regions. The relationship between crystalline and amorphous regions is particularly sensitive to the physical and chemical conditions used in the preparation of affinity adsorbents. Thus preferential reaction at the lower-order or more amorphous regions occurs, leading to microheterogeneity in ligand substitution, and hence creating a spectrum of affinity for the complementary macromolecule. This microheterogeneity in binding ability of the immobilised ligand can generate undesirable effects on the capacity of the adsorbent (§ 4.2) and on the adsorption and subsequent elution of the macromolecule.

The unique properties of cellulose have, nevertheless, contributed to its long and beneficial service to chromatography. Most commerically available cellulose preparations are microfibrous or microcrystalline, i.e., they contain rod or thread-like particles, and thus exhibit good flow rates. More recently, cellulose has become available in beaded form (see Appendix for details), although its potential in affinity chromatography has yet to be demonstrated. Conventional microfibrous cellulose has been used in the preparation of specific adsorbents for the purification of antibodies and enzymes. Thus, in the pioneering work of Lerman (1953) a phenylazophenol derivative of cellulose was used for the purification of mushroom tyrosinase whilst other specific cellulose derivatives were exploited to enrich flavokinase (Arsenis and McCormick, 1964) and avidin (McCormick, 1965) from extracts. Cellulose derivatives have found particularly widespread application in the field of nucleotide chemistry; DNA physically adsorbed onto cellulose ('DNA–cellulose') is particularly popular for the isolation and characterisation of DNA-binding proteins (Alberts et al., 1968) and likewise 'RNA–cellulose' (Smith et al., 1972) for the corresponding RNA-binding proteins. Furthermore, immobilised polynucleotides have been used as primers and templates for nucleotide-polymerising enzymes (Jovin and Kornberg,

1968). However, the use of cellulose as a matrix for affinity chromatography presents several fundamental practical problems, in addition to its heterogeneity. Because of its fibrous nature, it is readily compressed and thus easily clogged with particulate matter. This clogging leads to slow flow rates particularly with large columns. Furthermore, the heterogeneity of cellulose accentuates non-specific adsorption phenomena.

2.1.2.2. The cross-linked Dextrans

Dextran is an α-1,6-linked glucose polymer produced by *Leuconostoc mesenteroides* in sucrose-containing media. A commercial product Sephadex (see Appendix) is prepared by cross-linking refined Dextran fractions with epichlorohydrin and is available in bead polymerised form. The significant complement of hydroxyl groups on the polysaccharide backbone make a highly hydrophilic matrix which swells in water and electrolyte solutions. The process of swelling and drying in the Sephadex beads is reversible with no significant changes in the chromatographic properties of the gel after repeated drying and re-swelling cycles. Furthermore, Sephadex is chemically very stable, with the gel surviving intact after 2 months in 0.25 M NaOH at 60 °C or 6 months in 0.02 M HCl (Cruft, 1961). Furthermore, wet Sephadex can be heated to 110 °C without altering its properties although prolonged exposure to oxidising agents may cause a rise in the content of carboxyl functions.

Such considerations imply that the cross-linked Dextran derivatives possess many of the desirable features of the ideal matrix for affinity chromatography. Indeed this is so, except for their low degree of porosity. Sephadex is commercially available in eight different types which differ in their degree of cross-linkage and thus their swelling properties. The Sephadex series includes gels with fractionation ranges covering molecular weights in the region 0–700 (Sephadex G-10) to 5,000–800,000 (Sephadex G-200). Nevertheless, activation of the dextran gels by any of the common methods (§ 3.1.1) leads to a considerable degree of further cross-linking and thus makes the functionalised gels relatively ineffective as adsorbents for affinity purification, even for enzymes of low molecular weight.

However, for some types of macromolecule–ligand interactions, the porosity of the bead may not be a limiting factor. Thus, for example, supramolecular structures such as polysomes, ribosomes, intact cells, viruses, organelles or membrane fragments may not reasonably be expected to penetrate the pores of the matrix. In such systems it should suffice to immobilise one partner of the interacting system on the *surface* of the beads to permit adequate interaction with the supramolecular complex in the space between the beads. Thus Sephadex G-200 is ideally suited to application in affinity chromatography involving the separation of whole cells. Thus, intact B- and T-lymphocytes have been resolved on immuno-adsorbent columns comprising anti-globulin attached to Sephadex G-200 (Schlossman and Hudson, 1973). The adsorbed cells are conveniently recovered by digestion of the Sephadex beads with dextranase. Alternatively, a digestible gelatin bridge may be covalently attached to cyanogen bromide-activated (§ 3.1.1.1) Sephadex G-200 whence the anti-globulin fraction is subsequently attached to the gelatin using glutaraldehyde. The adsorbed cells may be released by digestion of the gelatin bridge with collagenase (Thomas and Phillips, 1973).

2.1.2.3. Agarose

The superior chromatographic properties of granulated agar gels were first appreciated by Polson (1961). The gels combined fractionation ranges at very high molecular weights, and thus beyond those obtainable with the cross-linked dextrans, with good mechanical stability under operational conditions.

Agar is obtained from various species of sea weed and consists of two groups of polysaccharides, agaropectin and agarose (Araki, 1937). The polysaccharide agaropectin contains sulphate and, to a lesser extent, carboxyl groups which impart ion-exchange properties to the agar and thus generate undesirable side effects in chromatography. Agaropectin and agarose may be separated by a number of procedures including acetylation, extraction of the agarose into a chloroform phase and subsequent regeneration of the poly-

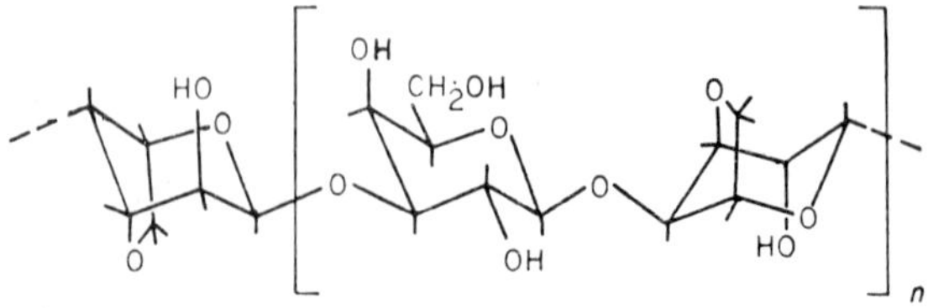

Fig. 2.1. The primary structure of agarose, an alternating copolymer of 3-linked β-D-galactopyranose and 4-linked 3,6-anhydro-α-L-galactopyranose residues. Reproduced with permission from Arnott et al. (1974), J. Mol. Biol., *90*, 269–284.

saccharides (Araki, 1937), by fractional precipitation with polyethylene glycol (Russel et al., 1964) or by selective precipitation of the agaropectin with cetylpyridinium chloride (Hjertén, 1961). Purified agarose yields stable gels at very low concentrations (<0.5%) on cooling from solution in boiling water. The resulting agarose gels fractionate in the high molecular weight range as did agar gels but are free of their associated ion-exchange properties.

The practical difficulties associated with the granulated forms of agarose prohibited the widespread application of this chromatographic material. It was not until Hjertén (1964) and Bengtsson and Philipson (1964) succeeded in preparing these gels in beaded form, and that Joustra (1969) had shown that the beaded agarose gels had greater resolving power than the corresponding granulated gels, that they received widespread attention.

Agarose is a linear polysaccharide consisting of alternating 1,3-linked β-D-galactopyranose and 1,4-linked 3,6-anhydro-α-L-galactopyranose residues (Araki, 1956) (Fig. 2.1). In contrast to the cross-linked dextran gels, the polysaccharide backbones of agarose are not bound together by covalent bonds, but are believed to interact via hydrogen bonds. Agents known to disrupt hydrogen bonds, such as urea or guanidine hydrochloride, reduce the mechanical stability of agarose gels. The decrease in stability is, however, considerably less than one would anticipate, and suggests that forces other than hydrogen bonding may be holding the polysaccharide chains together.

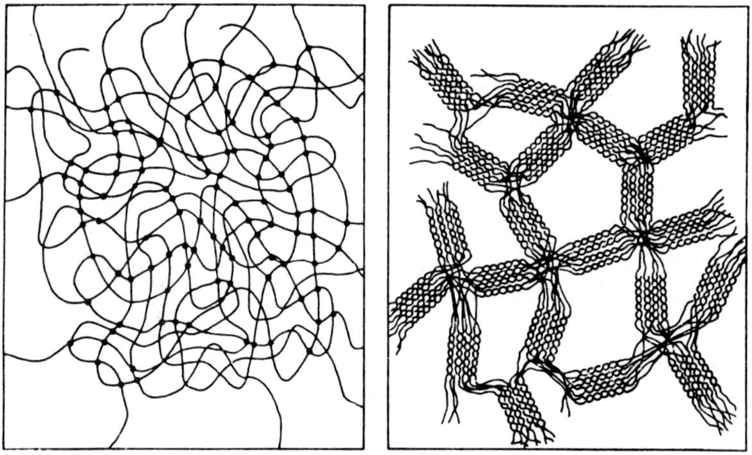

Fig. 2.2. A comparison between an agarose gel matrix (right) with a cross-linked dextran (Sephadex) matrix at equivalent polymer concentration. The aggregates in agarose gels may contain 10–10^4 bundles of polysaccharide helices rather than the smaller numbers shown above. Reproduced with permission from Arnott et al. (1974), J. Mol. Biol., *90*, 269–284.

The gel-forming fibres of agarose are believed to be relatively stiff bundles of polysaccharide helices (Arnott et al., 1974) and not flexible single chains as in the Sephadex gel filtration media. The accumulation of agarose chains into a separate 'network phase' in a gel which may contain up to 100 times more water than agarose means that the structure contains relatively large voids through which large macromolecules can diffuse. In contradistinction, a gel network comprising a comparable concentration of cross-linked soluble polymer, such as the cross-linked dextrans, would lead to a lattice in which the mean pore size would be considerably smaller. These relationships are shown diagrammatically in Fig. 2.2 and suggest that agarose should exhibit special properties as a chromatographic medium.

The lack of covalent cross-linkages in agarose compromises the stability of the gels under adverse conditions. Thus, for example, in aqueous solution prolonged exposure of the gels to temperatures above 50 °C leads to a gradual loss of stability and thus dissolution. It is thus recommended that temperatures below 0 °C and above 40 °C and heat sterilisation, be avoided. Furthermore, agarose gels are less stable to extremes of pH than the corresponding dextran gels and should not be employed outside the range 4–9. The gels will adequately tolerate exposure to 0.1 M NaOH or 1 M HCl for 2–3 h (Cuatrecasas, 1970). In contrast, however, agarose gels are remarkedly resilient to eluants containing high concentrations of salt, urea, guanidine hydrochloride, detergents or selected water-miscible organic solvents. Eluants containing 1–2 M NaCl, 6 M guanidine hydrochloride, 8 M urea and even 6 M guanidine hydrochloride containing 1 M acetic acid are tolerated by the gels, although agarose is not stable indefinitely in these strongly dissociating media (Glenner et al., 1972). In gels containing a low agarose content ($\leqslant 2\%$) structural disruption of the gel beads is more apparent. Agarose gels also withstand dissociating media containing up to 0.4% sodium deoxycholate (DOC), 3% sodium dodecylsulphate (SDS) and 0.5% Triton X-100. Low concentrations of water-miscible organic solvents such as ethylene glycol, ethanol, methanol, acetone, butanol, aqueous pyridine (up to 80% v/v) and aqueous dimethylformamide (50% v/v) are tolerated whilst dimethylsulphoxide is known to disrupt the structure of agarose.

It is important to realise, however, that despite the obvious limitations in the stability of underivatised agarose gels, activation of the gels prior to coupling ligands suitable for affinity chromatography, considerably improves the stability under adverse conditions. This is because activation of agarose (§ 3.1.1) leads, in most cases, to a degree of covalent cross-linking of the polysaccharide chains.

The structure of agarose makes it inadvisable to dry and re-swell the gels. When agarose is not in use it should be stored in the wet or moist state and protected from microbial growth by means of a suitable bacteriostat. A number of antimicrobial agents are in common use; 0.02% sodium azide, 0.5% butanol, trichlorobutanol and saturated toluene. Other bacteriostatic agents should only be used if they are known to be innocuous to the structure of agarose. In general, agarose gels should be stored for long periods in the presence of a suitable bacteriostat in a refrigerator below 8 °C but without freezing. Freezing results in irreversible structural disruption of the gel beads.

Despite some limitations in stability, the beaded derivatives of agarose exhibit many of the properties of the ideal matrix and have been widely acclaimed as the matrix of choice. Their loose structure allows ready penetration by large macromolecules and their hydrophilic polysaccharide nature and nearly complete absence of charged groups precludes the adsorption or denaturation of sensitive biochemical substances. Furthermore, the polysaccharide backbone can readily undergo substitution reactions to yield products with a moderately high capacity for further derivatisation.

Refined agarose suitable for chromatographic purposes is available commercially. Technical data for such agarose gels is collated in Table 2.1 and is taken from the manufacturers' technical information bulletins.

2.1.2.4. Cross-linked agarose

The beaded forms of commercially available agarose have long been acclaimed as superior media for the chromatography of sensitive biochemical substances. Nevertheless, their widespread application has been limited to some extent by the stability of the gels under extreme conditions of pH or temperature. Pharmacia Fine Chemicals now market a cross-linked agarose gel with substantially the same porosity as the parent gel, Sepharose®, but with greatly enhanced thermal and chemical stability. Sepharose CL is derived from Sepharose by treatment with 2,3-dibromopropanol under strongly alkaline conditions. The cross-links primarily involve the polysaccharide chains in a single gel fibre with inter-fibre cross-links not occurring to any significant extent. The exclusion limits of the gels are thus approximately the same as for the corresponding types of Sepharose (Table 2.1).

TABLE 2.1
Technical data for commercial agarose preparations.*

Designation or trade mark	Particle size (U.S. standard wet mesh size)	Wet particle diameter (μ)	Fractionation range	Approximate % agarose in gel	Maximum rec. hydrostatic pressure (cm H_2O)	Minimum flow rate ml/h/cm^2 column cross-section
1. *Bio-Rad Laboratories*						
Bio-Gel A – 0.5 M	50–100	150–300	10,000–500,000	10	100	110
	100–200	75–150				35
	200–400	37–75				15
Bio-Gel A – 1.5 M	50–100	150–300	10,000–1,500,000	8	100	90
	100–200	75–150				30
	200–400	37–75				10
Bio-Gel A – 5 M	50–100	150–300	10,000–5,000,000	6	100	70
	100–200	75–100				20
	200–400	37–75				9
Bio-Gel A – 15 M	50–100	150–300	40,000–15,000,000	4	90	50
	100–200	75–150				15
	200–400	37–75				6
Bio-Gel A – 50 M	50–100	150–300	100,000–	2	50	30
	100–200	75–150	50,000,000			10
Bio-Gel A – 150 M	50–100	150–300	1,000,000–	1	30	15
	100–200	75–150	7,150,000,000			4
2. *Pharmacia Fine Chemicals*						
Sepharose® 2B		60–250	2,000,000–	2		30
			40,000,000			
Sepharose 4B		40–190	300,000–	4		50
			20,000,000			
Sepharose 6B		40–210	10,000–4,000,000	6		65
Sepharose CL-2B		60–250	40,000,000	2		
Sepharose CL-4B		40–190	20,000,000	4		
Sepharose CL-6B		40–210	4,000,000	6		
3. *PL Biochemicals Inc.*						
Agarose				4		

* Other Suppliers of Agarose: Serva Feinbiochemica GmbH & Co., Aldrich Chemical Co., Industrie Biologique Française, Sigma Chemical Co., Marine Colloids Inc.

Cross-linked Sepharose has a considerably enhanced stability. It can be used in aqueous solutions in the pH range 3–14, in concentrated solutions of chaotropic ions, such as 3 M KSCN, in 6 M guanidine hydrochloride and at elevated temperatures. Sepharose CL may be repeatedly autoclaved at 110–120 °C without alteration in chromatographic properties and may be utilised for chromatography at temperatures up to 70 °C. More importantly, replacement of water in the gel by other miscible solvents exerts a relatively small effect on pore size. Thus the gels tolerate solvents such as ethanol, acetone, dimethylformamide, tetrahydrofuran, dimethylsulphoxide, chloroform, dichloromethane and dichloroethane. The stability of the cross-linked gels in such solvents extends the range of synthetic reactions permissible in the preparation of affinity adsorbents and permits elution of bound materials under more extreme conditions.

2.1.2.5. Polyacrylamide gels

In contrast to the polysaccharide gels, the neutral hydrophilic cross-linked polyacrylamide gels are entirely synthetic. They are produced by co-polymerisation of acrylamide ($H_2C = CH\,CONH_2$) with the bifunctional cross-linking agent, N,N^1-methylenebisacrylamide ($H_2C = CH\,CONH\,CH_2NHCOCH = CH_2$) (Hjertén and Mosbach, 1962) to form, under suitable conditions, a gel comprising cross-linked polyacrylamide chains (Fig. 2.3). By regulating the concentration of monomer and the proportion of cross-linker, a series of covalently-bonded gel products differing in pore size and hence swelling and chromatographic properties, may be obtained. *It should be emphasised, however, that the monomers used in the synthesis of polyacrylamide gels are highly toxic and thus should be handled with care.*

Beaded polyacrylamide gels can be purchased from Bio-Rad laboratories under the trade name, Bio-Gel P. Technical data for these products are given in Table 2.2. The beads are available in various pore sizes commencing at the highly cross-linked P-2 with an exclusion limit of 1,800 up to sparsely cross-linked P-300 which can include molecules with molecular weights up to 400,000. Com-

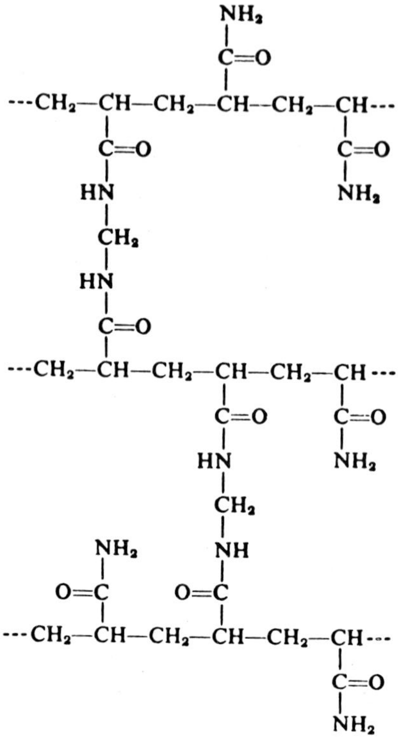

Fig. 2.3. The structure of part of a polyacrylamide matrix. Reproduced with permission from *A Laboratory Manual on Gel Chromatography* by Bio-Rad Laboratories.

mercial polyacrylamide beads are purchased in the dry state and are swollen by mixing with water or aqueous solutions for periods of 4–48 h depending on the porosity. Bio-Gel P products are stable to most eluants used in biochemical studies including dilute solutions of salts, detergents, urea and guanidine hydrochloride although high concentrations of these reagents may alter exclusion limits by up to 10%. The use of media with pH values outside the range 2–10 is to be avoided since some hydrolysis of the amide side groups may occur with the consequent appearance of ion-exchange groups. The use of strong oxidising agents such as hypochlorites or hydrogen peroxide is also inadvisable. The polyacrylamide gels are biologically inert and are not subject to enzymatic or microbial degradation.

Non-specific adsorption to the matrix backbone is restricted to very acidic, very basic and aromatic compounds and is evidenced by delayed emergence from the chromatographic bed. Ionic groups on the matrix are almost non-existent and ion exchange is of no practical significance unless ionic strengths less than 0.02 are used. The principal advantage of polyacrylamide gel is that it possesses an abundant supply of modifiable groups which, coupled with a considerable versatility in derivatisation technology, permits the covalent attachment of a variety of potential ligands (Inman and Dintzis, 1969). In principle, therefore, highly substituted derivatives may be prepared to assist in the purification of macromolecules that display poor affinity for the immobilised ligand.

The use of polyacrylamide matrices for affinity chromatography has nevertheless been rather limited. The reason for this is almost certainly the low degree of porosity of the beads currently available. This deficiency of polyacrylamide gels is emphasised by the fact that *Escherichia coli* β-galactosidase was unretarded on an adsorbent comprising a suitable galactoside inhibitor attached to polyacrylamide (Steers et al., 1971), but was retained on an analogous agarose adsorbent containing 50 times less ligand. The differences in chromatographic behaviour probably reflect differences in the concentration of ligand freely available to the macromolecule. Furthermore, they suggest that the capacity of an inert support for chemical substitution is not necessarily the best criterion for the design of an efficient adsorbent for affinity chromatography. The development of highly porous polyacrylamide beads as potential supports for affinity chromatography would certainly answer many of the present criticisms. Nevertheless the unique properties of polyacrylamide suggest that in certain circumstances polyacrylamide gels may prove superior to polysaccharide based gels.

For example, should degradation of the polyacrylamide matrix occur, there is unlikely to be confusion between the reaction products and sample molecules since polyacrylamide is an entirely synthetic

TABLE 2.2
Technical data for polyacrylamide gel filtration media.

Product designation	U.S. Standard wet mesh	Exclusion limit and fractionation range	Packed volume ml/g Xerogel	Approximate water regain
Bio-Gel P-2	50–100	100–1,800	3.5	1.5
	100–200			
	200–400			
	–400			
Bio-Gel P-4	50–100	800–4,000	5.0	2.4
	100–200			
	200–400			
	–400			
Bio-Gel P-6	50–100	1,000–6,000	8.0	3.7
	100–200			
	200–400			
	–400			
Bio-Gel P-10	50–100	1,500–20,000	9.0	4.5
	100–200			
	200–400			
	–400			
Bio-Gel P-30	50–100	2,500–40,000	11.0	5.7
	100–200			
	–400			
Bio-Gel P-60	50–100	3,000–60,000	14.0	7.2
	100–200			
	–400			

TABLE 2.2 *(continued)*

Product designation	U.S. Standard wet mesh	Exclusion limit and fractionation range	Packed volume ml/g Xerogel	Approximate water regain
Bio-Gel P-100	50–100 100–200 –400	5,000–100,000	15.0	7.5
Bio-Gel P-150	50–100 100–200 –400	15,000–150,000	18.0	9.2
Bio-Gel P-200	50–100 100–200 –400	30,000–200,000	25.0	14.7
Bio-Gel P-300	50–100 100–200 –400	60,000–400,000	30.0	18.0

Data taken from Bio-Rad Laboratories Catalogue B (April 1976).

polymer. These advantages of polyacrylamide gels are particularly apparent in applications involving the purification of carbohydrates or carbohydrate-binding macromolecules. Furthermore, polyacrylamide has special advantages over dextran and cellulose as an inert support in terms of improved stability and reduced background adsorption. Successful affinity chromatography has been achieved in the separation of whole, viable cells by virtue of the interaction of their membrane receptors with ligand attached to the outer surfaces of cross-linked polyacrylamide gels (Truffa-Bachi and Wofsy, 1970). Chemical derivatives of polyacrylamide may also be prepared by copolymerising a suitable acrylic or vinyl monomer with a bifunctional monomer such as N,N^1-methylenebisacrylamide. A number of polyacrylamide derivatives bearing useful functional groups may be made in this way and are available commercially (§ 3.1.2).

2.1.2.6. Polyacrylamide–agarose gels

The recent introduction of a new range of agarose–polyacrylamide copolymers, available under the trade name Ultrogel® from LKB products, has prompted speculation as to their suitability as inert matrices for affinity chromatography. These gels would seem to have all the advantages of each constituent polymer plus the availability of both amide and hydroxyl groups for functionalisation. LKB Ultrogel® is available in four types each comprising a three-dimensional polyacrylamide lattice enclosing an interstitial agarose gel. The gels are pre-swollen and calibrated within a narrow size range of 60–140 μm. The narrow size distribution of the beads reduces zone spreading and thus gives better resolution than other chromatography gels. Furthermore, Ultrogel® beads are more rigid and hence less compressible than conventional gel media and thus permit higher flow rates. The technical specifications of Ultrogel® are given in Table 2.3. Preliminary studies on the potential of Ultrogel® as a matrix for affinity chromatography have indicated differences in the selectivity of Ultrogel® and agarose adsorbents (Doley et al., 1976).

TABLE 2.3
Technical specifications of Ultrogel.

	AcA22	AcA34	AcA44	AcA54
Acrylamide concentration (%)	2	3	4	5
Agarose concentration (%)	2	4	4	4
Bead diameter (pre-swollen) (μm)	60–140	60–140	60–140	60–140
Fractionation range (globular proteins)	60,000–1,000,000	20,000–400,000	12,000–130,000	6,000–70,000

Data taken from LKB technical information.

2.1.2.7. Porous glass and ceramics

Controlled pore glass (CPG) is a material which fulfills many of the qualities of the ideal matrix and represents a new and unique carrier for application in affinity chromatography. It is produced when certain sodium borosilicate glass compositions are heat treated at 700–800 °C and subsequently leached with acid (Nordberg, 1944). During the heat treatment, two continuous and intertwined phases separate; one rich in silica and resilient to acid treatment and the other mainly boric oxide and thus readily etched by acids. The boric acid phase is leached out to leave a highly porous structure comprising anastomosing canals with pore diameters 30–60 Å. Subsequent treatment with mild caustic soda removes some siliceous material from the pore interiors and thus enlarges the pore diameter. Careful control of the various treatments can lead to a porous glass with an extremely narrow pore size distribution in the range 45–2500 Å. This range of pore diameters is adequate to include most biomolecules from substrates, enzymes and viruses to whole cells.

Controlled pore glass forms a rigid packing material which is insoluble and unaffected by changes in the eluant, pressure, flow rates, pH or ionic strength. The defined pore size produces sharp exclusion limits, good resolution and excellent reproducibility. The rigidity of the beads permit high flow rates and thus facilitates a fast and efficient

Subject index p. 519

separation. Glass beads are resistant to microbial attack and may be readily sterilised by disinfectants or autoclaving. The latter is a prime consideration in the purification of pyrogen-free enzymes destined for in vivo or clinical studies.

One potential disadvantage of porous glass beads is the non-specific adsorption of some proteins to the surface. Virgin glass will physically adsorb many enzymes, particularly those that are strongly basic, but also some neutral proteins and some viruses. The majority of acidic proteins, viruses, polysaccharides or nucleic acids elute without retardation or adsorption.

The non-specific adsorption of some proteins derives from the fact that like all silica glasses the surface of porous glass beads comprises silanol groups (Si–OH) which exhibit a slight negative surface charge in aqueous solution (Messing, 1969). Furthermore, the presence of boron in the glass results in the formation of surface Lewis acid sites and thus emphasises surface adsorption phenomena.

Recently, dextran-coated glass beads have been introduced as a novel way to eliminate non-specific adsorption (Regnier et al., 1974). Such beads could theoretically be activated by the cyanogen bromide technique and other techniques normally reserved for polysaccharide matrices (§ 3.1.1). One can envisage considerable potential for dextran-coated glass as a matrix for affinity chromatography of sensitive macromolecules.

Glass beads coated with antigen have been moderately successful in the separation of immune lymphoid cells (Wigzell and Makela, 1970). A number of enzymes have also been purified by affinity chromatography on porous glass matrices. Thus, for example, *E. coli* DNA and RNA polymerases were purified on DNA-glass (Scouten, 1974), and β-galactosidase on galactoside-glass (Woychik and Wondolowski, 1972).

One problem associated with glass beads is prohibitively slow flow rates often generated by clogging with particulate material. These problems were encountered in the purification of oestradiol receptor protein on oestradiol derivatised glass (Cuatrecasas and Anfinsen, 1971a), Scouten (1974) suggests that the extremely fine glass beads (200–325 mesh) used in the latter study may have exacerbated clogging and contributed to low flow rates and recommends the use of beads with mesh sizes 40–80 or 80–120.

The choice of pore diameter is a critical feature for optimal purification by affinity chromatography. The pore diameter determines to a considerable extent, in the same way as for agarose beads, the capacity for functionalisation. Thus, the capacity of glass with a nominal pore diameter of 2500 Å is considerably less than the smaller pore diameter beads, 1750 and 550 Å. Given the proper choice of matrix dimensions, however, porous glass constitutes an excellent affinity support.

Porous glass column packings are available from a number of suppliers. Electro-nucleonics Inc. market a series of porous quartz–glass granules with precisely controlled mean pore diameters in the range 40–3000 Å and with several mesh sizes. They are unaffected by significant changes in pressure, pH or temperature and are effective in separating molecular weights in the range 10^3–10^9 and beyond. Electro-nucleonics Inc. have also introduced a new type of glass packing which virtually eliminates non-specific adsorption phenomena.

Glyceryl–CPG is a controlled pore glass whose surface has been chemically modified to produce a hydrophilic, non-ionic coating which shares most of the same operating characteristics as conventional CPG. Glyceryl–CPG has distinct advantages for affinity chromatography; it may be activated by cyanogen bromide and other methods commonly employed for polysaccharide matrices. Table 2.4 lists some of the technical properties of controlled pore glasses taken from the technical bulletins of Electro-nucleonics Inc. and the other two major suppliers, Bio-Rad Laboratories and the Corning Biological Products Group.

2.1.2.8. Other inert supports for affinity chromatography

A number of other matrices are potentially available for functionalisation as supports for affinity chromatography. Many, however, require considerable chemical modification to make their use in

TABLE 2.4
Controlled pore glass (CPG).

Product description	Mesh size	Mol.wt exclusion limit	Average pore diameter (Å)
Bio-Rad Laboratories			
Bio-Glas 200	50–100	30,000	200
	100–200		
	200–325		
	minus 325		
Bio-Glas 500	50–100	100,000	500
	100–200		
	200–325		
	minus 325		
Bio-Glas 1000	50–100	500,000	1000
	100–200		
	200–325		
	minus 325		
Bio-Glas 1500	50–100	2,000,000	1500
	100–200		
	200–325		
	minus 325		
Bio-Glas 2500	50–100	9,000,000	2500
	100–200		
	200–325		
	minus 325		
Corning Biological Products Group			
G20–3900	20–80	350,000	550
G20–7900	20–80		1350
Electro-Nucleonics Ltd.			
Controlled pore glass	80–120	30,000	75
	120–200		
	200–400		
	80–120	130,000	120
	120–200		
	200–400		

TABLE 2.4 *(continued)*

Product description	Mesh size	Mol.wt exclusion limit	Average pore diameter (Å)
Controlled pore glass	80–120	400,000	170
	120–200		
	200–400		
	20–80		
	80–120	1,200,000	240
	120–200		
	200–400		
	20–80		
	80–120	4,000,000	350
	120–200		
	200–400		
	20–80		
	80–120	10,000,000	500
	120–200		
	200–400		
	20–80		
	80–120	30,000,000	700
	120–200		
	200–400		
	80–120	100,000,000	1000
	120–200		
	200–400		
	80–120	300,000,000	1400
	120–200		
	200–400		
	20–80		
	80–120	900,000,000	2000
	120–200		
	200–400		
	20–80		
	80–120	1,900,000,000	3000
	120–200		
	200–400		
	20–80		

affinity chromatography a practical proposition. Thus, for example, the polystyrene gels, cross-linked with divinylbenzene are too hydrophobic and display poor communication between the aqueous and solid phases. Such supports exhibit good swelling properties but, along with the acrylic gels, display low porosity and significant non-specific adsorption effects. This criticism may be relieved to some extent by use of the hydrophilic hydroxylalkylmethacrylate gels (Turková, 1974), available commercially under the trade mark, Spheron, and whose chemical structure is depicted in Fig. 2.4a. The hydroxyl groups attached to the matrix backbone exhibit properties analogous to those of agarose and other polysaccharides and are thus potentially activatable by cyanogen bromide. Furthermore, the number of reactive groups, porosity and bead size of these gels may be varied over a significant range during their production. Thus commercial Spheron gels are available with exclusion limits ranging from 100,000 (Spheron 100) to 10^8 (Spheron 10^5) and have been used for affinity chromatography (Turková, 1974).

By analogy to immobilised enzymes, inert supports such as nylon, metal oxides, starch and the copolymer of ethylene and maleic anhydride may also find application in affinity chromatography. The ethylene-maleic anhydride copolymer (Levin et al., 1964) (Fig. 2.4b) in particular, finds application in the binding of proteins to the anhydride groups via their amino groups. However, during attachment of proteins there is a concomitant liberation of carboxyl groups and the carrier acquires a polyanionic character. Nevertheless, the copolymer has been used to considerable advantage in the preparation of immuno-adsorbents. The requirements of these adsorbents are, however, somewhat less stringent than those of affinity adsorbents. This is because of the number of antigenic determinants borne by most proteins, their ready accessibility when such large proteins are coupled to matrices and the high avidity of antibodies for antigens.

The greater rigour required for affinity chromatography has necessitated optimalisation in the choice of matrix for the desired application. Unfortunately, there is no ready rule as to which inert

(a)

(b)

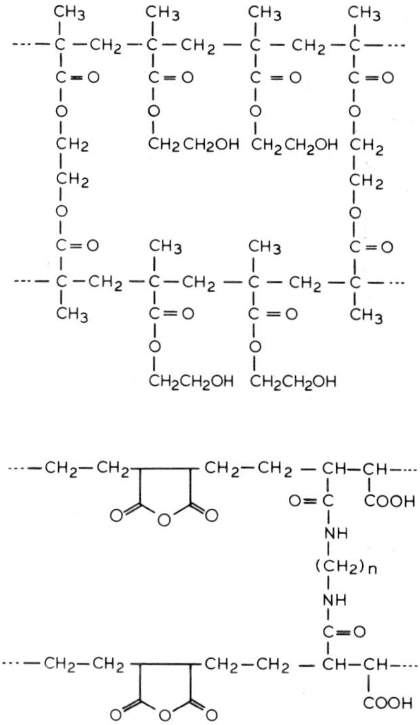

Fig. 2.4. The chemical structures of two synthetic support matrices for affinity chromatography: (a) poly(hydroxyethyl methacrylate) and (b) cross-linked ethylene–maleic anhydride copolymer.

matrix will give optimal results for the individual system under investigation. A largely empirical approach is still necessary, although experience has shown that the beaded derivatives of agarose are ideally suited to the purification of enzymes and other proteins by affinity chromatography.

2.2. Problems associated with the matrix

It has already been emphasised that the success of affinity chromatography depends on how closely the experimental conditions chosen permit the ligand–macromolecule interaction that is characteristic of the components in free solution. Ideally, therefore, there should be no matrix to interfere with this interaction. The enforced presence of an inert gel matrix to which the affinity ligand is attached generates two major problems for successful interaction with the complementary macromolecule. The first relates to the porosity of the matrix which determines the initial penetration of the protein into the vicinity of the ligand whilst the latter relates to the influence of the matrix backbone on the microenvironment of the immobilised ligand.

2.2.1. Macroenvironmental or exclusion effects of the matrix

When a specific ligand is immobilised to a gel lattice, the porosity of the beads will determine the overall accessibility of the ligand towards interactions with its complementary protein. It is the exclusion effects of the matrix that govern the effective ligand concentration, i.e., that concentration of the bound ligand freely available to the macromolecule, and thus determines the behaviour of the system under operational conditions. Thus immobilisation of a specific ligand within a gel lattice of low porosity will significantly impair the accessibility of the bound ligand and thus the effectiveness of the resulting adsorbent. The problem is largely circumvented by choice of a highly porous matrix such as derivatives of agarose or controlled pore glass.

In some applications, however, the porosity of the bead or matrix may not become a limiting factor. Thus, in cases where 'functionally homogeneous' cell populations are required, the affinity ligand, which recognises a surface receptor on the relevant cell type, is attached to the surface of the beaded matrix. Such applications require large spherical beads with correspondingly large inter-bead spaces such that the cells are not physically entrapped. Pharmacia Fine Chemicals AB market a product which is specially designed for the affinity chromatography of cells, Sepharose 6MB. These large beads of Sepharose 6B (200–300 μm) have a narrow range of bead sizes to eliminate physical entrapment of cells and thus exhibit excellent flow properties. Furthermore, they are available pre-activated by the cyanogen bromide technique to facilitate the efficient and safe coupling of ligands containing free amino groups. The applications of affinity chromatography to the resolution of cells and viruses are discussed in § 6.4.

2.2.2. *Microenvironmental or steric effects*

The second limitation imposed by the matrix relates to the effect of the matrix backbone on the microenvironment of the immobilised ligand. Thus, irrespective of the porosity of the matrix material, if the affinity ligand is coupled directly to the matrix backbone, the interaction with the complementary protein will be sterically hindered. This means that for successful purification by affinity chromatography the immobilised ligand must be placed away from the lattice backbone and this is generally achieved by locating the ligand at the terminus of a long chain or 'arm' which is anchored to the matrix backbone (Cuatrecasas et al., 1968; Cuatrecasas, 1970). Several general procedures have been devised for the preparation of adsorbents comprising ligands attached to matrices via *spacer arms* of various lengths or types (Cuatrecasas, 1970).

2.2.3. *Spacer molecules*

The importance of interposing a spacer arm between the ligand and the lattice backbone is amply justified by the relative ineffectiveness of an adsorbent comprising D-tryptophan methyl ester attached directly to agarose compared to the same ligand attached to agarose via a hexamethylene spacer arm for the purification of α-chymotrypsin (Cuatrecasas et al., 1968). Similarly, direct attachment of the relatively weak inhibitor, *p*-aminophenyl-β-D-thiogalactoside, to agarose generated an adsorbent that was incompetent in binding *E. coli* β-galactosidase (Fig. 2.5a), (Steers et al., 1971). By inserting a short spacer molecule (~1.0 nm) between the inhibitor and the matrix, β-galactosidase was retarded and appeared in a volume slightly greater than that of the void volume (Fig. 2.5b). In contrast, the use of a longer spacer molecule (~2.1 nm), resulted in strong retention of the enzyme from several sources (Fig. 2.5c).

Many other examples of the use of spacer molecules to enhance the steric availability of the ligand are to be found in the literature. The near universal use of spacer molecules prompts two important considerations: how long should the arm be to maximize the interaction between the ligand and the macromolecule and to what extent does its nature affect the interaction with the ligand?

2.2.3.1. *The length of the spacer arm*

The most general and extensively employed technique for the introduction of spacer molecules is to couple ω-aminoalkyl compounds of the general type $NH_2(CH_2)_nR$ to the matrix, where R may be a carboxyl group, an amino function or the ligand itself and $n = 2$–12 (Cuatreasas, 1970). Systematic studies on the effect of the length of the spacer arm have been performed with a homologous series of Sepharose-bound 8-(ω-aminoalkyl) derivatives of adenosine 5′-monophosphate in which the number of methylene groups in the spacer arm was increased from 2–8 (Lowe, 1977). Figure 2.6 shows

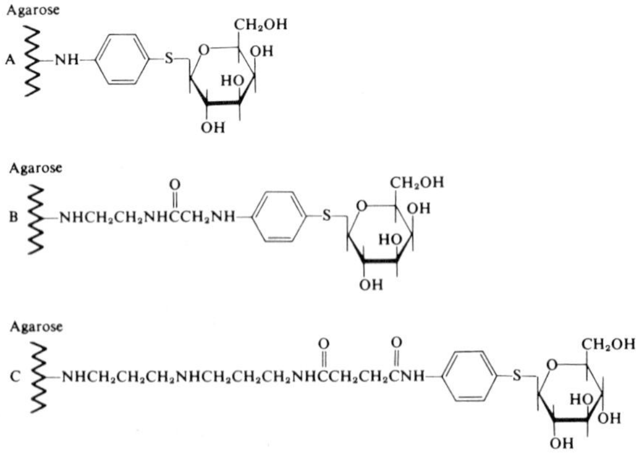

Fig. 2.5. The agarose adsorbents used for the purification of β-galactosidase by affinity chromatography. Reproduced with permission from P. Cuatrecasas (1972), Adv. Enzymol., *36*, 29.

the effect of increasing the length of the spacer arm on the binding of lactate dehydrogenase to the homologous series of adsorbents. The binding of the enzyme increased significantly from $n = 2$ to $n = 6$ whence further increases in the length of the spacer arm produced a smaller increase in the strength of the interaction.

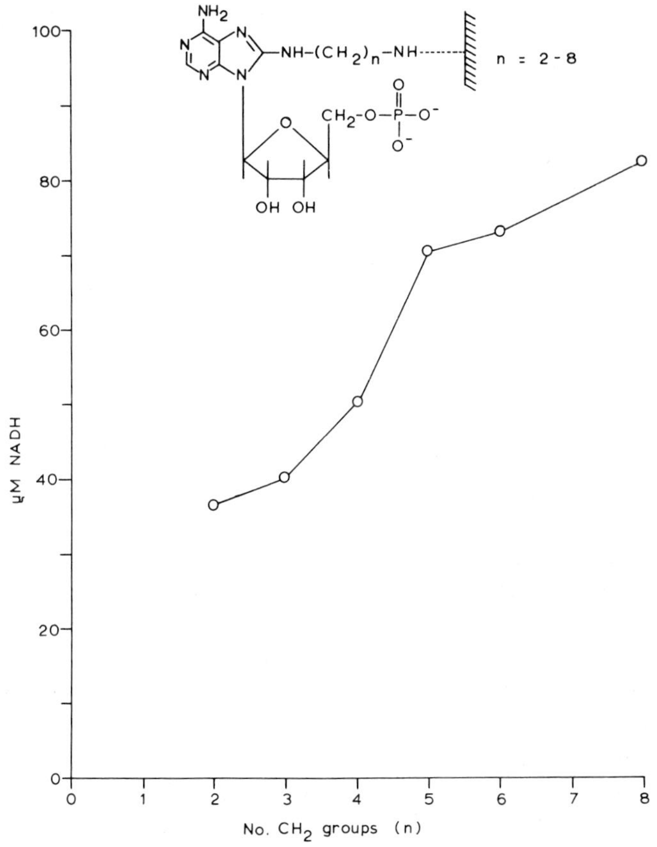

Fig. 2.6. The interaction of rabbit muscle lactate dehydrogenase with agarose-bound 8-(ω-aminoalkyl)-adenosine-5′-monophosphate derivatives containing polymethylene spacer arms of increasing length.

Subject index p. 519

Similar results were obtained with several other dehydrogenases, with a comparable homologous series of N^6-(ω-aminoalkyl)-adenosine-5′-monophosphate adsorbents (Hipwell et al., 1974) and were independent of whether the matrix was Sepharose 2B or 4B. These results are in general agreement with a study conducted with an adsorbent for Staphylococcal nuclease (Cuatrecasas, 1970). *It is generally accepted that a bridge containing at least 4 to 6 methylene groups must be interposed between the ligand and lattice backbone in order to achieve optimal interaction with the complementary molecule.* However, it should be pointed out that where the complementary macromolecule has a low apparent molecular weight or a high affinity for the immobilised ligand, the length of the spacer arm is not as critical as in the case of large proteins or with systems of low affinity.

The use of macromolecular spacer molecules has recently been suggested. Poly(lysyl-alanine), for example, possesses a polylysine backbone whose terminal ε-amino functions are substituted with oligoalanine peptides containing 14–16 residues. This polymer has a molecular weight of approximately 260,000 and may be attached to an activated matrix (§ 3.1) by more than one point. The remaining unattached amino terminals are then potentially available for the coupling of affinity ligands by, for example, carbodiimide promoted reactions (§ 3.4.1.1). The polymer containing all D-amino acids may be synthesised and should be resistant to proteolysis by proteases in a crude enzyme extract. Other macromolecular spacer arms may be constructed by coupling denatured albumin (Sica et al., 1973), polylysine, polyornithine or polyvinylamine (Wilchek, 1973) to a suitable matrix. The resulting derivatives may subsequently be activated or functionalised with suitable ligands for the affinity chromatography of a number of sensitive biochemical substances. For example, the general utility of polymeric spacer molecules may be exemplified in the purification of the oestradiol receptor (Sica et al., 1973). A conventional multimeric spacer arm composed of several smaller units linked together (§ 3.2) permitted a purification of 27-fold, whilst use of albumin increased the purification to 4,400-

fold and a poly(lysyl-alanine) arm allowed a purification in excess of 100,000-fold.

2.2.3.2. *The nature of the spacer arm*

With the exception of the recently introduced macromolecular spacer arms, the compounds most commonly employed as spacer molecules are linear aliphatic hydrocarbons with α,ω-terminal functional groups providing the points of attachment to the matrix and ligand. Until recently it was generally considered that these hydrocarbon molecules fulfilled their role as chromatographically inert spacer molecules with no undesirable side effects. However, it has been pointed out that spacer arms may not only generate local steric interference with the ligand–macromolecule interaction but may also accentuate non-specific adsorption phenomena (O'Carra et al., 1973). Thus, for example, in the purification of β-galactosidase, 'control' adsorbents prepared by coupling either a biologically inactive substrate analogue or no ligand at all to suitable spacer molecule–agarose conjugates, behaved remarkably similar to supposedly authentic adsorbents containing the galactoside ligand. Likewise, glycogen phosphorylase *b* was retarded by, or adsorbed to, a homologous series of alkyl–agaroses (agarose–$NH(CH_2)_nH$, $n = 2$–8) which did not contain the specific ligand, glycogen, attached to their termini (Er-El et al., 1972). Since, in the latter study, all the alkyl–agaroses were similar in structure, were neither substrate analogues nor effectors of the enzyme and differed only in the number of methylene groups in the hydrocarbon chain, it seems reasonable to conclude that retention of the phosphorylase *b* occurs, at least partly, by hydrophobic interactions. These observations suggest that hydrophobic interactions may significantly contribute to the tightness of binding between an immobilised ligand and a complementary macromolecule and thus supplement or even adumbrate those involved in recognition of the ligand. This 'compound' affinity is often detrimental to the efficiency of a biospecific adsorbent since it permits a degree of non-specific adsorption of protein and thus compromises the purity of the desired macromolecule on subsequent

Subject index p. 519

elution. The exclusive use of hydrophobic chromatography, where the ligand is deliberately omitted, for the purification of proteins with accessible hydrophobic binding sites has nevertheless shown considerable promise as a general technique for the purification of proteins (§ 8.2). Furthermore, it should be stressed that 'compound' affinity is not always detrimental but may be beneficial in supplementing the binding forces of ligand–macromolecule systems of low affinity.

Since there is a likelihood that the presence of hydrophobic spacer molecules will generate serious problems of non-specific adsorption the construction of more hydrophilic arms has been recommended. A number of prototype hydrophilic arms have been synthesised and tested. Thus, O'Carra et al. (1974) have interposed polar groups such as secondary amino, hydroxyl and peptido along the length of the spacer arm to break up hydrophobic enclaves. It was claimed that interference by non-biological adsorption was drastically reduced by replacement of the hydrophobic spacer molecules by more hydrophilic ones. However, in some cases a useful operational chromatography was only achieved when non-specific adsorption phenomena were present. In such cases, careful control and balancing of specific and non-specific effects may offer considerable advantage over complete eradication of non-specific effects.

Sesqui-peptides of glycine have also been interposed between a ligand and the matrix backbone (Lowe and Dean, 1974) although with somewhat disappointing results. Thus *N*-glycyl–glycine or *N*-(*N*-glycyl–glycyl)glycine in amide linkage to glucosamine and attached to Sepharose 4B proved relatively ineffectual in the purification of rat hepatic glucokinase, whilst the same derivative coupled to 6-aminohexanoate as spacer proved satisfactory for the adsorption of the enzyme (Holroyde et al., 1976). Interestingly, both the diglycine and 6-aminohexanoate derivatives of glucosamine were equally effective at inhibiting the reaction catalysed by glucokinase in free solution. When attached to an agarose support however, the glucosamine derivative coupled via a hydrophobic spacer molecule, 6-aminohexanoate, proved considerably more effective as an affinity adsorbent. These observations have been confirmed by the synthesis of a series of 8-substituted adenosine 5′-monophosphate (AMP) derivatives bearing spacer arms of the same length but differing degrees of hydrophobicity/hydrophilicity (Lowe, 1977). These derivatives exhibited similar competitive inhibition constants in free solution although the AMP derivative attached to Sepharose 4B via a hydrophobic hexamethylene arm tightly bound a series of NAD^+-dependent dehydrogenases and the use of a hydrophilic arm containing a peptide linkage and a carbinol grouping was totally ineffective. It is suggested that whilst the stronger binding to the more hydrophobic derivatives at equivalent ligand concentrations may be explained by a reinforcement of the ligand–macromolecule interaction with a non-specific hydrophobic interaction with the spacer arm, there is an alternative plausible explanation. The hydrophilic derivatives of glucosamine and AMP may not be physically available for interaction with the complementary enzymes, possibly as a result of interactions such as hydrogen bonding between the ligand–spacer molecule assembly and the matrix backbone. This phenomenon may thus lead to an adsorbent which behaves as one would anticipate for a lower effective concentration of ligand. Figure 2.7 shows the structures of some typical spacer arms.

2.3. Problems associated with spacer molecules

In general terms, the effects relating to the length and nature of a prospective spacer molecule cannot be divorced either from each other or from the nature of the ligand. It has been suggested that the effective length of the spacer molecule depends on the nature of both ligand and spacer arm since it is known that hydrophobic ligands attached to matrices via hydrophobic spacer arms give very disappointing results (O'Carra et al., 1973). It is conceivable that under these circumstances the hydrophobic ligand 'folds' back onto the hydrophobic arm and thus in such cases a hydrophilic spacer arm may be preferable. Furthermore, the optimal length of a suitable spacer arm does not appear to bear a simple relationship to known

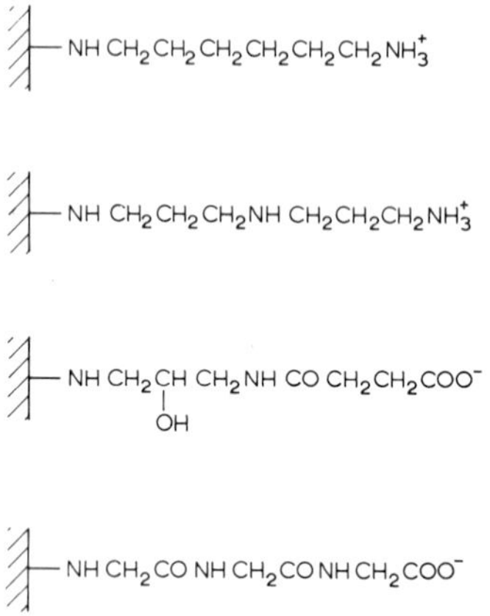

Fig. 2.7. The structures of some typical spacer arms.

parameters of the system and may vary from ligand to ligand for a particular enzyme or from enzyme to enzyme for a particular ligand. Thus a spacer arm of nominal length 0.8 nm is optimal for the interaction of lactate dehydrogenase and an immobilised pyruvate analogue, whilst 1.1 nm is required for AMP linked through the exocyclic N^6-amino of the adenine moiety and 1.5 nm for a similar ligand linked to the 8-position of the adenine nucleus. These observations reflect the need for a more rational strategy in the design of spacer arms for affinity chromatography. Indeed, the selection of a spacer molecule seems to be based either on precedent or on preparative simplicity. Each purification should be considered independently and if a spacer molecule is necessary, the length to achieve optimal separation efficiency should be determined. This is important because unnecessarily increasing the spacer length may enhance non-specific adsorption (§ 4.4) or decrease the effectiveness of the column by virtue of folding or coiling of the flexible spacers and thus decreasing the availability of ligand to the solute molecules. The use of hydrophilic spacers should also be investigated. Bearing these considerations in mind, a largely empirical approach is generally adopted with a hexamethylene spacer molecule being that most commonly employed.

2.4. *Considerations relating to the selection of the ligand*

The ligand that is to be covalently attached to the inert matrix must be one that exhibits specific and unique affinity for the solute to be purified. The ligand is usually, but not always, chosen from a biological source which has a natural pairing tendency with the complementary solute. Ideally the physiochemical basis for the interaction should be at least partially understood. Table 2.5 illustrates the diversity of ligands that have been immobilised and used for affinity chromatography. Such diversity of potential ligands for immobilisation prompts consideration of the principles underlying the selection of a competent ligand and its immobilisation for an affinity adsorbent.

2.4.1. *The nature of the ligand–macromolecule interaction*

Careful consideration should be given to the nature of the macromolecule–ligand interaction when prospective ligands for immobilisation are evaluated. In most cases, the choice of the ligand for immobilisation is limited to a substrate, a reaction product, a suitable analogue or any other ligand that interacts directly with the macromolecule to be purified. In more complex multi-substrate reactions considerable latitude in the choice of ligand is permissible. In random order bi- or multisubstrate reactions the choice of ligand for attachment to the matrix will be determined largely by the relative affinities and ease of immobilisation of the ligands. For ordered bisubstrate reaction mechanisms, however, in which ligand A binds compulsorily to the enzyme before ligand B can interact with the binary complex, advantage can be taken of the mechanism of interaction. For example, immobilisation of ligand A will follow the usual rules of affinity chromatography, whilst immobilisation of ligand B will generate an adsorbent which is incompetent for the complementary enzymes unless ligand A is included in the irrigant in

Subject index p. 519

TABLE 2.5
Some ligands that have been immobilised and used for affinity chromatography.

Allosteric effectors
Antibiotics
Antibodies
Chromophores and dyestuffs
Coenzymes
Enzymes
Hormones
Hydrophobic ligands
Inhibitors
Lectins
Nucleic acids
Nucleotides
Plant hormones
Polynucleotides
Protease inhibitors
Steroids
Substrate and substrate analogues
Sugars and polysaccharides
tRNA

suficient concentrations to generate the binary complex. Thus the binding of the complementary enzyme to an adsorbent comprising immobilised B depends on the presence of ligand A in the irrigant and its subsequent removal will lead to prompt elution of the enzyme from the adsorbent.

Many pyridine nucleotide-dependent dehydrogenases display such compulsory ordered kinetic mechanisms in which the pyridine nucleotide binds first. O'Carra and Barry (1972) have demonstrated that in the presence of 100 μM NADH, lactate dehydrogenase is strongly retarded by the presence of an immobilised analogue of pyruvate, oxamate, and that subsequent removal of the reduced nucleotide effected prompt elution of the enzyme. Clearly, a greater degree of specificity is inherent in the 'negative' elution achieved by removal of the ligand required for formation of the binary complex, than would be effected by 'positive' elution with a suitable eluant ligand. Thus the purification of lactate dehydrogenase on an immobilised oxamate adsorbent is equivalent to two independent affinity chromatography steps since it is dependent, not only, on the dual affinity of the enzyme for NADH and oxamate, but also on the nature of the kinetic interaction between them. Other examples of this dual affinity chromatography are available in the literature (Andrews, 1970) and demonstrate the advantage to be gained in exploiting the nature of the ligand–macromolecule interaction.

2.4.2. *The affinity of the macromolecule for the ligand*

The affinity of the ligand for the complementary macromolecule is an important consideration relating to the potential efficacy of a prospective ligand for immobilisation. However, whilst a fair number of empirical guidelines have been developed for affinity chromatography, little in the way of theory has appeared to place the technique on a more rigorous footing. A simple mathematical model describing affinity adsorption in terms of a few measurable parameters could thus prove useful in assessing prospective ligands (Lowe and Dean, 1974). If the concentration of enzyme is denoted by E and that of the ligand by L, then

$$E + L \overset{K_L}{\rightleftharpoons} EL$$

Where the dissociation constant (K_L) of the enzyme-immobilised ligand complex (EL) is given by:

$$K_L = \frac{[E][L]}{[EL]} = \frac{[E_0 - EL][L_0 - EL]}{[EL]} \qquad (2.1)$$

Where E_0 and L_0 are the initial concentrations of enzyme and immobilised ligand respectively. In most cases $L_0 \gg E_0$ and hence $L_0 \gg EL$, whence:

$$K_L = \left(\frac{E_0 - EL}{EL}\right) L_0$$

and rearranging,

$$\frac{[EL]}{[E_0]} = \frac{\dfrac{[L_0]}{K_L}}{1 + \dfrac{[L_0]}{K_L}} \tag{2.2}$$

This equation (2.2) defines the fraction of enzyme bound to the affinity adsorbent for a fixed amount of enzyme input (E_0) when the ligand concentration L_0 is treated as the variable.

Similar equations have been derived by O'Carra et al. (1973) and Nishikawa et al. (1976). This elementary treatment permits an approximate estimate of the maximum K_L between an immobilised ligand and an enzyme to effect an operationally useful binding of the enzyme from the raw material. Thus if the adsorbent comprises an immobilised ligand whose concentration in the matrix is, typically, about 5 μmol/g (or ml) Sepharose, i.e., 5 mM, then K_L for the enzyme-immobilised ligand must be at least 0.1 mM in order to effect near quantitative (98%) adsorption of the enzyme to the matrix. At $L_0 = 10$ mM, calculations based on eq. 2.1 suggest that to effect good binding of the enzyme, K_L should be at least 0.5 mM. With ligands exhibiting K_L values greater than 1 mM, the problem is that of coupling ligands to the gel at sufficiently high concentrations. Here there is a practical limit imposed by the matrix itself (§ 2.4.3).

Thus, this simplified theoretical treatment shows that at a typical immobilised ligand concentration of 5–10 mM an upper limit for the affinity of the prospective ligand for the enzyme to be purified would be 0.1–0.5 mM. Prospective ligands with $K_L > 0.5$–1 mM should be discarded in preference to those with higher affinity. However, it should be stressed that calculations based on eq. 2.2 should be used for guidance only since they assume that (a) the overall ligand concentration (L_0) is equivalent to the effective ligand concentration (L_{eff}) and (b) that K_L for the enzyme-immobilised ligand is comparable to that for the enzyme-free ligand system.

2.4.3. *The mode of attachment of the ligand to the matrix*

The particularly striking feature of enzymes that may be exploited

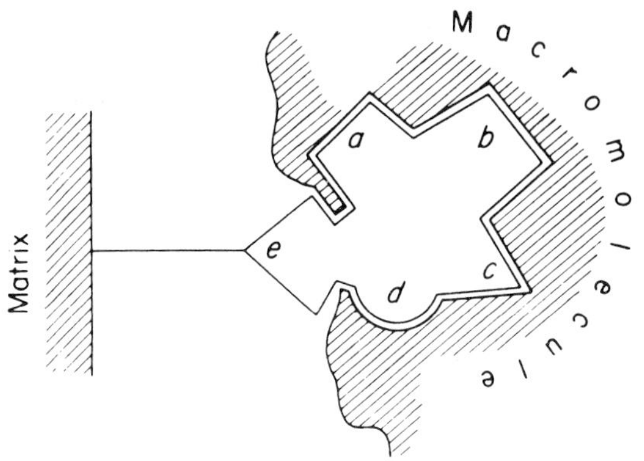

Fig. 2.8. The importance of the mode of attachment of the ligand to the matrix in affinity chromatography. Reproduced with permission from Lowe and Dean (1974), John Wiley and Sons Ltd.

to advantage in the preparation of affinity adsorbents is their high substrate specificity. The substrate to be insolubilised must possess functional groups that can be chemically modified for linkage to the solid support without impairing or abolishing the interaction with its complementary enzyme. It is this restriction that often makes the selection of ligand and its mode of attachment to the matrix somewhat empirical in practice. Figure 2.8 illustrates the importance of the mode of attachment of the ligand to the matrix by demonstrating that of the five potential binding points to a small ligand, only that denoted by *e* is free from interaction with the complementary enzyme. Thus covalent attachment of the prospective ligand to the matrix through this functional group should yield an effective adsorbent whilst attachment by any other point (*a* – *d*) could generate an adsorbent either only partially effective or entirely ineffectual. Needless to say, information is required to indicate which part of the ligand can be chemically modified without being detrimental to the interaction with the complementary enzyme. Rarely, however, is such data readily available to the researcher and, more often than not the collection of such data entails considerable screening of the literature. For example, information relating to substrate specificity or the synthesis of analogues of prospective ligand, affinity labels or active-site directed irreversible inhibitors can point to areas on the ligand where some tolerance to chemical modification is permissible. Under ideal circumstances, *X*-ray diffraction studies on enzyme–substrate complexes may be consulted to give a detailed picture of the way in which the substrate is orientated in the active site cleft. X-ray diffraction data on dog fish muscle lactate dehydrogenase and a number of related pyridine nucleotide-dependent dehydrogenases has revealed that the coenzyme NAD^+ is bound in a deep cleft, nicotinamide end innermost and adenine moiety outermost, with most functional groups on the coenzyme involved in binding to the enzyme (Chandrasekhar et al., 1973). The adenine moiety of NAD^+ is bound in a shallow hydrophobic crevice with the exocyclic N^6-amino group of the adenine nucleus protruding out of this pocket, and thus, potentially available as a point of attachment to the matrix backbone. Other functional groups on the coenzyme are not so fortunately placed for attachment to the matrix without impairing the interaction with the complementary enzymes. Studies by Lowe (1977), Harvey et al. (1974) and Trayer and Trayer (1974) on the effectiveness of several immobilised adenine nucleotides for the affinity chromatography of pyridine nucleotide-dependent dehydrogenases have shown that the point of attachment of the nucleotide to the matrix is of fundamental importance. Figure 2.9 shows the structures of four affinity adsorbents prepared by immobilising adenosine 5′-monophosphate (AMP) by different procedures. Adsorbent a (N^6-(6-aminohexyl)-AMP–Sepharose), comprises AMP attached to agarose via the N^6-amino group of adenine, whilst adsorbent b (C^8-(6-aminohexyl)-AMP–Sepharose), c (P^1-(6-aminohexyl)-P^2-(5′-adenosine)-pyrophosphate–Sepharose) and d (ribosyl-linked AMP) are linked via the 8-position of the adenine, the terminal phosphate and the ribose hydroxyls respectively. Ox heart lactate dehydrogenase exhibits preference for adsorbent a, binds to adsorbent c and to a lesser extent b but shows no affinity to adsorbent d (Trayer and Trayer, 1974). The concentration of NAD^+ required to elute lactate

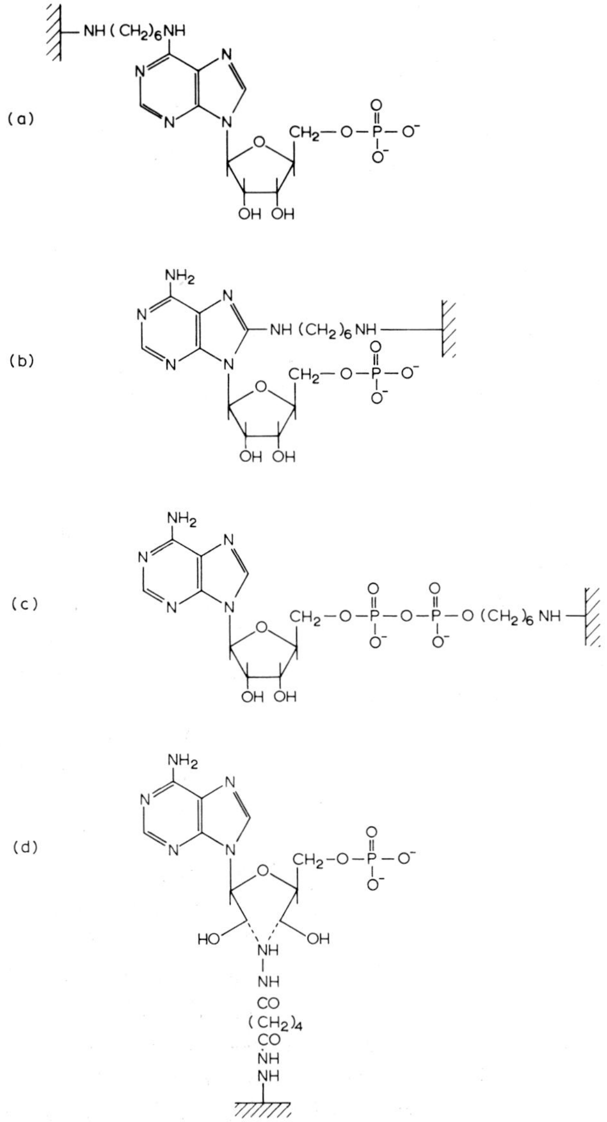

Fig. 2.9. The structures of several immobilised-AMP adsorbents. (a) N^6-(6-aminohexyl)-AMP–agarose [N^6-AMP]. (b) 8-(6-aminohexyl)-AMP–agarose [C^8-AMP]. (c) P^1-(6-aminohexyl)-P^2-(5′-adenosine)-pyrophosphate–agarose [P-ADP]. (d) Ribosyl-linked AMP [R-AMP].

TABLE 2.6
The elution of ox heart lactate dehydrogenase from immobilised adenosine phosphate derivatives by NAD^+

Sepharose-bound adenosine phosphate derivative	Concentration of NAD^+ required for elution (mM)
N^6-AMP	9–9.5
P-ADP	3.50
C^8-AMP	1.40
R-AMP	No binding

Data from Trayer and Trayer (1974). The abbreviations for the various immobilised nucleotide adsorbents listed above refer to the structures given in Fig. 2.9.

dehydrogenase from the various Sepharose-bound adenosine phosphate derivatives at equivalent immobilised ligand concentrations can be compared directly. Table 2.6 shows that the tightness of binding of lactate dehydrogenase to the four adenosine phosphate derivatives is entirely consistent with the X-ray crystallographic analysis (Rossmann et al., 1972). The exocyclic 6-amino group of the adenosine moiety is orientated away from the enzyme and attachment of the alkyl 'spacer' arm at this point on the AMP molecule (Fig. 2.9a) might be expected to permit maximal interaction of the AMP with the enzyme. The C^8 position on the adenine nucleus is orientated towards the surface of the hydrophobic crevice in which the adenosine moiety of NAD^+ binds (Chandrasekhar et al., 1973) and thus attachment through this position (Fig. 2.9b) might be expected to impair the interactions with the enzyme. The second phosphate of ADP does not appear to play a significant role in binding ADP to lactate dehydrogenase and, indeed, the ADP difference electron-density map shows two partially occupied sites for the second phosphate (Chandrasekhar et al., 1973). Attachment of the 'spacer arm' through this second phosphate (Fig. 2.9c) therefore provides another ligand capable of interacting strongly with the complementary enzyme. However, linkage through this point may weaken the inter-

TABLE 2.7
The binding of several AMP-dependent enzymes to Sepharose-immobilised N^6-(6-aminohexyl)-AMP and C^8-(6-aminohexyl)-AMP.

Enzyme	Source	Concentration of NADH required for elution (μM)	
		N^6-AMP	C^8-AMP
Alcohol dehydrogenase	Horse liver	75	30
Alanine dehydrogenase	*B. subtilis*	8	80
Malate dehydrogenase	Pig heart	35	0
Malate dehyrogenase	*B. subtilis*	0	10
Lipoamide dehydrogenase	Pig heart	0	0
D-galactose dehydrogenase	*Ps. fluorescens*	0	0
Phosporylase *b*	Rabbit muscle	3.0*	2.0*

* mM AMP.
Ligand concentration in each case: 2.5 μmol AMP/g moist weight Sepharose 50 mM KH_2PO_4–KOH, pH 7.5.

actions between the 5′-phosphate(s) and the enzyme and thus this immobilised ligand is not as effective as the immobilised N^6-substituted AMP. In contrast, however, the 2′,3′-vicinal diols of the adenosine ribose are buried within the active site crevice and thus not surprisingly the ribose-immobilised AMP derivative (Fig. 2.9d) is ineffectual in binding the enzyme.

The order of effectiveness of the immobilised-AMP derivatives depicted in Fig. 2.9 for the binding of ox heart lactate dehydrogenase (N^6-AMP > P-ADP > C^8-AMP > R-AMP) is also found for the enzyme from other sources. However, the preference for the N^6-immobilised AMP derivative is by no means universal for other AMP-dependent enzymes. Table 2.7 compares the effectiveness of immobilised 6- and 8-substituted AMP derivatives for the binding of a typical selection of enzymes from several sources. This table demonstrates the importance of attaching the ligand to the 'spacer arm' through more than one functional group on the ligand in order to exploit small differences in the binding specificities of the enzymes. For example, whilst immobilised N^6-substituted AMP is preferred by several dehydrogenases from mammalian sources, considerable enrichment of alanine and malate dehdrogenases from a crude extract of *B. subtilis* may be achieved using immobilised C^8-substituted AMP.

In cases where the prospective ligand has a number of potential sites for attachment of the 'spacer arm' and the mode of binding to the enzyme is unknown, the reader is advised to synthesise several immobilised derivatives in order to assess their relative binding merits for the enzyme under consideration. Where two or more points of attachment appear to be equally favourable the preferred site would probably be dictated by the relative susceptibilities to chemical modification. Thus, in the above example of immobilised-AMP derivatives, 8-substituted ligands may be preferred to N^6-substituted since the preparation of these ligands follows a relatively simple and straightforward procedure from inexpensive starting materials. In contrast, the synthesis of N^6-substituted and phosphate-substituted derivatives (Fig. 2.9a,c) proceeds by a lengthy and tortuous route requiring considerably more expertise in organic chemistry. As a general rule, the larger the ligand, the more points of attachment between it and the complementary protein and thus the greater the degree of latitude in preparing the affinity adsorbent. This is particularly true where the prospective ligand is a protein or other macromolecule.

The utilisation of protein–protein interactions has been widely used for the purification by affinity chromatography of antigens and antibodies, proteolytic enzymes and a number of binding and transport proteins (Lowe and Dean, 1974). In each case, effective adsorbents were obtained by linking the proteinaceous ligand directly to Sepharose by the cyanogen bromide technique (§ 3.1.1.1). However, it is important to realise that the immobilised biologically active protein should retain its native tertiary structure and hence its ability to specifically and reversibly bind the complementary molecule. For this reason, the macromolecule should be covalently attached to the matrix by the smallest number of linkages. Proteins react with cyanogen bromide-activated Sepharose through the unprotonated form of their free amino groups. Therefore, since proteins are richly endowed with surface lysyl residues, such molecules will be attached to the matrix by multiple linkage if the coupling reaction is performed at pH $\geqslant 9.5$. This problem of multiple attachment may be lessened by coupling at less favourable pH values. Thus, for example, the capacity of an immuno-adsorbent for antigen could be considerably enhanced by coupling the antibodies to the matrix at pH 6.0–6.5 (Cuatrecasas, 1969).

2.4.4. The ligand concentration

Having selected a suitable ligand for immobilisation and decided on its mode of attachment and scheme of synthesis, an important consideration, and perhaps one of the most neglected, is the concentration of ligand on the gel. Relatively few reports have included data of sufficient quality to document its importance. A simple model will suffice to demonstrate the relevance of the gel-bound ligand concentration. In § 2.4.2 the dissociation constant (K_L) of the enzyme-immobilised ligand complex (EL) was given by:

$$K_L = \frac{[L_0 - EL][L_0 - EL]}{[EL]}$$

where E_0 and L_0 are the total concentrations of enzyme and ligand respectively. Rearranging and denoting $[E_0] - [EL]$ by the concentration of free enzyme, $[E]$

$$[EL] = \frac{[E]\,L_0}{K_L} \Bigg/ \left(1 + \frac{[E]}{K_L}\right) \qquad (2.3)$$

This equation is similar to that derived by Nishikawa et al. (1976) and demonstrates that the concentration of bound enzyme, $[EL]$, expressed in units of moles per unit volume (or weight) of gel, is clearly limited by the concentration of the immobilised ligand, $[L_0]$. In other words, if the concentration of enzyme is increased, the capacity of the adsorbent, $[EL]$, is determined by the immobilised ligand concentration. For a fixed concentration of added enzyme, $[E_0]$, the percentage of enzyme bound $[EL]/[E_0]$, will be given by eq. 2.2,

$$\frac{[EL]}{[E_0]} = \frac{[L_0]}{K_L} \Bigg/ \left(1 + \frac{[L_0]}{K_L}\right)$$

and is a hyperbolic function of L_0 in the same way that the percentage saturation of an enzyme is determined by the substrate concentration. This is illustrated in Fig. 2.10a for the binding of three enzymes to N^6-(6-aminohexyl)-AMP–Sepharose in which the immobilised ligand concentration is varied from 0–750 nmol/ml Sepharose. Figure 2.10b shows theoretical curves for several K_L values based on eq. 2.2.

It is doubtful, however, that the simple model described above has any real operational significance and serves merely to emphasise the importance of ligand concentration in affinity chromatography. Indeed, the capacity of affinity adsorbents is much smaller than that expected from the concentration of the immobilised ligand, L_0. The apparent or 'effective' concentration of ligand (L_{eff}) within the gel matrix is approximately 1% or less of the chemically determined value (Lowe et al., 1973; Harvey et al., 1974; Nishikawa et al., 1976). In addition, lack of information on the effective affinity (K_L^{eff}) often precludes any rational strategy and the highest attainable ligand concentration is generally opted for. This is obviously more important for ligand–macromolecule systems of low affinity.

These considerations relating to ligand concentration also have an important practical consequence for interacting systems of high affinity. In cases where elution of an adsorbed protein is difficult without using such drastic conditions that denaturation is experienced, reduction of the ligand concentration will permit a more facile elution under milder conditions. This may be achieved simply by diluting the derivatised gel with unmodified Sepharose.

A second problem associated with unnecessarily high ligand concentrations may generate chromatographic aberrations if the ligand is charged. Adsorbents containing covalently linked cholinergic

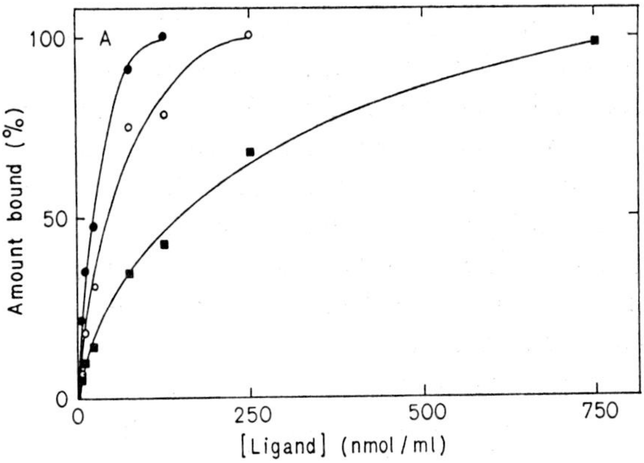

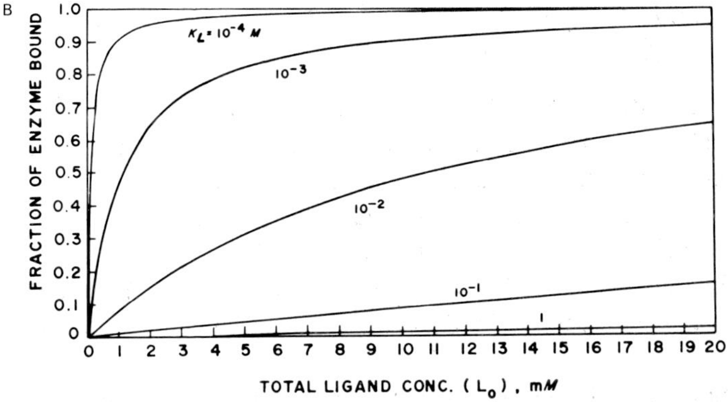

Fig. 2.10. The effect of ligand concentration on the binding of enzymes to affinity adsorbents. (a) The binding of rabbit muscle lactate dehydrogenase (●), pig heart lactate dehydrogenase (○) and glycerokinase (■) to N^6-(6-aminohexyl)-AMP-agarose as a function of ligand concentration. Enzyme (5 IU) and bovine serum albumin (1.5 mg) were applied to a 1 g column of adsorbent. Reproduced with permission from Harvey et al. (1974), Eur. J. Biochem., *41*, 335. (b) Fractional enzyme binding (EL/E_0) for low enzyme concentrations calculated from eq. 2.2 at several values of L_0 and K_L. Reproduced with permission from Graves and Wu (1974) Methods Enzymol., *34*, 145.

ligands (Kalderon et al., 1970) and ostensibly specific for acetylcholinesterase were found to lose specificity when the immobilised ligand concentration was increased from 1.5×10^{-4} M to 1.6×10^{-3} M. A ligand concentration of approximately 10^{-3} M was found to give optimal purification of the enzyme. This behaviour was attributed to the fact that at ligand concentrations below 10^{-3} M, the ligands may be spaced sufficiently distant to prevent non-specific proteins from interacting with more than one charged ligand at once. At ligand concentrations above 10^{-3} M, the charged groups are close enough to permit non-specific binding of proteins. Thus, only at ligand concentrations below 10^{-3} M, is the complementary macromolecule capable of interacting specifically with the immobilised ligand. Furthermore, the low capacity of the adsorbents can be compensated by scaling up the chromatographic procedures.

2.5. *Other considerations relevant to the design of affinity adsorbents*

As intimated in § 2.4.4, ligands which are ionic can generate non-specific ion-exchange problems in the adsorbent. Ideally therefore, ionic ligands should be avoided. Where ionic ligands must be utilised, however, experience suggests that the inclusion of salt (at least 0.15 M) in the column irrigants be employed to optimise the specificity of binding (§ 4.4). Higher salt concentrations often weaken the affinity between the immobilised ionic ligand and the complementary enzyme.

Similar problems are often experienced with hydrophobic ligands. These ligands, expecially aromatic ones, can yield adsorbents which exhibit non-specific binding of proteins. For example, a prospective chymotrypsin-specific adsorbent comprising 4-phenylbutylamine (PBA) attached to agarose (Fig. 2.11) not only binds chymotrypsin strongly (Stevenson and Landman, 1971) but also a number of acidic proteins such as ovalbumin and β-lactoglobulin (Hofstee, 1973). Similar non-specific adsorption phenomena have been observed with other aromatic adsorbents. In most cases, however, the ad-

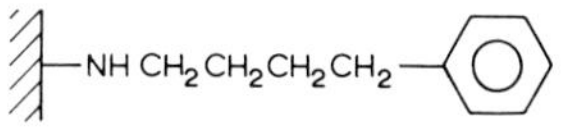

Fig. 2.11. The structure of the 4-phenylbutylamine adsorbent.

sorption of a complementary enzyme to an affinity adsorbent involves both ionic and hydrophobic interactions. Under such conditions these interactions can mutually reinforce each other to yield a significant degree of non-specific binding. Thus, strongly ionic or hydrophobic ligands should be avoided if possible. Section 4.4 reviews the factors generating non-specific adsorption, how to recognise the phenomenon and how to eliminate or minimise it.

2.6. *Examples of the rationale involved in the preparation of affinity adsorbents*

A few examples of the logic underlying the selection of the ligand for the preparation of selective adsorbents will be given to illustrate the principles discussed above.

(a) *Penicillinase.* Penicillinase is secreted into the medium by cultures of *Bacillus licheniformis*. Crane et al. (1973) have devised a rapid and mild procedure for isolating microgram amounts of the enzyme from the Bacillus employing affinity chromatography on agarose-immobilised cephalosporin C. The antibiotic and

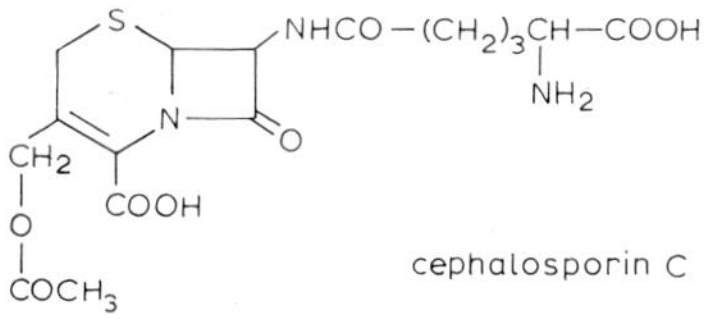

substrate analogue, cephalosporin C, is a particularly competent ligand for affinity chromatography for several reasons. It has a high affinity for the penicillinase ($K_m \sim 1$–2×10^{-5} M) coupled with a relatively slow rate of enzymic hydrolysis. Furthermore, unlike penicillin analogues, cephalosporin C is stable to non-enzymatic hydrolysis and the aminoadipoyl side chain provides a non-essential and distal amino function for covalent attachment to the matrix. The ligand was coupled directly to CNBr-activated Sepharose 4B (§ 3.1.1.1) to yield an adsorbent containing 0.6–0.8 μmol cephalosporin C/g Sepharose 4B. This ligand concentration was adequate to adsorb the exo-penicillinase.

(b) *Staphylococcal nuclease*. The extracellular nuclease of *Staphylococcus aureus* hydrolyses DNA and RNA and is competitively inhibited by 3′-phosphoryl-deoxythymidine 5′-phosphate (pdTp). It is the presence of a free 5′-phosphoryl group that often endows synthetic derivatives with strong inhibitory properties. Thus 3′-(4-aminophenylphosphoryl)-deoxythymidine 5′-phosphate (Cuatrecasas et al., 1968) is an ideal ligand for affinity chromatography since it has a high affinity for the nuclease ($K_i \sim 10^{-6}$ M), is stable in the pH

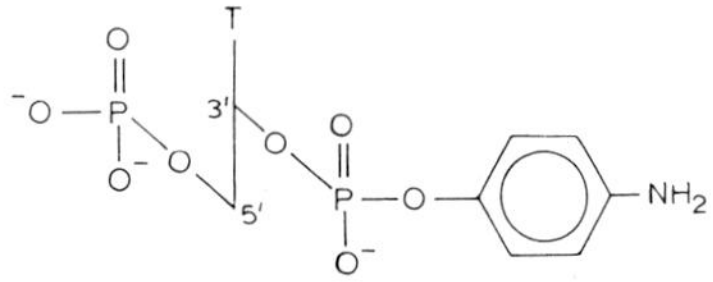

range 5–10 and has an amino group removed from the basic unit (pdTp) recognised by the enzymatic binding site. The inhibitor was coupled directly to CNBr-activated Sepharose (§ 3.1.1.1) and an adsorbent containing 1–1.5 μmol inhibitor/g Sepharose proved effective in binding the nuclease.

(c) *Thymidylate Synthetase*. Thymidylate synthetase catalyses the reductive methylation of deoxyuridine 5′-monophosphate (dUMP) to thymidylate with the concurrent conversion of N^5,N^{10}-methylenetetrahydrofolate to dihydrofolate. The tetrahydrofolate analogue, tetrahydromethotrexate, bound to ω-aminoalkyl–Sepharose via its carboxyl groups did not adsorb the enzyme from *E. coli* (Slavik et al., 1976).

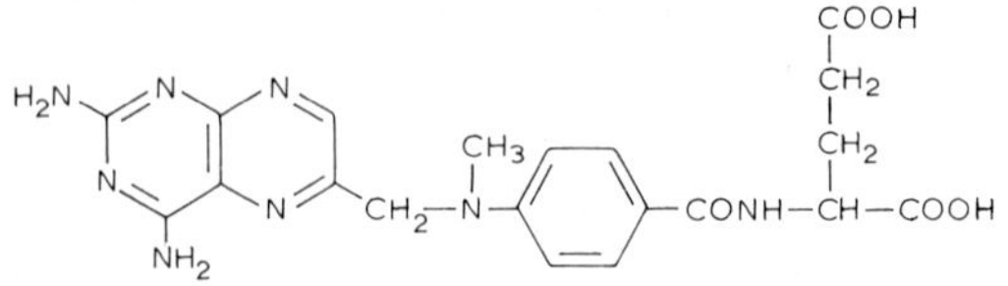

However, the binary complex of thymidylate synthetase with dUMP binds tetrahydrofolate and its analogues with greater affinity than just the enzyme alone. This fact may be exploited in the purification of the enzyme by affinity chromatography. Thus, the presence of 2×10^{-5} M dUMP in the crude enzyme preparation was sufficient to generate the binary complex and enhance the affinity of the enzyme towards the immobilised tetrahydromethotrexate. The enzyme was quantitatively adsorbed to the matrix and subsequently desorbed on removal of the dUMP from the elution buffer. This example illustrates the value in knowing the mechanism of the enzyme for the design of efficient adsorbents for affinity chromatography.

The chemical technology of affinity chromatography

The preceeding chapter has described features important in the design of biospecific adsorbents for affinity chromatography. It should permit the preliminary selection of a suitable ligand, matrix and spacer arm. This chapter describes some chemical procedures commonly employed in the covalent attachment of the prospective ligand to suitable matrices.

Two distinct approaches are distinguishable for the synthetic chemistry involved in the preparation of affinity adsorbents. The complete ligand–spacer arm conjugate may be preassembled by conventional organic chemistry prior to covalent attachment to the matrix. Whilst this approach almost invariably involves tedious chemical fractionation, purification and characterisation steps, it is to be recommended since the precise chemical nature of the immobilised ligand is known. Such confidence is essential for the design and interpretation of experiments in affinity chromatography. Furthermore, such investment of considerable effort in the careful preparation of biospecific adsorbents may well be rewarded by significantly enhanced resolution from inert proteins on subsequent chromatography.

On the other hand, some workers claim that attachment of a prospective ligand to a pre-coupled spacer arm–matrix conjugate has a number of advantages. For example, all excess reactants and by-products may be removed following every coupling step simply by washing the derivatised gel on a sintered glass funnel. Furthermore, the convenience of the solid-phase approach permits considerable flexibility in the synthesis of derivatives for a particular problem. This approach is therefore that most commonly adopted in the preparation of affinity adsorbents. Indeed, many derivatised gels comprising a matrix-spacer molecule assembly are available commercially and suitable for the direct attachment of appropriate ligands.

Nevertheless, it is important to realise that the difficulty in detecting side reactions and of reducing immobilised by-products is a serious drawback which could compromise the usefulness of adsorbents prepared by this method. Much of the non-specific adsorption of inert proteins to affinity adsorbents may be directly attributable to such drawbacks in the synthetic chemistry. However, in the interests of preparative simplicity, unwelcome side effects are often tolerated.

Irrespective of the synthetic route adopted, the conditions under which the ligand is coupled to the matrix must be sufficiently mild to be innocuous to both. This is particularly important where labile ligands such as coenzymes or proteins are attached to the matrix.

3.1. The activation and functionalisation of matrices

Much of the chemistry involved in coupling ligands to matrices has not been developed specifically with affinity chromatography in mind, but is an adaptation of conventional approaches. It is important to realise, therefore, that the chemistry employed in the activation of inert support matrices is largely dictated by the nature and stability of the matrix itself. Thus, for example, conditions applicable to the functionalisation of glass would totally destroy agarose. Furthermore, the chemistry of assembly of the biospecific adsorbent should preserve the 'ideal' characteristics of the matrix insofar as is possible. Thus, in most cases the relatively mild conditions required to maintain the 'ideality' of matrices seriously limit the number of chemical procedures that can be exploited.

3.1.1. Polysaccharide matrices

The hydrophilic polysaccharide supports have been widely used as chromatographic supports for the resolution and purification of sensitive biological compounds. It is important that the activation procedure for the polysaccharide matrix be sufficiently mild as to be innocuous to its structure.

In chemical terms the hydroxylic polysaccharide matrices are activated by introducing electrophilic groups into the gel in order to make them more reactive towards nucleophilic groups present on the ligand or spacer molecule. The following groups are of especial interest:

Imidocarbonate 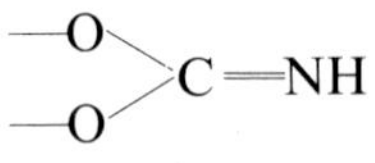

Oxirane 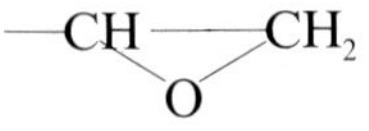

Aziridine 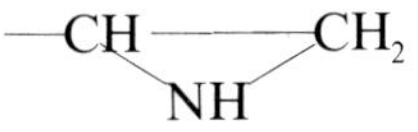

Activated double bond 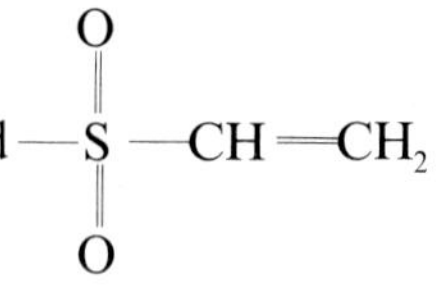

Activated halogen —CO—CH_2—Br

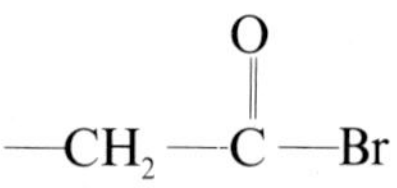

and will be described along with some other methods in the following sections. In almost all cases the derivatisation procedures follow two steps; a preliminary activation step of the matrix resulting in the introduction of reactive groups, and the subsequent coupling step for the ligand.

3.1.1.1. Cyanogen halides

By far the most general and widely employed technique for attaching ligands to agarose, dextran or cellulose involves activation of the

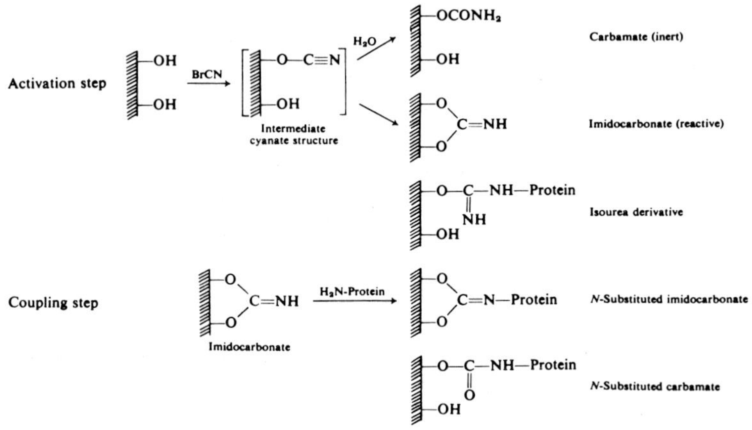

Fig. 3.1. Chemical activation of polysaccherides by means of cyanogen halides and coupling of protein ligands to cyanogen halide-activated polysaccharides. Reproduced with permission from Axen and Ernback (1971), Eur. J. Biochem., *18*, 351.

carrier with cyanogen halides and subsequent coupling of primary aliphatic or aromatic amines to the activated matrix. The activation of polysaccharides with cyanogen bromide (CNBr) at alkaline pH was introduced to affinity chromatography by Axén et al. (1967). The precise chemistry of the activation process and the coupling reaction has only been partially elucidated. Figure 3.1 shows the probable course of events; the initial 'activation' step involves the formation of a labile cyanate intermediate which can interact with an adjacent hydroxyl to form both cyclic and acyclic imidocarbonates. A hypothetical structure for CNBr-activated Sepharose is shown in Fig. 3.2. The presence of the acyclic imidocarbonates, generates by reaction of the cyanate intermediate with an adjacent polysaccharide chain, and thus producing interchain cross-links, accounts for the decrease in swelling capacity of the gel and its enhanced thermal stability which accompanies activation (Axén et al., 1967). The intermediate cyanate may also be hydrolysed to an inactive carbamate; for example, prolonged alkaline hydrolysis under mild conditions converts the activated gel into the completely unreactive

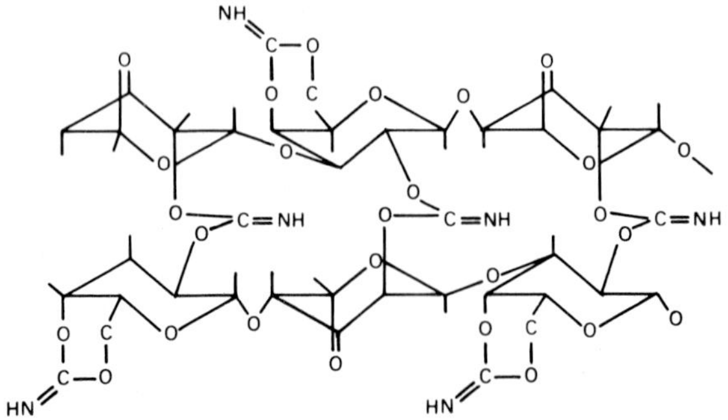

Fig. 3.2. Hypothetical structure of CNBr-activated agarose illustrating both cyclic and acyclic imidocarbonates. Reproduced with permission from *Affinity Chromatography: Principles and Methods* by Pharmacia Fine Chemicals.

carbamylated polymer. It is thus essential to conduct the activation step in a relatively short time period, typically 8–12 min.

Cyanogen bromide (CNBr) is available commercially as a white crystalline solid which is sufficiently volatile at room temperature to generate highly poisonous and irritant vapours. Consequently all manipulations should be performed in a well ventilated hood. Furthermore, some aged batches of CNBr have a distinct yellow-orange colour and are liable to further explosive decomposition. Such samples should be destroyed by complexing with alkaline ferrous sulphate to produce the harmless and easily disposable ferrocyanides. Alternatively, pure cyanogen bromide may be recovered by sublimation.

3.1.1.1.1. The activation step. A number of procedures are available for the activation of polysaccharides by cyanogen bromide with the choice dictated largely by the individual preference of the reader. Two such approaches are described in detail below. The polysaccharide matrix must be thoroughly washed prior to activation or derivatisation to remove bacteriostatic agents and other preservatives present in commercial preparations. This operation is conveniently performed on a large sintered funnel. The gel is measured either by weighing a quantity of gel sucked moist on the funnel or by suspending the gel in a measuring cylinder in distilled water and reading the settled volume. The wet weight of the gel is approximately equivalent to its settled volume. The precise conditions used for estimating the gel should be given in order to assist reproducibility.

Pure cyanogen bromide is sparingly soluble in water (20 mg/ml) and as such reacts relatively slowly. The problem may be circumvented either by grinding the crystals to a powdered state or by solubilising the CNBr in organic solvents (March et al., 1974). The latter technique not only facilitates the handling of CNBr but also reduces its precipitation during the activation process. Cyanogen bromide is readily soluble in dimethylsulphoxide, acetonitrile or 10% aqueous *N*-methylpyrrolidone. Where possible, acetone should be avoided since bromoform can be generated in this solvent under some circumstances. Grinding old discoloured cystals of CNBr should also be avoided in view of the potential explosion risk.

The most universally employed procedure for the activation of polysaccharides, especially agarose, by CNBr is probably the so-called titration–activation method. In this procedure, say 20 g exhaustively washed moist agarose cake is added to 20 ml H_2O in a 100 ml beaker equipped with a 0–110°C thermometer, a pH meter and a 25 mm magnetic stirring bar. The suspension is stirred slowly, the temperature lowered to about 10–15°C by the addition of crushed ice and the pH adjusted to 10.8 ± 0.1 by the addition of 1–2 drops 4 N NaOH. The activation procedure is initiated by the addition of the CNBr and the pH of the reaction maintained at 10.8 ± 0.1 by manual titration with the 4 N NaOH. The CNBr (100 mg/g moist weight gel) may be added as a crystalline solid, a crushed solid, an aqueous solution or by adding an aliquot of a stock solution. The latter may be prepared by dissolving CNBr in acetonitrile (1 g/ml) and storing in a tightly stoppered vial at −20°C. The relatively low initial temperature allows the experimenter time to become accustomed to the manual addition of alkali without 'overshooting' the maximal pH of 11.0. The temperature is subsequently allowed to rise to 18–20°C. Alternatively, an autotitrator coupled to a magnetic valve (radiometer TT3 and MM1) may be used. The autotitrator is preset to titrate upscale with a null point at 10.9 such that when the pH in the reaction vessel falls below 10.9, the magnetic valve is opened and 4 N NaOH from a reservoir is injected into the suspension. When the pH rises above 10.9, the magnetic valve is closed. In the author's experience, however, the manual titration method is facile and with careful control of the pH and temperature, the activation is virtually complete within 10–12 min. The end point is apparent when the addition of alkali is no longer necessary and when this is so the gel is washed rapidly according to the procedure given below (§ 3.1.1.1.2).

Despite the relative simplicity of the titration method a faster and technically simplified version is available (March et al., 1974). The activation procedure is preformed in concentrated carbonate buffer. The required amount of washed gel is suspended in an equal volume of 2 M $NaHCO_3$–Na_2CO_3 buffer (pH 10.9) in a beaker equipped with a thermometer and magnetic stirring bar. The slurry is cooled to approximately 4–5 °C in an ice bath and the reaction initiated by the addition of CNBr (100 mg/g gel dissolved in acetonitrile). After 10 min at 4–5 °C, the activated gel is transferred to a sintered funnel and washed. It is important to maintain the low temperature during the activation since at room temperature the subsequent coupling of the ligand may be reduced by as much as 35% (March et al., 1974).

The concentration of CNBr recommended in the procedures above is satisfactory for moderate levels of ligand substitution. When lower or higher levels of activation are required, 50 mg and 200–300 mg CNBr/g moist weight gel respectively may be employed together with 2 M and 8 M NaOH for the titration.

The amount of a small ligand, glycine, bound to activated agarose was linearly proportional to the concentration of CNBr employed for activation, up to 80 mg CNBr/ml agarose, as shown in Fig. 3.3 similar effects are apparent when the ligand is a protein.

3.1.1.1.2. Washing the activated gel. It is generally recognised that the CNBr-activated intermediate (Fig. 3.2) displays limited stability

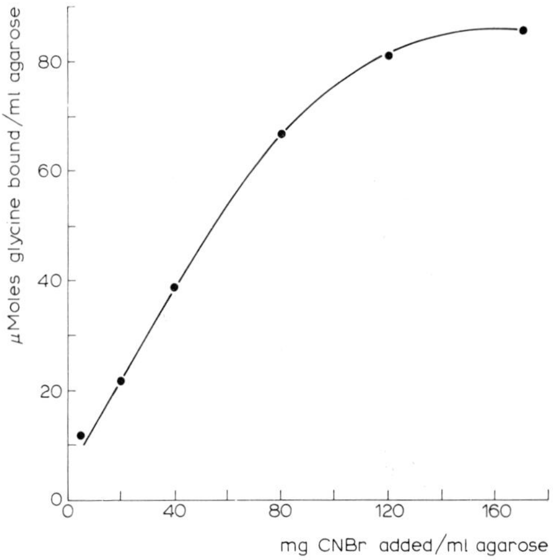

Fig. 3.3. The relationship between the concentration of CNBr used in activating agarose to the subsequent binding of glycine. Aliquots (3 ml) of agarose were activated for 6 min with the indicated amount of CNBr/ml agarose. After washing, 111 μmol of [2-^{14}C]glycine were added/ml agarose, the gels tumbled for 16 h at 4 °C and then thoroughly washed. The bound glycine was determined by hydrolysing gel samples in 1 M HCl at 100 °C for 20–30 min and counting. Reproduced with permission from Stage and Mannik (1974), Biochim. Biophys. Acta, *343*, 382.

and it is therefore essential that the washing procedure be performed as rapidly as possible prior to transferring the gel to the coupling-reaction medium. At the end of the activation step the gel is rapidly cooled by the addition of crushed ice and poured into a large sintered glass funnel which has been pre-cooled with crushed ice. The suspension is rapidly filtered into a Büchner flask (2 litre) containing solid ferrous sulphate to remove unreacted CNBr and cyanides as harmless ferrocyanide. The gel is subsequently washed under suction with 1 litre ice-cold distilled H_2O and 1 litre of the buffer to be used in the coupling stage, typically ice-cold 0.1 M $NaHCO_3$–Na_2CO_3 buffer (pH 8.5–9.5).

3.1.1.1.3. CNBr-activated Sepharose 4B. CNBr-activated Sepharose 4B is available commercially from Pharmacia and obviates the hazardous manipulation of CNBr. The activated gel is freeze dried in the presence of dextran and lactose to preserve the beaded form and supplied in 15 g air-tight packs. The product can be stored unopened in a refrigerator below 8 °C for at least 18 months but once the package is opened it should be stored dry below 8 °C.

The required amount of freeze-dried powder is swollen in 1 mM HCl on a glass filter and washed with at least 200 ml of the same solution per gram of powder. As a rough guideline, 1 g freeze-dried material is equivalent to 3.5 ml final gel volume. The ligand-binding capacity of the gel is conserved more effectively by washing with solutions of low pH than with solutions of pH > 7. The gel is then ready to couple ligand as soon as the washing is completed.

Pharmacia also market CNBr-activated Sepharose 6MB for use in cell biology and immunology for the separation of 'functionally homogeneous' cell populations. It is produced by activation of Sepharose 6MB macrobeads (diameter 200–300 μm) with cyanogen bromide and is handled in a manner analogous to CNBr-activated Sepharose 4B.

3.1.1.1.4. The coupling step. Unprotonated primary amines couple to CNBr-activated agarose and dextran, often with considerable efficiency, to yield products of the type shown in Fig. 3.1. Whilst the exact nature of the coupled product remains unclear, there is some evidence for the formation of N-substituted isourea derivatives (Svensson, 1973).

The coupling capacity of CNBr-activated polysaccharides may be evaluated by means of a model compound such as alanine (Cuatrecasas, 1970) or glycine (Stage and Mannik, 1974). Ligands such as alanine which carry an α-amino group, exhibit an optimum coupling pH of 9.5–10.0 (Fig. 3.4). This optimum in the coupling efficiency is a reflection of two competing phenomena, the charge on the α-amino group and the stability of the activated complex. The ascending limb of Fig. 3.4 is determined by the reactive unprotonated

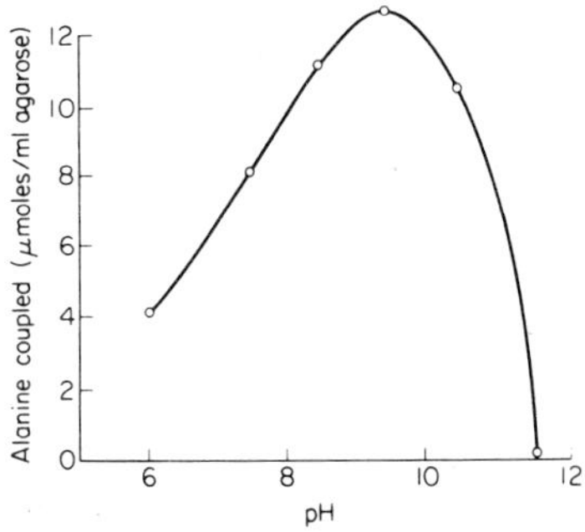

Fig. 3.4. The effect of pH on the coupling of [^{14}C]alanine to CNBr-activated agarose. Packed Sepharose 4B (60 ml) was mixed with 60 ml water and treated with 15 g CNBr. To the cold, washed, activated agarose were added 2.2 mmol [^{14}C]alanine (0.1 μCi per μmol) in 60 ml cold distilled water, and 20 ml samples of the mixed suspension were added rapidly to beakers containing 5 ml of cold 0.5 M buffer. The final concentration of alanine was 15 mM. After 24 h the suspensions were thoroughly washed, aliquots were hydrolysed by heating at 110 °C in 6 N HCl for 24 h and the amount of [^{14}C]alanine released was determined. Buffers were: 0.1 M sodium citrate (pH 6.0), phosphate (pH 7.5), borate (pH 8.5 and 9.5) and carbonate (pH 10.5 and 11.5). Reproduced with permission from Cuatrecasas (1970), J. Biol. Chem., *245*, 3059.

form of the ligand and the coupling pH should be chosen to be above the pK_a of the ligand but less than 10. Thus the coupling pH should be 7–8 for aromatic amines (pK_a ~ 5), 9.5–10 for amino acids (pK_a α-NH_2 ~ 8) and 10 for aliphatic amines (pK_a ~ 9–10). Tris and other amino-containing buffers should be avoided during the coupling procedure, since these amino groups compete with the amino functions on the ligand for the activated groups. Borate or bicarbonate buffers are commonly used. The decreased coupling efficiency at pH values above 9.5–10.0 (Fig. 3.4) reflects the sharp decline in stability of the activated complex. CNBr-activated Sepharose is also unstable at elevated temperatures and the coupling reaction is generally performed at +4 °C. Under such conditions the coupling capacity for glycine declines exponentially with a half-life of approximately 22 h (Stage

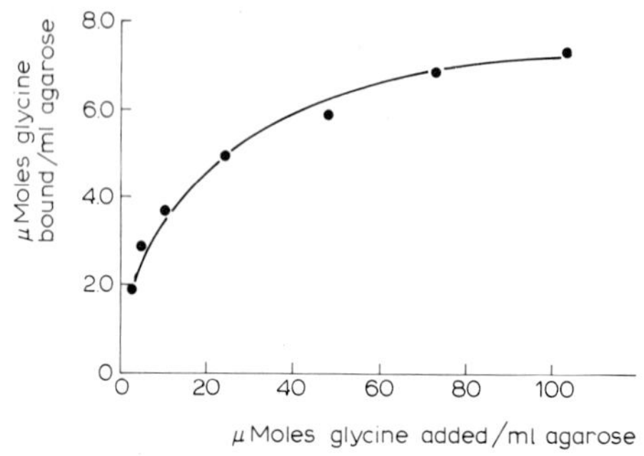

Fig. 3.5. The effect of glycine concentration on the binding of glycine to CNBr-activated agarose. Agarose was activated with 40 mg CNBr/ml agarose at pH 10.9 ± 0.1 and 23 °C for 6 min. Following washing, the indicated amount of [2-^{14}C]glycine was added/ml agarose and the gels tumbled at 4 °C for 16 h and then thoroughly washed. The bound glycine was determined by hydrolysing in 1 M HCl at 100 °C for 20–30 min and counting. Reproduced with permission from Stage and Mannik (1974), Biochim. Biophys. Acta, *343*, 382.

and Mannik, 1974). However, if as is usual, the activated gel is brought into contact with the small ligand immediately following activation, the coupling reaction is usually complete within 2–3 h. It is advisable, however, to allow the reaction to stand overnight at +4 °C to ensure complete loss of reactive groups on the matrix. Even under such conditions, however, residual activated groups sometimes remain, particularly with bulky ligands. The residual active groups may be neutralised by reaction with 1-amino-propan-2,3-diol, D-glucosamine or ethanolamine.

The amount of ligand bound per gram of activated gel is a function of the amount added in the coupling step. The effect of glycine concentration on the binding of glycine to activated agarose is shown in Fig. 3.5. Similar curves were observed with immunoglobulin G (Stage and Mannik, 1974). Cyanogen halide-activated polysaccharides have a remarkable capacity to covalently fix proteins under slightly alkaline conditions. For example, chymotrypsin and papain may be attached to Sephadex G-200, cellulose or agarose to yield conjugates containing as much as 30% protein (Axén and Ernback, 1971). However, exposure of enzymes to high-density activated polymers under pH conditions where most of the reactive amino groups of the protein are unprotonated can lead to multi-point attachment and consequently a reduced activity for the fixed enzyme. This problem can be circumvented by coupling at less favourable pH values, such that fewer residues are unprotonated.

3.1.1.1.5. The uniformity of distribution of coupled ligand to beaded supports. It is relevant to know whether large bulky ligands such as proteins are uniformly distributed within activated agarose beads or whether they are attached more or less exclusively to the surface. Originally it was suggested that all parts of an activated Sepharose 4B matrix were available for binding macromolecules. For example, the use of immunofluorescent techniques has shown that leucine aminopeptidase (mol. wt. 326,000) binds throughout the gel (Lasch et al., 1972) as does IgG (mol. wt. 150,000) and IgM (mol. wt. 900,000) (Stage and Mannik, 1974). More recently, however, it has been suggested that the distribution of the immobilised ligand in agarose beads may be related to the level of CNBr used in the activation step (Lasch et al., 1975).

3.1.1.1.6. The stability of CNBr-coupled ligands – the problem of leakage. As a general rule, affinity adsorbents may be reused repeatedly without appreciable loss of binding potency. Nonetheless, recent applications of affinity chromatography to very high affinity systems have revealed a significant and continuous 'leakage' of the ligand from CNBr-activated supports (Tesser et al., 1974). Such solubilisation of the coupled ligand will not seriously interfere with the interaction between the complementary protein and the immobilised ligand, provided that the affinity between the reactants is not high. If the affinity of the system under study is very high, i.e., $K_d \sim 1$ nM as is frequently found in hormone–receptor systems, and if the solvolysed free ligand is more effective in complex formation than the immobilised ligand, then the complementary protein may bind the free ligand in an essentially irreversible manner. In such cases, the protein may not be adsorbed to the gel. Furthermore, the presence of the soluble ligand–protein complex in the column eluants may pass unnoticed since it is 'inactivated' by the solubilised ligand. This may lead to the conclusion that irreversible adsorption to the matrix had occured when in fact no adsorption was experienced in the first place. These problems are particularly acute in the purification of steroid binding and receptor proteins (Sica et al., 1973).

The rate of leakage of [^{14}C]alanine, or ^{125}I-labelled albumin from substituted agarose deviates from first order kinetics over a 30-day

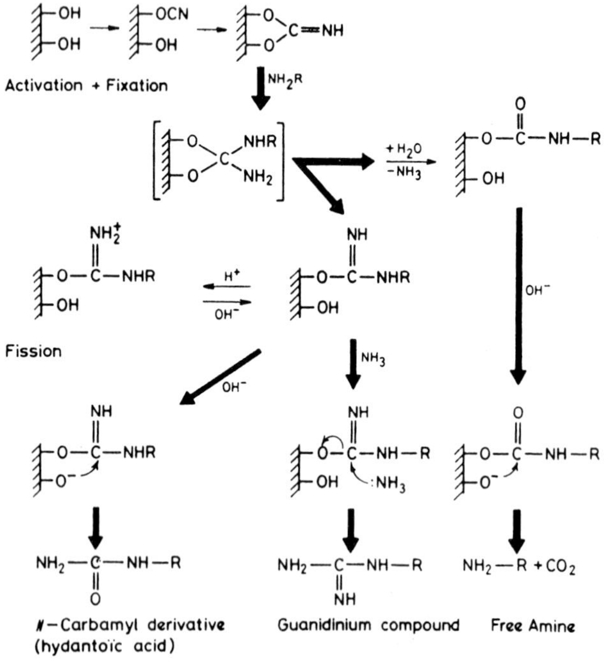

Fig. 3.6. Solvolysis of ligands from CNBr-activated polysaccharide gels: a tentative scheme based on the coupling mechanism proposed by Axen and Ernback (1971). Reproduced by permission from a lecture by G. I. Tesser at the 8th FEBS Conference, Amsterdam, 1972.

period. Simple monoamines are solvolysed at an initial rate of about 0.1% per day but decreasing by about 6-fold after 30 days. Protein ligands are released about 5 times slower. The rate of leakage is enhanced at pH values greater than 5 and by elevated temperatures; the products vary with the conditions. Thus, in Tris buffer (pH 8) [^{14}C]alanyl–agarose liberates a mixture of *N*-alanyl guanidine and alanine (1 : 4); in Tris buffer (pH 9) the ratio was 2 : 1. At pH 11 in aqueous triethylamine, only *N*-carbamyl alanine and a small amount of 4-methyl-hydantoin could be detected. A tentative scheme rationalising these observations is shown in Fig. 3.6. Attempts to stabilise the susceptible bonds by reduction with $NaBH_4$ have so far proven unsuccessful.

The leakage problem is currently controlled by exhaustive washing and manipulation of such variables as the degree of ligand substitution, flow rates, temperature, pH, etc. Other approaches include the use of soluble polyfunctional spacer molecules (§ 3.2.3) or the use of alternative methods of matrix activation.

3.1.1.1.7. Other effects of CNBr-activation. CNBr-activated polysaccharide gels exhibit enhanced stability as a result of cross-linking. Thus, CNBr-treated agarose, in contrast to underivatised agarose, can withstand 1 N HCl and 1 N NaOH and also demonstrates a decreased swelling capacity. Activation of agarose with CNBr leads to incorporation of up to 6% nitrogen and a trace of bromine. The low but significant bromine content of the activated gel probably reflects the presence of a low concentration of tribromotriazine, the product of trimerisation of CNBr in slightly alkaline solution, in the CNBr. The triazine will participate in coupling reactions under the same conditions as for CNBr (§ 3.1.1.2). The incorporation of these two elements should be borne in mind if elemental analysis is employed as a method of assessing the immobilised ligand concentration (§ 3.6.5).

Perhaps the most serious potential drawback in the use of CNBr for the activation of polysaccharide matrices lies in the introduction of ion exchange groups. The N-substituted isourea derivatives formed

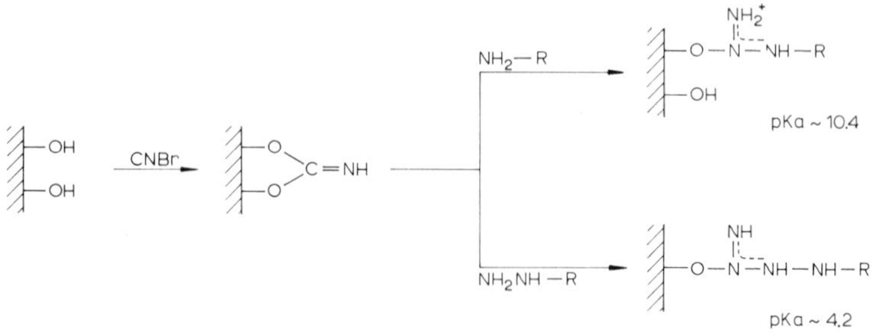

Fig. 3.7. The nature of the linkage between CNBr-activated agarose and hydrazide and amino-containing ligands.

on coupling primary amines to CNBr-activated agarose (Fig. 3.1) retain the basicity of the amino group and exhibit a $pK_a \sim 10.4$. The problem may be circumvented by coupling bishydrazides rather than bisamines to the CNBr-activated agarose. Hydrazide derivatives of Sepharose are easy to prepare and exhibit some of the properties of polyacrylamide gels in that they lack charged groups. Figure 3.7 shows the differences between hydrazide and amino derivatives of Sepharose with particular reference to the pK_a values of the terminal amino groups, 4.2 and 10.4, respectively. Details of the preparation of these derivatives are given in § 3.2.

3.1.1.2. Triazines

Polyhydroxylic supports may be activated with dichloro-*sym*-triazines or with trichlorotriazine (cyanuric chloride) in a manner reminiscent of the CNBr-activation procedure. Soluble triazines, prepared by substituting with carboxymethoxy or carboxymethylamino functions, react rapidly with suitable matrices at pH 9–11 and 20 °C to yield a monochloro-*sym*-triazinyl–polysaccharide complex (Fig. 3.8). The triazinyl-activated matrix subsequently reacts slowly (16–20 h) at pH 7–9 and 0–20 °C with strong nucleophiles such as primary amino groups (Kay and Crook, 1967). Similarly, cyanuric chloride reacts with hydroxylic matrices to form a dichloro-*sym*-triazinyl-derivative, eminently suitable for coupling proteins at pH values around 7. In

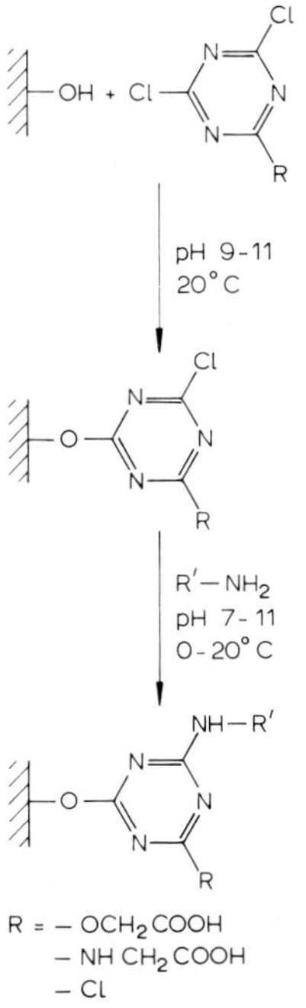

Fig. 3.8. The activation of polyhydroxylic support matrices with dichloro-*sym*-triazines.

a typical coupling procedure, 1 g of chloromethyoxy-*sym*-triazinyl-support was allowed to couple overnight at 5 °C to several millilitres of protein solution in 0.1 M phosphate buffer (pH 8).

3.1.1.3. Periodate oxidation

The activation of polysaccharide matrices by the oxidation of the vicinal diol groups with sodium metaperiodate ($NaIO_4$) to generate aldehydic functions (Sanderson and Wilson, 1971) represents a rapid, simple and safe alternative to the use of cyanogen halides. Sodium metaperiodate is soluble in aqueous media and may be readily removed by washing after reaction. The 'activated' gel can be stored at +4 °C for at least a month without loss of coupling potential. The

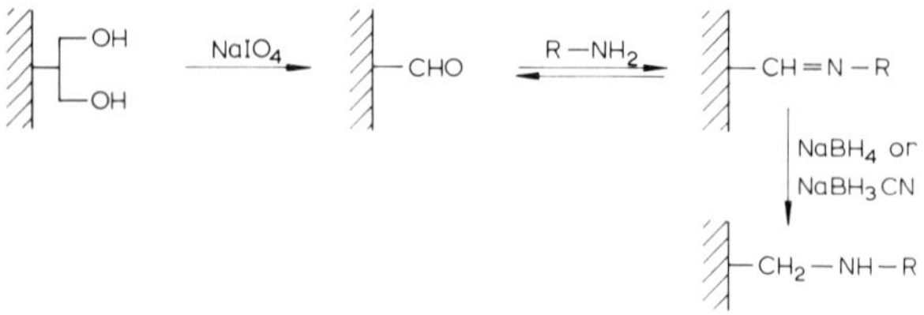

Fig. 3.9. The reductive amination of periodate-oxidised agarose.

aldehydic functions react with primary amines at pH 4–6 to form Schiff bases and subsequent stabilisation may be achieved by reduction with sodium borohydride ($NaBH_4$) or preferably sodium cyanoborohydride ($NaBH_3CN$). At pH 6–6.5 the latter reagent preferentially reduces the Schiff bases without reducing the aldehyde functions (Borch et al., 1971) and thereby drives the overall reaction shown in Fig. 3.9 to completion. In practice, the gel to be activated is mixed with an equal volume of 0.2 M $NaIO_4$, placed in a tightly closed polyethylene bottle and gently shaken on a mechnical shaker for 2 h at room temperature. The periodate-oxidised matrix is filtered, thoroughly washed on a coarse sintered funnel with distilled water and subsequently allowed to react with the amino-containing ligand. For example, ω-aminoalkyl–agarose may be prepared by adding oxidised-agarose (100 ml) to an equal volume of 2 M aqueous diaminodipropylamine at pH 5.0 and gently shaking for 6–10 h at 20 °C. The pH is subsequently raised to 9.0 by the addition of solid Na_2CO_3, whence 10 ml of freshly prepared aqueous 5 M $NaBH_4$ is added in small portions at 4 °C with gentle stirring to avoid foaming over a period of 12 h. The reduced gel is exhaustively washed with 1 M NaCl until the filtrates contain no free diamine. Substitutions up to 2–3 μmol diamine/ml agarose may be achieved by this procedure.

The Schiff base resulting from the reaction of oxidised agarose with an amino ligand may also be reductively aminated with sodium cyanoborohydride ($NaBH_3CN$). In some cases the use of the latter is to be preferred; for example, when the supply of the amino-containing ligand is in short supply. The use of $NaBH_3CN$ favours complete coupling of the amino ligand. Furthermore, the latter reagent is preferable when the ligand to be coupled is labile at the pH values (pH 9–10) required for reduction with $NaBH_4$. Equal volumes of periodate-oxidised agarose, which was previously washed with 0.5 M phosphate buffer (pH 6), and 0.5 M phosphate buffer (pH 6) containing 1–50 mM amino ligand and 0.5 mM $NaBH_3CN$ are gently shaken for 3 days at room temperature. The gel is exhaustively washed and unreacted aldehyde groups subsequently reduced with $NaBH_4$ using 1 ml 1 M $NaBH_4$/ml agarose gel for 15 h at 4°C. The gel is extensively washed again.

Solid supports such as the cross-linked dextrans, agarose, cellulose and Ultrogel® are well suited for activation and coupling by the periodate oxidation technique. Indeed, the stability of cellulose under the coupling conditions permits a very high degree of substitution. In contrast, the more porous Sephadex gels (G-100, G-200) tend to shatter after periodate treatment and are therefore not recommended.

3.1.1.4. Oxirane coupling

Highly strained three-membered electrophilic ring systems such as are found in bisoxiranes (bisepoxides) or in the halohydrins are particularly useful reagents for introducing low molecular weight ligands containing amino or hydroxyl functions (Porath and Sundberg, 1970). As in the case of CNBr activation, the reaction takes place in two steps: an activation step in which one of the oxirane groups of the reagent (bisglycidyl ether of 1,4-dihydroxy-*n*-butane) reacts with the hydroxyl on the polymer matrix and a coupling step whereby the remaining functional group reacts with the ligand to be coupled. The intermediate oxirane-containing polymer is attacked by nucleophiles

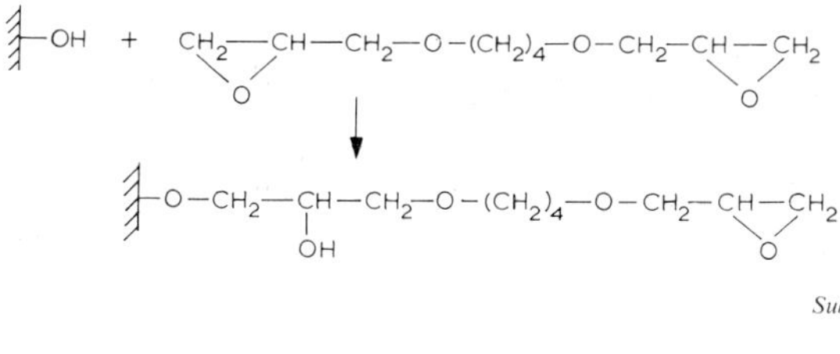

Subject index p. 519

whose reactivity follows the usual order SH > NH > OH. Aliphatic hydroxyl groups react at strongly alkaline pH (~ 11) whilst primary amines react at lower pH values and thiol groups even lower. The coupling of a primary amine, R-NH_2, generates an alkylamine:

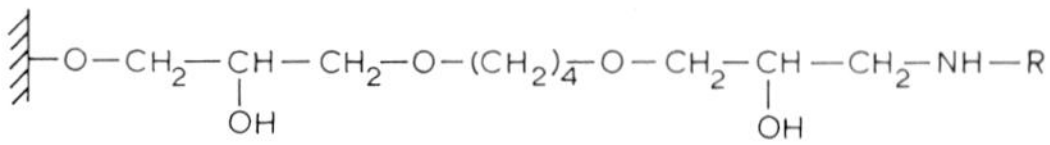

whilst hydroxyl and thiol groups generate ether and thioether linkages respectively.

Optimal conditions for the oxirane activation of agarose have been reported (Sundberg and Porath, 1974). Accordingly, the preparation of a highly activated gel may be acomplished as follows. About 7 ml washed Sepharose 6B are suspended in 5 ml 1 M NaOH containing $NaBH_4$ (2 mg/ml) and 1 ml 1,4-butanediol-diglycidyl ether (1,4-bis-(2,3-epoxy propoxy)-butane). The reaction is allowed to proceed at 25 °C for 5 h with gentle stirring whence the activated gel is thoroughly washed with distilled water. The gel may be stored at 4 °C in neutral solution for at least 1 week without significant loss of oxirane groups. Furthermore, no decrease in coupling potential was observed after storage of the activated gel in dimethyl-sulphoxide or acetone for 1 month.

The synthesis of gels with differing degrees of oxirane substitution may be achieved by varying the reaction time or bisoxirane concentration. The procedure outlined above generates about 200 μmol oxirane/g dry gel.

Optimal conditions for coupling low molecular weight amino-containing ligands have also been established (Sundberg and Porath, 1974). The coupling efficiency increases continuously in the pH range 8–12 and is considerably enhanced by raising the temperature from 20–40 °C. The relatively slow rate of reaction below room temperature may be compensated for by permitting the reaction to proceed for several days or even weeks.

Hydroxylic polymers may also be converted to oxirane derivatives by reaction with epichlorohydrin (Porath and Fornstedt, 1970) in hot alkaline solution. However, in extremely alkaline media (pH $\geqslant$ 13) oxirane-containing gels, derived either from bisoxirane or epichlorohydrin activation, can undergo cross-linking as a side reaction:

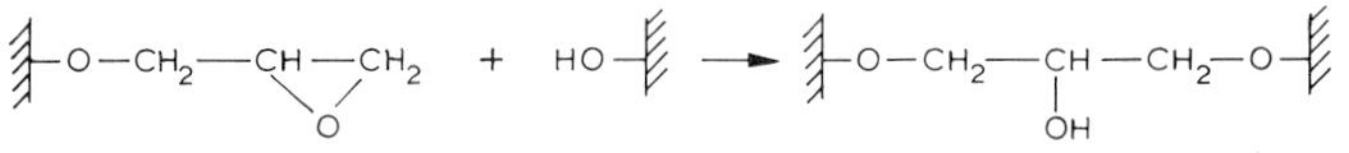

This cross-linking reaction renders the matrix insoluble in boiling water and considerably more stable in acid media.

The oxirane activation method offers some advantages over other coupling procedures including imido carbonate coupling. Thus, for example, the O–C, N–C and S–C bonds generated on coupling ligands with suitable functional groups are extremely stable. This is advantageous whenever treatment of the adsorbent under drastic alkaline conditions is necessary. Furthermore, the use of a long-chain bisoxirane reagent introduces a long (equivalent to 12 carbon atoms) hydrophilic spacer molecule between the ligand and the matrix backbone. The spacer molecule is coupled to the matrix via a stable uncharged ether linkage, in marked contrast to the nature of the imido carbonate linkage.

Epoxy-activated Sepharose 6B, formed by reacting Sepharose 6B with 1,4-bis-(2,3-epoxypropoxy)-butane, is available commercially from Pharmacia Fine Chemicals AB as a freeze-dried powder in packs of 15 g. The required amount of freeze-dried powder is swollen in distilled water (1 g freeze-dried material gives about 3,5 ml final gel volume) and extensively washed with water on a filter to remove the additives required to preserve the swelling characteristics of the gel. The ligand to be coupled is dissolved in the coupling solution and allowed to couple to the epoxy-activated gel for 16 h at 25–40 °C with gentle shaking. The coupling solution may comprise distilled water, buffers such as carbonate, phosphate or borate, sodium hydroxide solution or contain organic solvents such as dimethylformamide or dioxane in concentrations up to 50% of the final mixture. The coupling solution should not contain Tris, glycine or any other nucleophile.

After coupling, excess ligand is washed away with distilled water, 0.1 M borate buffer (pH 8.0), 0,5 M NaCl, 0.1 M acetate buffer (pH 4.0) and 0.5 M NaCl whence excess oxirane groups may be neutralised by reaction with 1 M ethanolamine overnight. Extensive washing with distilled water renders the ligand-spacer–Sepharose conjugate ready to use. Alternatively, the product may be stored at 4 °C. Full details of epoxy-activated Sepharose 6B are found in the manufacturers technical literature.

3.1.1.5. Other bifunctional reagents

Other bifunctional reagents, such as the bisaziridines, may be used for the attachment of ligands containing amino or hydroxyl functions to hydroxylic matrices. Unfortunately such reagents are ionogenic in that they form amines on coupling the ligand and thus convert the matrix into an anion exchanger.

Reaction of hydroxyl-containing polymers with divinylsulphone in alkaline solution leads to the introduction of reactive vinyl-sulphonyl groups (Porath and Sundberg, 1972).

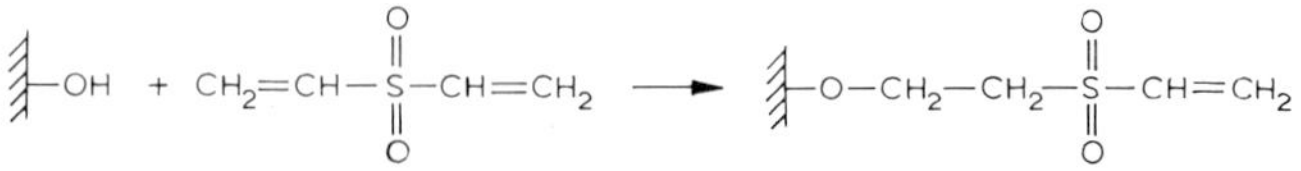

The rates of nucleophilic attack by ligand borne thiols, amines or hydroxyls follow the same order as for oxiranes except that the reactions proceed at about 1 pH unit lower. Reaction with primary amines produces an alkylamino linkage:

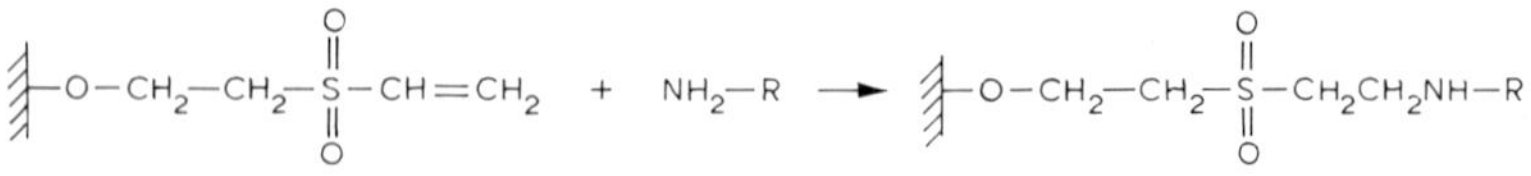

A slow release of low molecular weight ligands occurs at pH values > 8. Protein ligands are more tightly bound.

Quinones such as benzoquinone are even more reactive than divinylsulphone in attacking hydroxylic matrices. Studies with model compounds have indicated the following sequence of reactions for benzoquinone (Brandt et al., 1975).

Activation:

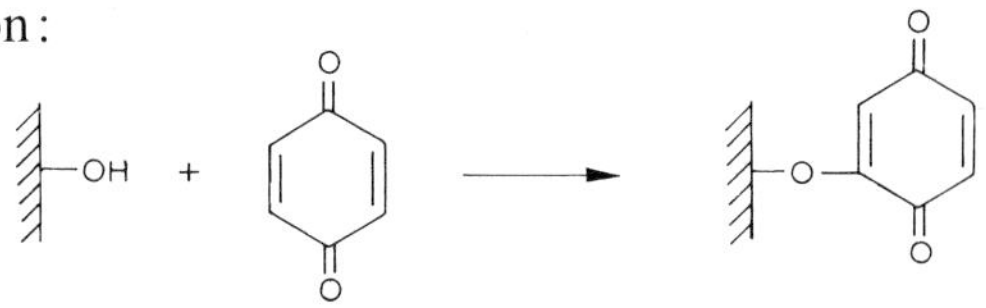

Coupling:

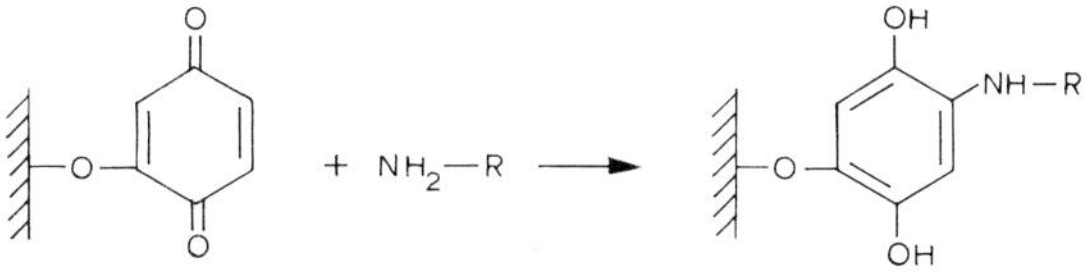

The procedure is facile, gives high coupling yields and can be used over a wide pH range. For example, serum albumin has been coupled to Sepharose 4B at pH 3–10. Typically, Sepharose 4B is suspended in 0.1 M sodium phosphate buffer (pH 8.0) containing 20% ethanol and 100 mM benzoquinone. The reaction is allowed to proceed for 1 h at 20 °C with gentle stirring whence the activated gel is thoroughly washed with 20% ethanol, water, 1 M NaCl, water and finally the solution is used for coupling. The coupling reaction is allowed to proceed for 24 h at 20 °C. De-activation of residual groups is effected by adding glycine and the gel finally washed as for the oxirane-coupled gels. The quinone coupling procedure appears promising for protein ligands.

3.1.1.6. Other methods

Highly reactive halogens may be introduced into hydroxylic matrices to facilitate ligand binding. Thus, polysaccharide hydroxyl groups can be acylated with bromoacetyl bromide followed by alkylation of a suitable amino group (Jagendorf et al., 1963). The matrix is stirred with a solution of bromoacetic acid in dioxane for 16 h at 30 °C whence bromoacetyl bromide is added and stirring continued for

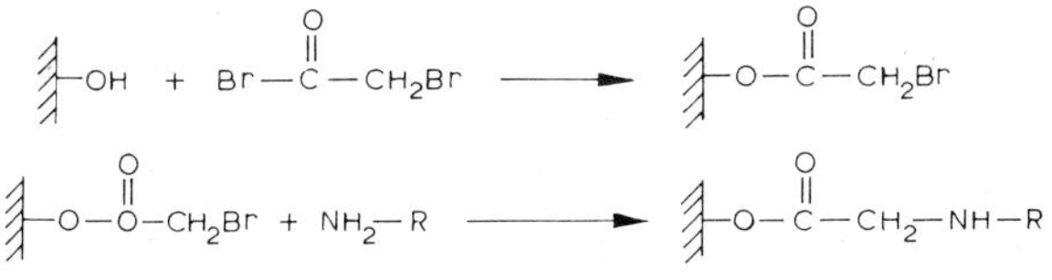

a further 7 h. Reaction with the ligand can generate high levels of ligand substitution. The lability of the ester linkage even at neutral pH can, however, be a disadvantage. Bifunctional dibasic acid dichlorides, such as glutaric acid dichloride, may also prove useful for ligand immobilisation:

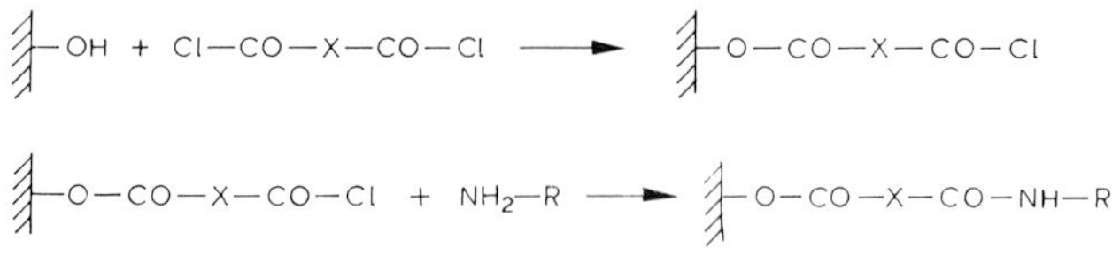

3.1.2. Polyacrylamide

The cross-linked polyacrylamide gels consist mainly of segments of linear polyethylene with alternate backbone carbon atoms bearing primary amide groups ($-CONH_2$). The plentiful supply of these amide groups endows the polymer with a marked hydrophilic character and accounts for its low adsorption of macromolecules.

Chemical derivatisation of polyacrylamide may be achieved by (1) modification of unadulterated preformed polyacrylamide gels (Inman and Dintzis, 1969) or (2) by co-polymerisation of acrylamide and cross-linker with a functional group-bearing acrylic or vinyl monomer (Ohno and Stahmann, 1972; Schnaar and Lee, 1975).

3.1.2.1. Direct activation of polyacrylamide

The commercial availability of polyacrylamide beads (Table 2.2) permits the preparation of affinity adsorbents with independent selection of functional group, degree of substitution, porosity and particle size. Furthermore, the abundance of modifiable carboxamide groups permits a high degree of substitution. The carboxamide side groups are resistant to hydrolysis over a wide pH range. The amide nitrogen is, however, readily displaced by certain nitrogen compounds with the liberation of free ammonia (Inman and Dintzis, 1969).

Polyacrylamide may be aminoethylated by adding the dry beads slowly to anhydrous ethylene diamine preheated to 90 °C. The incorporation of aminoethyl groups into the polyacrylamide gel is linearly related to the time of heating at 90 °C over a 7-h time period. Under these conditions, substitution densities as high as 1.2 mmol/g polymer may be achieved. The reaction rate is dependent on the water content of the ethylenediamine, increasing with increasing concentration of water. Carboxyl groups are formed by alkaline hydrolysis of the carboxamide groups in increasing amounts relative to aminoethyl groups at high moisture contents. Ideally, therefore, ethylene diamine should be used fresh from the supplier.

The hydrazide derivative of polyacrylamide is prepared by gradually adding the dry beads to an aqueous solution of hydrazine (1–6 M) and stirring gently at constant temperature for the desired period. The degree of substitution may be regulated by the concentration of hydrazine and the time and temperature of the reaction. Thus, for example, almost 4 mmol hydrazide/g polymer may be introduced by heating the beads for 8 h at 47 °C. The proportion of carboxyl groups generated during hydrazinolysis (2–3%) was less than that during aminoethylation (8%).

Partial alkaline hydrolysis of polyacrylamide leads to deamidation and the appearance of titratable carboxyl functions. The dry beads are gradually added to 0.5 M $NaHCO_3$–Na_2CO_3 buffer (pH 10.5) and the degree of hydrolysis controlled by the time of heating at 60 °C. Alternatively, the functional group density may be controlled by heating for 3 h at 60 °C with bicarbonate–carbonate buffers of different pH values.

These reactions for the primary derivatisation of polyacrylamide are summarised in Fig. 3.10. The resulting functional derivatives of polyacrylamide may be formed at any predetermined level of substitution and subsequently converted into a wide range of other derivatives by suitable reactions (Inman and Dintzis, 1969) prior to coupling the desired ligand. Protein ligands can be coupled directly to polyacrylamide with glutaraldehyde (Weston and Avrameas, 1971). Underivatised Bio-gel P-300 beads are activated by incubation at 37 °C for 16–40 h in 5% (v/v) glutaraldehyde in 0.1 M potassium phosphate buffer (pH 6.9). The activated beads are washed with 0.1 M potassium phosphate buffer (pH 6.9 and pH 7.7) and proteins

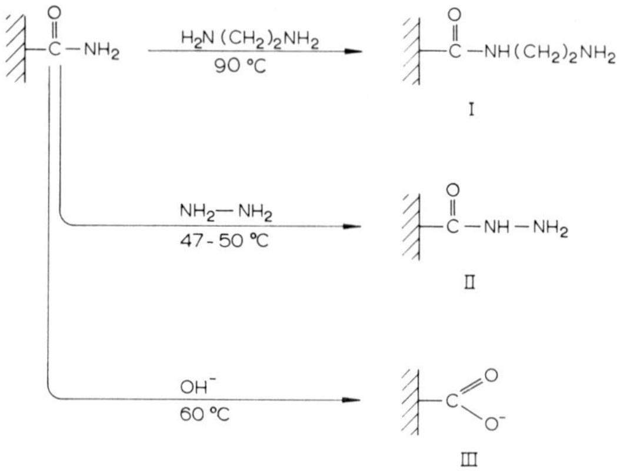

Fig. 3.10. Three reactions for the primary derivatisation of polyacrylamide by aminoethylation (I), hydrazinolysis (II) and alkaline hydrolysis (III) respectively.

coupled by incubation for 18 h at 4°C in the pH 7.7 buffer. This procedure represents a very facile approach to the preparation of immobilised proteins for affinity chromatography or as immunoadsorbents.

3.1.2.2. Co-polymerisation techniques

As an alternative to chemically modifying native polyacrylamide, polyacrylamide gels containing the desired ligands may be formed by coupling acrylic acid to the amino terminal of the ligands and then co-polymerising them with acrylamide and *N,N*¹-methylenebisacrylamide. This approach permits not only easy control of gel porosity but also the level of ligand substitution. However, the necessity of synthesising individually modified ligands containing double bonds discourages general application of this method.

In an attempt to circumvent these objections, Schnaar and Lee (1975) have synthesised the acrylic acid esters of *N*-hydroxysuccinimide and *N*-hydroxyphthalimide and co-polymerised them with acrylamide and methylenebisacrylamide. The resulting 'active' gels react readily with ligands containing primary amino functions (Fig. 3.11). The

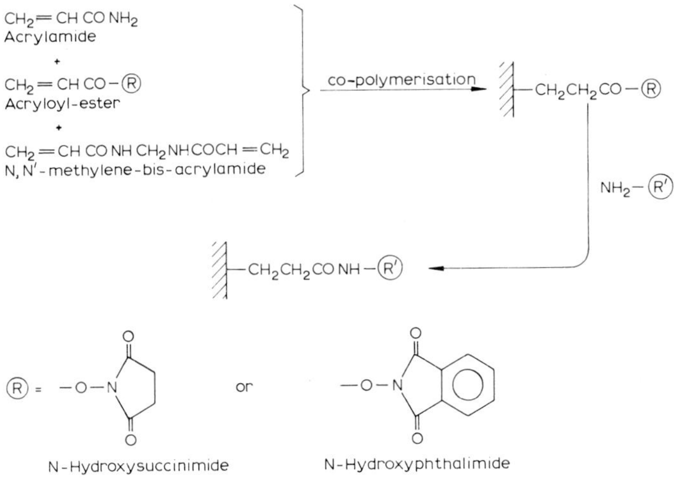

Fig. 3.11. Polyacrylamide gel copolymerised with active esters. Active esters of acrylic acid (*N*-succinimidyl acrylate and *N*-phthalimidyl acrylate) were synthesised according to Schnaar and Lee (1975) and copolymerised with acylamide and *N,N'*-methylene-bisacrylamide. Affinity adsorbents were generated by displacing the active ester in the gel by ligands containing aliphatic amino groups.

versatility of this system has been demonstrated by the efficient purification of wheat germ haemagglutinin on polyacrylamide gel incorporating 6-aminohexyl-2-acetamido-2-deoxy-β-D-gluco-pyranoside as ligand (Schnaar and Lee, 1975).

3.1.3. Porous glass and ceramics

The most common approach to chemically bonding reactive organic groups to inorganic materials such as glass or ceramics is through reaction with silane coupling agents. These silane reagents have dual functionality, with an organic functional group at one end and a silylalkoxy group at the other. Typical organosilanes commercially available include epoxy-, vinyl-, thiol-, alkylamine-, alkylchloro- and phenyl-silanes. The organosilane, γ-aminopropyl-triethoxysilane is commonly used to introduce primary amino groups onto porous

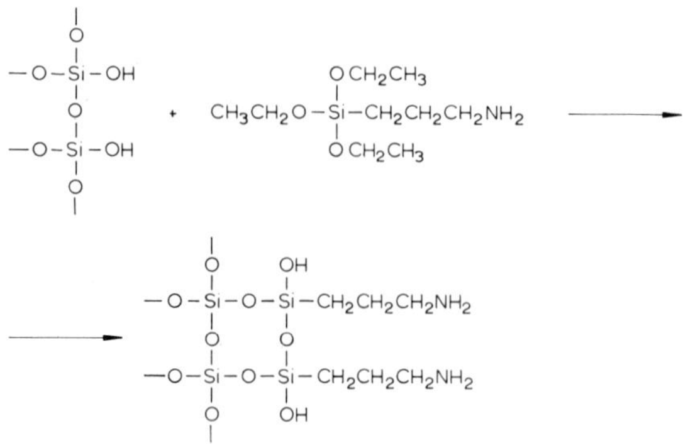

Fig. 3.12. The derivatisation of porous glass with organo functional silanes such as γ-aminopropyl-triethoxysilane.

glass to permit subsequent secondary derivatisation into a number of other functional groups. In aqueous media, the ethoxy groups hydrolyse to hydroxyls and condense with the silanols on the glass to generate a monolayer of silane across the carrier surface (Fig. 3.12). Prior to silanisation the controlled pore glass carrier is washed with 5% HNO_3 at 80–90 °C for 1 h followed by rinsing with distilled water in order to hydrate and clean the carrier surface. To 1 g of the clean glass is added 18 ml distilled water plus 2 ml 10% (v/v) γ-aminopropyl-triethoxy silane (available commercially from Union Carbide) and the pH adjusted to 3–4 with 6 N HCl. The reactants are heated at 75 °C for 2 h, washed with distilled water and finally dried in 115 °C oven for at least 4 h. The product, alkylamine-glass, is stable and may be stored for later use.

The silanization technique may also be performed in organic solvents. This approach gives higher alkylamine substitutions but the distribution is patchy and the beads are not quite as durable (Weetall, 1976). Typically, 1 g of clean glass is refluxed overnight in 50 ml 10% (v/v) γ-aminopropyl-triethoxysilane in toluene, washed with toluene and acetone and finally dried as above. The resulting alkyl-amine-glass can subsequently be used to immobilise ligands and proteins by the techniques discussed in § 3.4. Similar procedures may be employed to silanize, and hence derivatise, alumina, silica, nickel oxide, titania and several metal silicates. Furthermore, the silanol residues on the surface of glass and the metal oxide groups on ceramics are susceptible to activation with CNBr. The procedure is identical to that given for polysaccharide matrices (§ 3.1.1.1).

3.1.4. Other support matrices

The hydroxylalkyl methacrylate gels (Turková, 1974) contain neutral hydroxyl groups and are thus susceptible to activation by CNBr and all the procedures used for derivatising polysaccharide carriers (§ 3.1.1). Alternatively, the gels may be formed by co-polymerisation with monomers containing reactive groups by procedures analogous to those used in the synthesis of polyacrylamide gels (§ 3.1.2.2). On many accounts, nylon is also an attractive support matrix for the immobilisation of ligands, particularly proteins, since it is readily available in a wide range of physical forms. Furthermore, it displays hydrophilic character by virtue of its secondary amide linkages and is thus suitable for the fractionation and manipulation of cells (Edelman et al., 1971). A range of chemical reactions are now available for the functionalisation of nylon (Hornby and Goldstein, 1976).

3.2. Spacer arms

The necessity for a spacer group has been extensively corroborated by numerous reports in the literature describing successful purifications using adsorbents in which a spacer molecule has been interposed between the ligand and the matrix backbone. This seems to be particularly important for small ligands, for protein-ligand complexes of low affinity ($K_i \geqslant 10^{-4}$ M) and for those involving very high molecular weight proteins.

3.2.1. Hydrophobic spacer arms

The most general and extensively used technique for the introduction of spacer molecules is to couple ω-aminoalkyl compounds of the

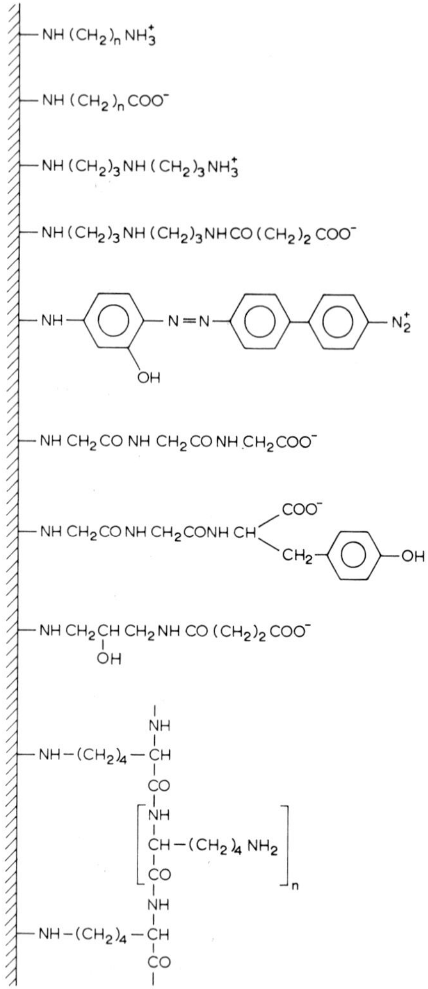

Fig. 3.13. Some common matrix–spacer arm conjugates used in affinity chromatography. Agarose bound: (1) α,ω-diaminoalkanes ($n = 2$–12); (2) α-amino, ω-carboxylalkanes ($n = 2$–10); (3) 3,3′-diaminodipropylamine; (4) 3,3′-diaminodipropylamine succinylated with succinic anhydride; (5) *m*-aminophenol reacted with bis-diazotised benzidine; (6) glycyl–glycyl–glycine; (7) glycyl–glycyl–tyrosine; (8) 1,3-diaminopropan-2-ol succinylated with succinic anhydride; (9) poly(lysine).

general type $NH_2(CH_2)_nR$ to an activated matrix, where R may be an amino or carboxyl function and $n = 2$–12 (Fig. 3.13). Typically, adsorbents comprising spacer arms containing 6 carbon atoms with amino or carboxyl termini may be prepared by covalently attaching 1,6-diaminohexane and 6-aminohexanoic acid respectively to CNBr-activated Sepharose. Significant cross-linking of the gel with the bifunctional amine is avoided by using a large excess of the diamine during the coupling stage.

A number of these agarose-coupled spacer molecules are available commercially. Pharmacia Fine Chemicals market 6-aminohexyl–agarose and 6-aminohexanoyl–agarose under the designations AH-Sepharose 4B and CH-Sepharose 4B respectively. P.L. Biochemicals Inc. market similar products under the trademarks, AGHEXAMINE and AGHEXANOIC respectively. Miles Laboratories offer a range of ω-aminoalkyl–agaroses containing 2, 4, 6, 8 and 10 methylene groups. This is particularly helpful since it permits the experimenter to couple his ligand to a number of spacer molecules of differing lengths and thus assess the value of each in the purification of his particular protein. He could well find that the 6-aminohexyl-spacer molecule most commonly employed may be superceded by either a longer or a shorter spacer molecule to achieve optimal purification. Technical data for these derivatised gels are given in Table 3.1. The data are taken from the manufacturers' technical information.

For longer spacer molecules, 3,3'-diaminodipropylamine may be attached to the gel matrix with cyanogen bromide (Fig. 3.13). The terminal amino group can subsequently be succinylated by treatment with succinic anhydride in aqueous solution (§ 3.4.3) to generate an ω-carboxyl group, some 21 Å distant from the matrix skeleton (Steers et al., 1971). However, flexible hydrophobic polymethylene spacer arms have a tendency to fold back on themselves in aqueous media. For this reason the use of long hydrocarbon arms containing 12 or so carbon atoms is not recommended. Not only is there no increase in the strength of the ligand–protein interaction but the use of such long hydrocarbon arms has attendant solubility problems and non-specific denaturation of proteins at hydrophobic interfaces.

TABLE 3.1
Technical data for commercial derivatised adsorbents

Designation	Chemical nature	Spacer molecule concentration (μmol/ml swollen gel)	Availability	Supplier
AH-Sepharose 4B	CNBr-activated Sepharose 4B plus 1,6 diaminohexane	6–10	Freeze-dried powder equivalent to 60 ml swollen gel	Pharmacia Fine Chemicals Sigma Chemical Co.
AGHEXAMINE™	CNBr-activated Sepharose 4B plus 1,6-diaminohexane	5–10	10 ml, 50 ml and 100 ml packs	PL Biochemicals
ω-Aminoalkyl–agarose ($n=2, 4, 6, 8, 10$)	CNBr-activated agarose plus α,ω-diaminoalkane	10–25	Hydrophobic chromatography kit (1 ml of each) 20 ml and 50 ml packs	Miles Laboratories
ω-Aminoalkyl–agarose ($n = 2, 6$)				ICN
3,3′-diaminodipropyl-amino–agarose				ICN
Servachrom A-A1	2-Aminoethyl–agarose			SERVA
Servachrom A-A2	3,3′-Diamino-dipropyl-amino–agarose			
Affi-Gel 101	3-Aminopropyl–agarose		50 ml	Bio-Rad
Affi-Gel 102	3,3′-Diaminodipropyl-amino–agarose (Both linked via stable ether linkage)		50 ml	Bio-Rad

Aminohexyl-cellulose		100 μequiv./ml	10 g (moist)	Biochemica MERCK
Aminopropyl-controlled pore glass	Pore diameter 550 Å	0.17 mequiv./g		Corning Pierce
Aminohexyl-CPG		0.1 mequiv./g		
Servachrom-CPG aminopropyl	Amino-propyl-CPG			SERVA
Aminoethyl-Bio gel P	P-2, P-150		P-2: 25 g P-150: 12 g	Bio-Rad
CH-Sepharose 4B	CNBr-activated Sepharose 4B plus ε-aminocaproic acid	10–14	Freeze-dried powder. 15 g dry powder equivalent to 60 ml swollen gel	Pharmacia Fine Chemicals Sigma Chemical Co.
AGHEXANOIC™	Agarose–hexanoic acid	5–10	10, 50 and 100 ml packs	PL Biochemicals
ε-Aminocaproic acid substituted-agarose				ICN
Succinylated aminohexyl and aminododecyl-cellulose		100 μequiv./ml bed volume	10 g (moist)	Biochemica MERCK
Affi-Gel 201 Affi-Gel 202	Succinylated aminopropyl and 3,3′-diamino-dipropylamino–agarose		50 ml packs	Bio-Rad
Servachrom A-C1	Succinylated 2-aminoethyl–agarose			SERVA

TABLE 3.1 *(continued)*

Designation	Chemical nature	Spacer molecule concentration (μmol/ml swollen gel)	Availability	Supplier
Servachrom A-C2	Succinylated 3,3′-diamino-dipropylamine–agarose			
Albumin-agarose				ICN
Poly(L-lysine)–agarose, poly(DL-alanine)–poly(L-lysine)–agarose and succinylated analogues		10 μmol/ml	50 ml	Miles Laboratories

3.2.2. *Hydrophilic spacer arms*

The recent emphasis on hydrophobic chromatography, particularly with residues containing aliphatic polymethylene or benzenoid hydrocarbons prompts the use of more hydrophilic spacer molecules. A typical polar or hydrophilic spacer molecule may be prepared by coupling 1,3-diamino-2-propanol to CNBr-activated agarose, bromoacetylating (§ 3.4.4) and addition of further 1,3-diamino-2-propanol. Oligoglycine arms may also be interposed between the ligand and matrix (Lowe and Dean, 1974). Such oligopeptides, containing up to 4 glycine units, serve the dual purpose of being both more hydrophilic and more rigid than their polymethylene counterparts. The terminal ω-amino group of the peptide is coupled to CNBr-activated Sepharose and the free carboxyl group subsequently utilised for linkage to the ligand. Furthermore, oligopeptide sequences terminating in tyrosine, such as glycyl–glycyl–tyrosine, or cysteine, serine or glutamic acid enable hydrophilic spacers with phenolic, thiol, hydroxyl or carboxyl functions to be incorporated into the gel. The oligopeptides are useful in combining the function of spacer with the introduction of new reactivities and thus enchance the versatility of this range of gels. One drawback of these methods, however, is that a terminal carboxyl function is also introduced along with the desired function and that this could impair the interaction with the complementary macromolecule.

3.2.3. *Multivalent macromolecular spacer molecules*

Extension of the above logic suggests that polypeptides or proteins may prove useful hydrophilic spacer molecules. Thus, Wilchek (1973) introduced the preparation and use of a number of polypeptide–agarose derivatives. These gels provide a multiple-linked hydrophilic spacer arm which reduces simultaneously both spacer–ligand leaching (§ 3.1.1.1.6) and non-specific hydrophobic interactions. Poly(L-lysyl)–agarose (Wilchek, 1973) is prepared by coupling poly(L-lysine) to CNBr-activated agarose through the ε-amino groups of the lysyl residues (§ 3.1.1.1). Similarly, poly(DL-alanyl)–poly(L-lysyl)–agarose comprises the multichain polymer poly(DL-alanine)–

poly(L-lysine) attached to beaded agarose. Poly(DL-alanyl) side chains branch off the poly(L-lysine) backbone and are bound to the support matrix. Both the poly(L-lysyl)– and the poly(DL-alanyl)–poly(L-lysyl)–agarose bear free amino groups suitable for subsequent derivatisation with ligand and are available commercially from Miles Laboratories. This company also market the corresponding succinylated derivatives which provide free carboxyl functions for coupling to reactive amino groups on the ligand (§ 3.4.3).

Bovine serum albumin may also be coupled to CNBr-activated agarose in the presence of a high concentration of urea to promote unfolding and thus to increase the likelihood of multiple attachment to the protein to the solid support. The coupling to agarose is performed as described in § 3.1.1.1 except that the coupling buffer comprises 0.2 M $NaHCO_3$ (pH 9) containing 10 M urea. Alternatively native albumin may be coupled to CNBr-activated agarose in the absence of urea.

3.2.4. Charge-free spacer molecules

It has been mentioned in § 3.1.1.1.7 that the N-substituted isourea derivatives formed on coupling primary amines to CNBr-activated agarose exhibit strong ion-exchange properties. This problem may be alleviated by coupling adipic acid dihydrazide, succinic acid dihydrazide, polyglutamic acid hydrazide or polyacrylic acid hydrazide to CNBr-activated agarose. These conjugates represent stable, high capacity agarose derivatives free of ion-exchange groups. Polyacrylic hydrazide–agarose, for example, is relatively easy to prepare. Polyacrylic hydrazide is prepared by heating polymethylacrylate with 98% hydrazine at 100°C for 3 h (Kern et al., 1957) and is then coupled to the CNBr-actived agarose in the usual way.

Polyacrylic hydrazide–agarose and its succinylated derivative are available commercially from Miles Laboratories in packs of 50 ml. This company also markets adipic hydrazide–agarose, and its succinylated analogue, prepared by coupling the bishydrazide of adipic acid (C6) to agarose. However, hydrazide agarose derivatives, although charge free, do exhibit some leakage of the spacer from the agarose in common with all single point attachment spacers linked to CNBr-activated agarose. Bio-Rad Laboratories offer a number of spacer arm gels under the designation Affi-Gel. The matrix employed in Affi-Gel is Bio-Gel A15 m, a 4% beaded agarose, and the supports employ extremely stable neutral ether bonds to link the spacer arms to the agarose backbone rather than CNBr coupling. Each gel has a nominal terminal group concentration of 6–12 μmol/packed ml of gel. The company markets two amino gels, Affi-Gel 101 and Affi-Gel 201 comprising spacer arms of 4 and 10 atom lengths respectively. The former is an aminopropylated-agarose (agarose–$O(CH_2)_3NH_2$) whilst the latter contains 3,3'-diaminodipropylamine as part of its structure (agarose–O-CH_2-CO-NH-$(CH_2)_3$-NH-$(CH_2)_3$-NH_2) and as such is generally less prone to hydrophobic interference than purely methylenic spacer molecules such as hexanediamine. The succinylated derivatives of these two gels are also available as Affi-Gel 201 and Affi-Gel 202, respectively. Technical data for these gels is included in Table 3.1.

3.3. *Preparation of high capacity adsorbents*

Interacting systems of low affinity are often not amenable to affinity chromatography unless the immobilised ligand concentration can be made sufficiently high to retard the enzyme. Porath and Sundberg (1972) have coupled the polyhydric phenol, phloroglucinol (1,3,5-trihydroxy-benzene) to epichlorohydrin-activated Sepharose. The resulting phloroglucinol-substituted polymer has, compared to the parent gel, an increased number of hydroxyl groups which are not only more reactive phenolic hydroxyls but are also distant from the matrix backbone. This second generation chemisorbent thus has greatly increased capacity for ligand substitution and can be activated with cyanogen halides, bisepoxides, epihalohydrins, triazines or divinyl sulphones in the usual way prior to attachment of the ligand.

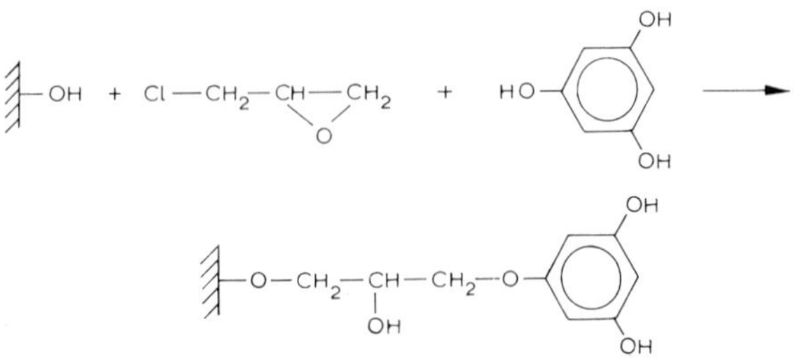

3.4. Reactions for coupling ligands to spacer arms

The aminoalkyl- and carboxyalkyl-derivatives of agarose discussed so far are versatile gels which can be further extended and modified in order to coordinate with a wide range of functional groups on prospective ligands. There are two main types of reactive groups which may be introduced into the support matrix in order to couple ligands:

(a) Nucleophilic groups: amino, thiol and hydroxyl groups.

(b) Electrophilic groups: cyclic imidocarbonates, oxirane groups, isocyanates, isothiocyanates, diazonium salts, phenols, aromatic amines, anhydrides and activated carbonyls such as acid chlorides and those activated by carbodiimides.

The synthesis of such secondary intermediates extends the applicability of affinity chromatography to almost all ligands. The successful development of a competent adsorbent will depend to some extent on the possibility of linkage through one particular group. Unfortunately, none of the methods so far devised is absolutely specific for an individual functional group, although with careful control of coupling conditions a degree of specificity may be achieved.

3.4.1. Ligands containing amino groups

Ligands containing aliphatic or aromatic amino groups can be coupled directly to CNBr-activated agarose (§ 3.1.1.1) or to matrices activated by any of the other methods cited in § 3.1.1. Thus, for example, if the ligand to be immobilised is a protein, the N-terminal α-amino group and the ε-amino groups of lysyl residues are the most likely to react. However, in most cases the ligand is a small molecular weight compound that must be attached to the terminus of a preformed spacer arm–matrix conjugate. A number of methods are available to link amines to matrices. These methods are summarised in scheme 3.1.

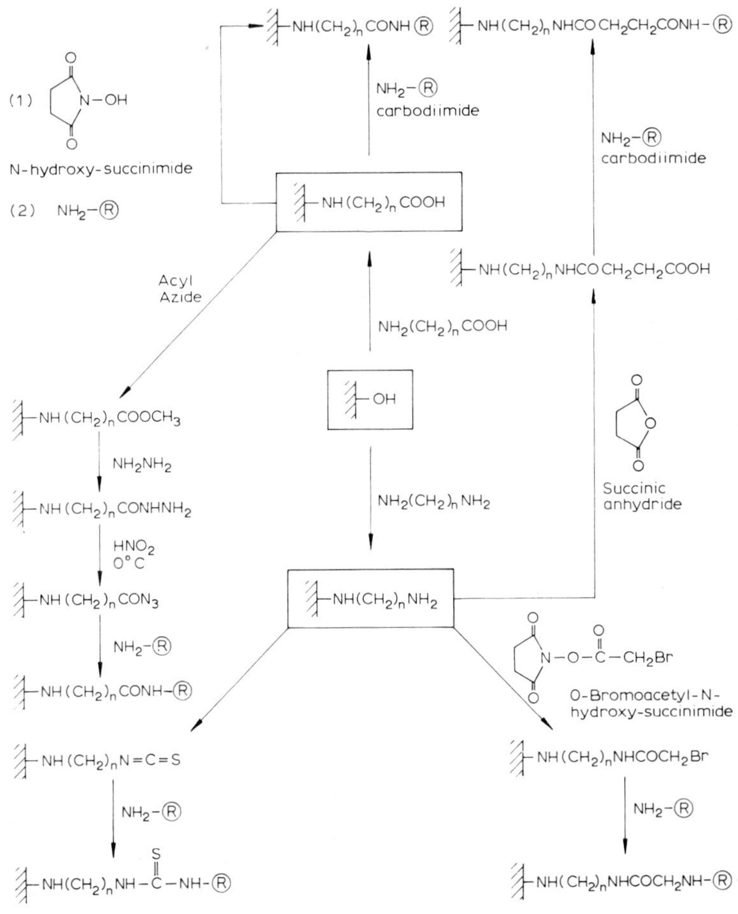

Scheme 3.1. Methods for coupling amino ligands.

3.4.1.1. Carbodiimide condensations

Ligands carrying primary amino functions may be coupled to ω-carboxyalkyl-derivatives by a carbodiimide promoted reaction. The *N,N*¹-disubstituted carbodiimides are some of the most versatile synthetic reagents at the disposal of the biochemist. They may be regarded as anhydrides of urea and thus their most important application in biochemistry has been in the synthesis of amide and

$$\underset{\text{urea}}{NH_2{-}\overset{\overset{\displaystyle O}{\|}}{C}{-}NH_2} \;\overset{-H_2O}{\rightleftharpoons}\; \underset{\text{carbodiimide}}{HN{=}C{=}NH}$$

peptide bonds by acid catalysed addition of water, i.e., hydration, to form the corresponding ureas. Figure 3.14 shows the mechanism of amide formation in the presence of an N,N¹-disubstituted carbodiimide. The carbodiimide (I) is protonated to give an intermediate (II) which is attacked by a carboxylate anion to generate an *O*-acyl-urea (III) which can undergo an intramolecular rearrangement to form the

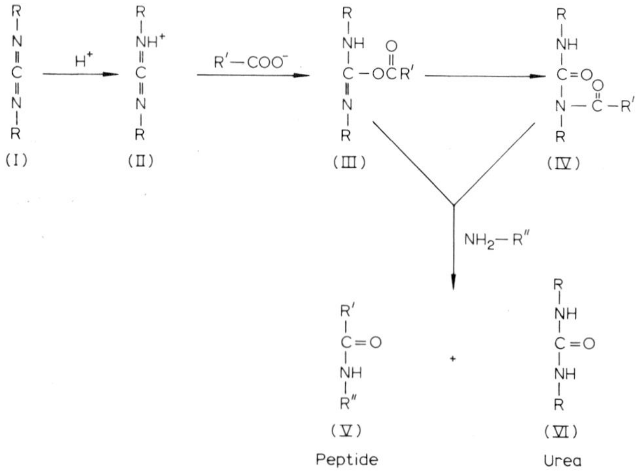

Fig. 3.14. Carbodiimide promoted condensations.

Dicyclohexyl carbodiimide (DCC)

1-Ethyl-3-(3-dimethylaminopropyl)-carbodiimide hydrochloride (EDC)

1-Cyclohexyl-3-(2-morpholinoethyl)-carbodiimide p-toluene methosulphonate (CMC)

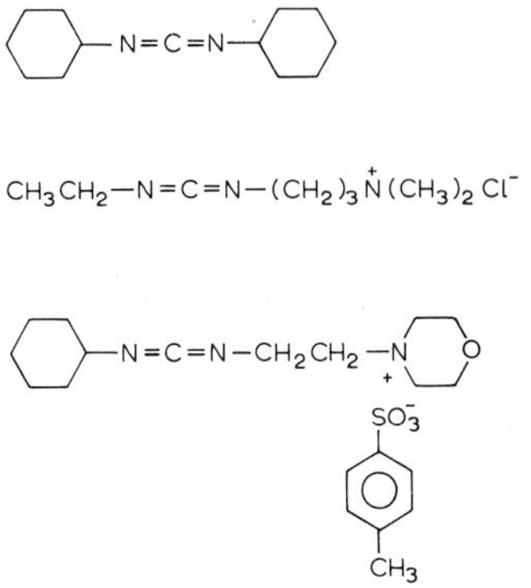

Fig. 3.15. The structures of the three recommended carbodiimides.

N-acyl-urea (IV). Both *O*- and *N*-acyl-ureas can react with the primary amine to form the peptide (V) and the corresponding urea (VI). Figure 3.15 gives the structures of three recommended carbodiimides.

Dicyclohexyl carbodiimide (DCC) is insoluble in water but may be used to couple amines to ω-carboxyalkyl-–agarose in 80% (v/v) aqueous pyridine or acetonitrile. However, the derived urea is also insoluble in water and must be removed from the gel by extensive washing in organic solvents such as ethanol and butanol. The problem may be circumvented by using one of two soluble carbodiimides, 1-ethyl-3-(3-dimethylaminopropyl)-carbodiimide hydrochloride (EDC) or 1-cyclohexyl-3-(2-morpholinoethyl)-carbodiimide *p*-toluene methosulphonate (CMC). These carbodiimides are convenient to use since their corresponding ureas are freely soluble in water and are hence easily removed from the gel product by washing with water. They have been used under a variety of experimental conditions and it is thus important that the pH, diimide concentration and reaction time be optimised for each individual application. In the authors own experience EDC often gives better and more consistent coupling yields.

The reaction is normally performed in distilled water with the pH maintained at 4.5–6.5 by the addition of acid or base as necessary.

The pH of the reaction will be determined largely by the stability of the ligand being coupled with the preferred pH being maintained at 4.5–5.0 if possible. Buffers may be used to maintain these low pH values but should not contain amino, carboxyl or phosphate groups which would compete in the reaction. A concentration of carbodiimide greater than stoichiometric is required since these reagents are unstable. Carbodiimides should be stored dry at −20 °C and used fresh from the bottle. The concentration used should be 10–100 times greater than the concentration of immobilised spacer groups.

A typical coupling procedure for linking a ligand bearing a primary amine to ω-carboxyalkyl–agarose might be as follows. A concentration of the ligand in excess of the concentration of bound spacer molecules is dissolved in distilled water and the pH adjusted to 4.5–4.7. The washed gel suspension is added and then the large excess of solid carbodiimide slowly in small portions continuously stirring and readjusting the pH to 4.5–4.7 when necessary. The pH is monitored and readjusted over a period of 2 h at room temperature and then the reaction is allowed to proceed with gentle agitation overnight either at room temperature or at 4 °C. The gel is thoroughly washed to remove all material not covalently attached. It should be stressed that the conditions of coupling should be optimised in each individual case.

3.4.1.2. Other methods for peptide bond formation

Other methods are available for linking amino functions to ω-carboxy-alkyl–agarose and vice versa. For example, carboxyl groups can be activated by mixed anhydride formation with iso-butyl chloroformate (Dean et al., 1972); Woodwards K reagent may effect a similar condensation (Patel and Price, 1967).

3.4.1.3. Anhydride reactions

Amino containing ligands may be coupled to succinylated ω-amino-alkyl-gels by the methods given above. The aminoalkyl-gel is treated for 12 h with 0.3 g succinic anhydride dissolved in 50 mM sodium phosphate buffer (pH 6.0) for each gram of alkylamine gel. The final product, the carboxyl derivative is exhaustively washed.

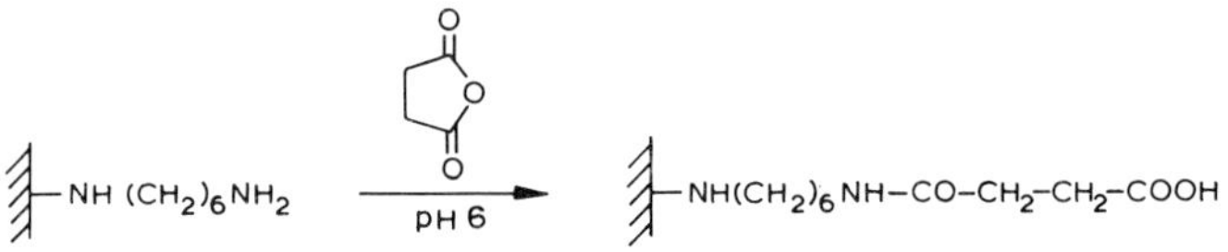

Likewise, the hydrazide derivatives of polyacrylamide, glass or agarose may be converted into the succinyl hydrazide by treatment with succinic anhydride at pH 4.

3.4.1.4. N-*Hydroxysuccinimide reactions*

Agarose derivatives containing stable 'activated' functional groups such as the *N*-hydroxysuccinimide ester (Cuatrecasas and Parikh, 1972) represent convenient alternatives for coupling amino-containing ligands. Furthermore, these reactions are performed under very mild conditions and are almost exclusively specific for amino functions. Typically, the *N*-hydroxysuccinimide ester of ω-carboxyalkyl-agarose is prepared by reacting the agarose derivative for 90 min at room temperature with 0.1 M dicyclohexyl carbodiimide (DCC) and 0.1 M *N*-hydroxysuccinimide in anhydrous dioxane. The resulting ester is washed thoroughly with dioxane and methanol to remove precipitated dicyclohexyl urea and may be stored, at 24 °C and protected from light, as a 1 : 1 suspension in anhydrous dioxane for several months with very little loss of coupling potential. Ligands carrying free unprotonated amino functions can be coupled in the pH range 5–8.5 at 4 °C under mild aqueous conditions with reaction times between 10 min and 6 h. Buffers such as citrate, phosphate, acetate or bicarbonate are tolerated but those carrying a nucleophilic group capable of reacting with the active ester, such as Tris, glycine and ammonium or guanidinium salts, should be avoided. At the end of the coupling reaction, residual active ester groups should be eliminated by incubating the gel with 1 M glycine at pH 9 for 2 h at room temperature. The adsorbent is finally washed exhaustively with the desired buffer. Figure 3.16 illustrates the reactions involved in the preparation and use of *N*-hydroxysuccinimide esters of agarose.

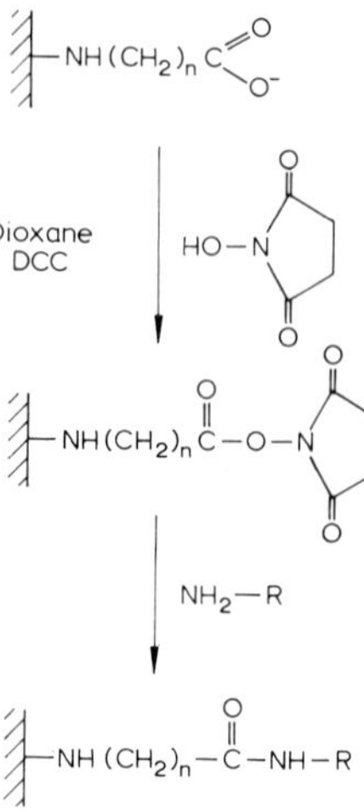

Fig. 3.16. Scheme of reactions involved in the preparation and use of *N*-hydroxy-succinimide esters of agarose.

These derivatives are available commercially. Pharmacia Fine Chemicals market the *N*-hydroxysuccinimide ester of 6-amino-hexanoyl–agarose under the designation Activated CH-Sepharose 4B. The activated gel is supplied as a freeze-dried powder in packs of 15 g equivalent to approximately 45 ml swollen gel. The freeze-dried powder is stable when stored dry below 8 °C and may be swollen in 1 mM HCl prior to coupling. Each ml of swollen gel will couple 5–7 μmol glycyl-leucine. PL Biochemicals market the *N*-hydroxy-succinimide ester of succinylated aminopropyl–agarose (agarose–*N*-hydroxysuccinimide coupler) whilst Bio-Rad Laboratories offer a similar product coupled via the stable uncharged ether linkage (Affi-Gel 10). The latter comes in 1 g packs of dehydrated material in a vial large enough to serve as the reaction vessel. The unique advantage of this type of product is that one simply adds the ligand and shakes; no other reactions are necessary.

Ligands containing primary aliphatic or aromatic amines, imidazole or phenolic groups also react with the bromoacetyl-derivatives of agarose. Bromoacetamidoalkyl–agarose may be prepared under mild aqueous conditions by reacting ω-aminoalkyl–

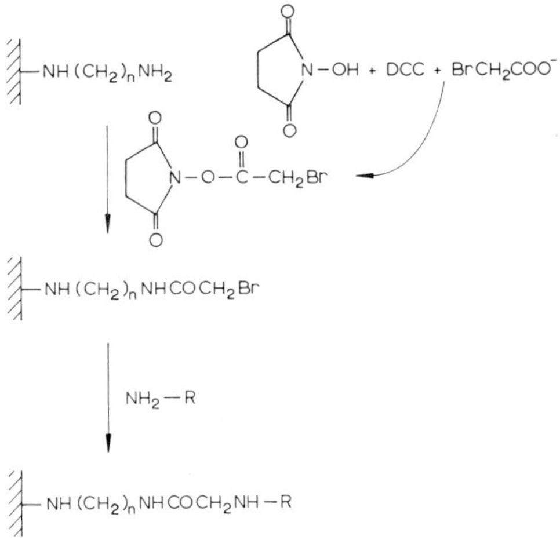

Fig. 3.17. The preparation and use of bromoacetamido derivatives.

agarose with *O*-bromoacetyl-*N*-hydroxysuccinimide. The latter is generally prepared in situ by coupling bromoacetic acid to *N*-hydroxysuccinimide with *N,N*¹-dicyclohexyl carbodiimide in dioxane. After filtration to remove the precipitated dicyclohexyl urea, the filtrate is added to a suspension of ω-aminoalkyl–agarose at pH 7.5 and incubated for 30 min at 4 °C. Figure 3.17 illustrates these reactions.

3.4.1.5. The acyl azide procedure

This procedure (Fig. 3.18) may be used to attach a ligand bearing a primary aliphatic amino group to a hydrazido carrier (§ 3.2.4). The hydrazide polymer is cooled to 0 °C in an ice bath in 0.25 M HCl, crushed ice added, followed by the rapid addition of 1 M sodium nitrite solution. The reaction mixture is stirred for 1.5 min, whence the amino ligand is added, the pH adjusted to 8.5–10.5 and the reaction allowed to proceed for up to 2 h. The advantage of this approach is that it involves no intermediate washings or transfers.

3.4.1.6. Isothiocyanate coupling

An alkylamine carrier may be converted to an isothiocyanate and thereby be used to covalently attach a primary amine (Wide and

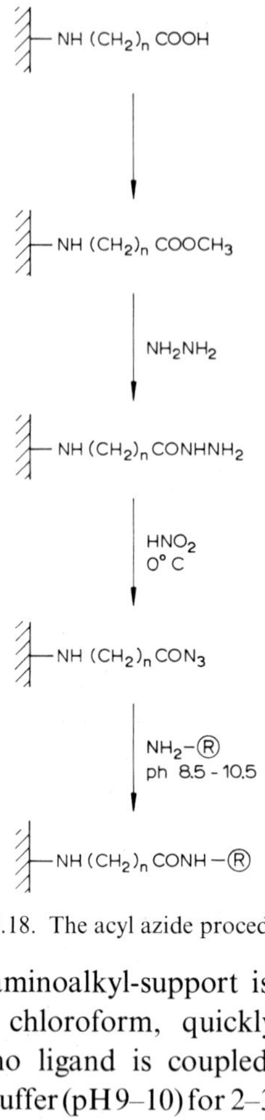

Fig. 3.18. The acyl azide procedure.

Porath, 1967). The ω-aminoalkyl-support is refluxed for 4 h with 10% thiophosgene in chloroform, quickly air dried and used immediately. The amino ligand is coupled to the isothiocyanate product in bicarbonate buffer (pH 9–10) for 2–3 h at room temperature. The product is washed with distilled water and stored as a wet cake at 4 °C.

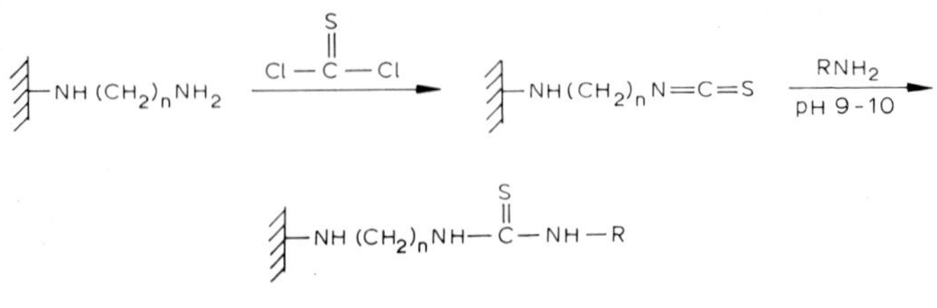

3.4.1.7. Bifunctional reagents

A number of bifunctional reagents may be used to link primary amines on the prospective ligand to alkylamino-support matrices. These reagents are summarised in Table 3.2. For detailed information on their application the reader is advised to consult the review by Wold (1967). The use of potential bifunctional reagents is, however, subject to two limitations; firstly, it is important that both functional groups on the molecule react in order that residual active groups are not left on the adsorbent. Secondly, a large excess of reactants is required to ensure that the bifunctional reagent does not cross-link the matrix.

3.4.2. Ligands containing carboxyl groups

Carboxyl-containing ligands may be coupled to primary-amine-bearing supports by carbodiimide promoted condensation reactions

TABLE 3.2
Bifunctional reagents reactive with amino groups

Glutaraldehyde	$CHO(CH_2)_3CHO$
Hexamethylene diisocyanate	$O{=}C{=}N{-}(CH_2)_6{-}N{=}C{=}O$
Dimethyladipimidate	$(NH_2^+)(OCH_3)C{-}(CH_2)_4{-}C(NH_2^+)(OCH_3)$
Dimethylsuberimidate	$(NH_2^+)(OCH_3)C{-}(CH_2)_6{-}C(NH_2^+)(OCH_3)$
Divinylsulphone	$CH_2{=}CH{-}S({=}O)_2{-}CH{=}CH_2$
Bis-oxiranes	$\overbrace{CH_2{-}CH}^{O}\,CH_2\,O(CH_2)_4O\,CH_2\,\overbrace{CH{-}CH_2}^{O}$

(§ 3.4.1.1). Alternatively, any of the other methods available for peptide bond formation may be used (§ 3.4.1.2).

The reaction of aryl isocyanides, carbonyl compounds and carboxylic acids to form arylamides is known as the Passerini Reaction. The reaction has been modified to include the reaction of amines and carbonyls with isocyanides and carboxylic acids. The so-called four-component Ugi Reaction involves the formation of an immonium ion (Schiff base) from the reaction of an amine with a carbonyl compound. The immonium ion subsequently reacts with an isocyanide to produce an intermediate which reacts with nucleophiles such as carboxylic acids. The final product, containing an amide linkage, is formed by a rearrangement:

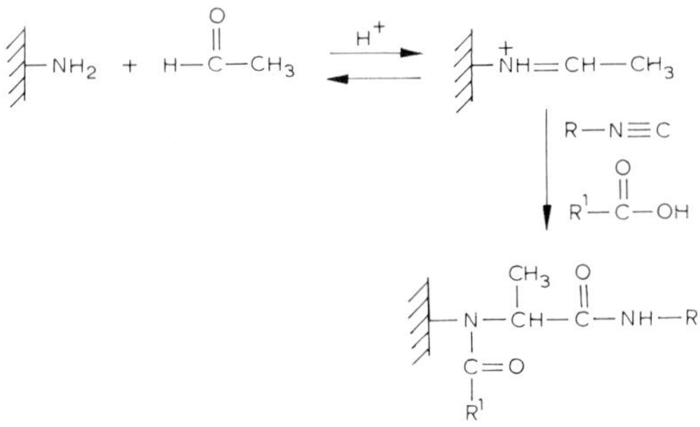

The four component Ugi reaction has been extended to the attachment of proteins, peptides and amino acids to amino and carboxyl-containing polymers (Axén et al., 1971).

3.4.3. Ligands containing aromatic functions

Diazonium derivatives of agarose, polyacrylamide or glass react with phenols, imidazoles or other compounds susceptible to electrophilic substitution and are readily prepared under mild aqueous conditions from suitable *p*-aminobenzamido-alkyl derivatives. The *p*-aminobenzamidoalkyl derivative of the matrix may be prepared by treatment of the aminoalkyl-gel with either *p*-nitrobenzoyl azide or *p*-nitrobenzoyl chloride in alkaline solution, typically 0.2 M sodium borate (pH 9.3) or 10% triethylamine. The derivatised gel is thoroughly washed with a suitable solvent, such as 40% (v/v) dimethylformamide, and the washed gel reduced with 0.1 M sodium dithionite for 40 min at 40–50 °C and subsequently diazotised by treatment with 0.1 M sodium nitrite in 0.5 M HCl for a few minutes at 0 °C. The resulting diazonium derivative may be used in situ without further washing simply by adding the phenol, imidazole or protein to be coupled in a strong buffer such as saturated sodium borate. The pH should be maintained for 8 h at 0 °C at 8 for imidazolyl residues and 10 for phenolic groups.

Ligands containing diazotisable aromatic amines may be coupled via azo linkages to immobilised tyrosyl derivatives in high yield. The tripeptide, glycyl–glycyl–tyrosine, containing a carboxyl terminal tyrosine residue, is particularly suitable for this approach. The entire reaction mixture containing the diazotised ligand is added to a suspension of tyrosyl–agarose in 0.2 M carbonate–bicarbonate buffer (pH 9.4) at 0 °C. Diazonium derivatives of this type have been used to purify Staphylococcal nuclease (Cuatrecasas et al., 1968) and neuraminidase from *Vibrio cholerae* (Cuatrecasas and Illiano, 1971).

The use of azo-bonded ligands has an important advantage: the bound ligand may be rapidly and quantitatively released by reduction of the azo linkage with 0.1 M sodium dithionite in 0.2 M sodium borate buffer (pH 9). This permits, not only the release of an intact protein–ligand conjugate under mild conditions but also allows an estimation of the amount of ligand bound to the gel.

3.4.4. Ligands containing aldehyde or ketone groups

Aliphatic aldehydes and ketones react rapidly and reversibly with primary amino groups under mild, slightly alkaline conditions, to yield adducts, Schiff's bases, which may be reduced with mild reducing agents such as sodium borohydride to yield stable alkyl-amino groups.

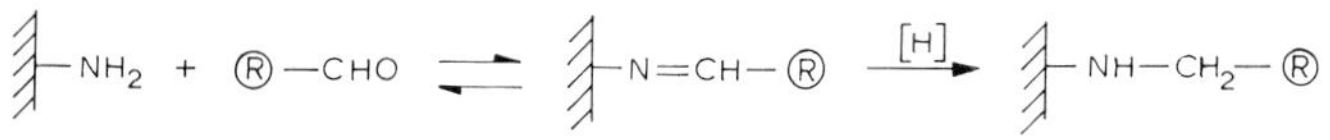

These reductive alkylation procedures are particularly useful for the immobilisation of nucleosides and nucleotides via their sugar moieties. Thus, for example, periodate oxidation of the ribose moiety of AMP, RNA, UMP, GTP, NAD^+, ATP, tRNA, etc., Schiff's base formation with the terminal amino of 6-aminohexyl–agarose or the corresponding hydrazide derivative and subsequent reduction with sodium borohydride at pH ~ 8 has been used to prepare a series of immobilised nucleotides for affinity chromatography. A typical procedure involves the prior preparation of adipic acid dihydrazide–agarose by coupling adipic acid dihydrazide to CNBr-activated agarose in 0.1 M $NaHCO_3$ pH 9.5 and subsequent reaction with periodate-oxidised $NADP^+$. The latter is prepared by oxidation of $NADP^+$ with a 4-fold molar excess of sodium periodate in 0.1 M potassium phosphate buffer (pH 7.0) for 3 h at 4 °C.

Immobilised pyridoxal-5'-phosphate derivatives may also be prepared by Schiff's base formation.

Ligands bearing ketone groups, particularly the steroid ketones, may be immobilised by prior reaction with *O*-carboxymethoxylamine hemihydrochloride. Thus, for example, 6-keto-oestriol or testosterone may be converted into the corresponding 6-(*O*-carboxymethyl)- or 3-(*O*-carboxymethyl)-oximes, respectively, be refluxing with *O*-carboxymethoxylamine hemihydrochloride in an appropriate solvent. The resulting *O*-carboxymethyloximes may be coupled to ω-amino-alkyl–agarose by any of the methods described in § 3.4.2.

$$R{-}C{=}O + NH_2{-}O{-}CH_2\text{-}COOH \rightarrow R{-}C{=}N{-}O{-}CH_2\text{-}COOH$$

3.4.5. Ligands containing hydroxyl groups

Ligands bearing hydroxylic functions may be coupled to ω-carboxyl-alkyl–agarose via carbodiimide promoted condensations. However, bis-oxirane activated matrices are undoubtedly the reagents of choice for coupling ligands through hydroxyl functions (§ 3.1.1.4). The reaction of agarose with 1,4-bis-(2,3-epoxy propoxy)-butane generates a hydrophilic spacer molecule linked to the matrix via a stable uncharged ether linkage with an active terminal oxirane group available for spontaneously coupling the hydroxyl groups of sugars and carbohydrates via stable ether linkages. Epoxy-activated Sepharose 6B is available commercially from Pharmacia Fine Chemicals.

3.4.6. *Ligands containing thiol groups*

Thiol-disulphide interchange is a particularly attractive method for linking thiol-bearing ligands to an activated thiol-matrix. Thiol groups may be introduced into polymers by coupling cysteine or glutathione directly to CNBr-activated agarose or by reacting the ω-aminoalkyl derivatives with *N*-acetyl homocysteine thiolactone for 24 h at 4 °C in 1 M Na HCO_3 (pH 9.7) containing a small amount of imidazole as catalyst.

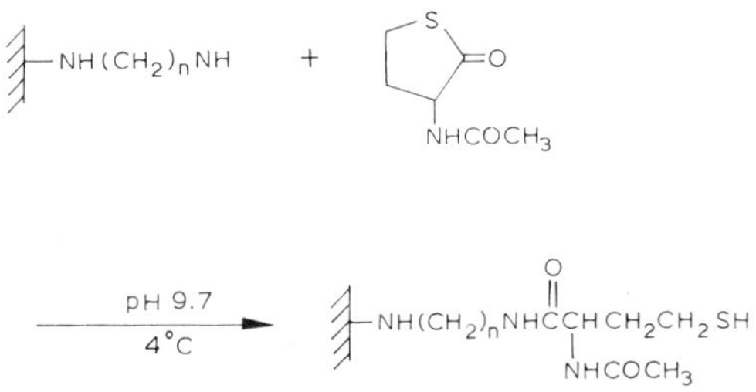

Disulphide bridges can be formed by the oxidative coupling of a thiol-containing ligand to a polythiol matrix in the presence of alkaline ferricyanide. The resulting linkage is readily cleaved by brief exposure of the disulphide to excess L-cysteine, β-mercaptoethanol or dithiothreitol.

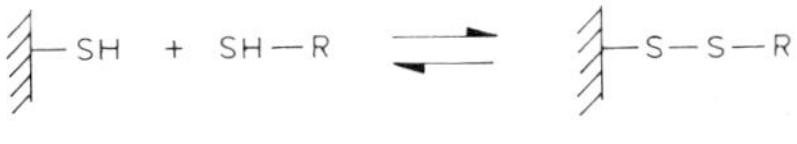

Alternatively, thiol–agarose may be activated with 2,2′-dipyridine disulphide or 5,5′-dithiobis-(2-nitrobenzoic acid) at pH ⩾ 8 and subsequently reacted with the thiol-bearing ligand (Brocklehurst et al., 1973).

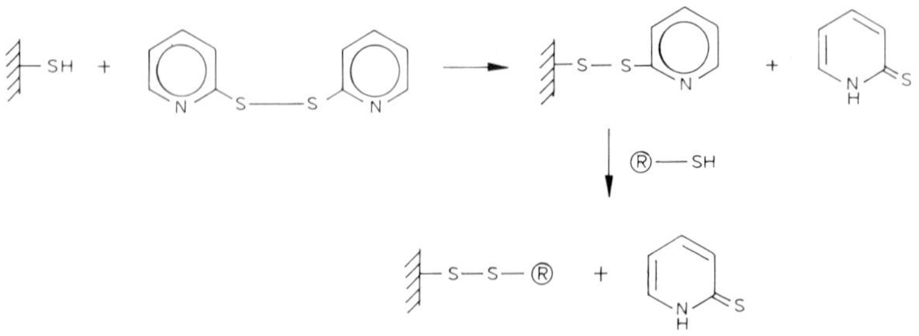

Thiol-containing ligands may also be coupled to ω-carboxylalkyl-derivatives via carbodiimide condensations. The resulting thiol ester linkage can be specifically cleaved by brief exposure to alkaline pH or to 1 M neutral hydroxylamine (Cuatrecasas, 1970). In contrast, a stable thiol ether linkage is formed on reaction with alkyl halides such as the bromoacetamidoalkyl derivative of agarose.

3.5. *Attachment of ligands via reversible linkages*

It is often desirable to provide a simple and gentle method for removing the intact ligand–complementary protein complex from the matrix. This may be achieved by inserting suitable linkages capable of specific cleavage between the ligand and the matrix and a number of methods are available:

(a) *Thiol esters.* The thiol ester linkage ($\mathrm{R{-}S{-}\overset{\overset{\displaystyle O}{\|}}{C}{-}R'}$) formed by a carbodiimide promoted reaction between a thiol and a carboxylic acid is susceptible to cleavage by brief exposure to alkaline pH or 1 M neutral hydroxylamine. This approach has been exploited to release *O*-succinyl-oestradiol and its serum binding protein from an appropriate adsorbent (Cuatrecasas, 1971).

(b) *Azo linkages.* Azo-bonded ligands may be quantitatively released by reduction of the linkage with 0.1 M sodium dithionite in 0.2 M sodium borate buffer (pH 9).

(c) *Disulphide linkages.* Ligands attached via disulphide linkages may be quantitatively removed by reaction with excess thiols such as 2-mercaptoethanol, L-cysteine, *N*-acetyl-L-cysteine, dithiothreitol or dithioerythritol.

3.6. *Methods for the measurement of the immobilised ligand concentration*

When the ligand is covalently attached to the inert matrix and the gel washed free of the unbound material, it is essential to determine the concentration of the immobilised ligand. Many methods are available for determining the concentration of covalently bound ligands.

3.6.1. *Difference analysis*

The amount of ligand coupled to a known weight or volume of gel is generally estimated by the difference between the total amount of ligand added to the coupling mixture and that recovered after exhaustive washing. In many cases this method is sufficiently accurate for most purposes and especially if the ligand may be assayed by a sensitive spectrophotometric, fluorimetric or radiometric method. In the authors laboratory, difference analysis is used as a preliminary guideline only, prior to application of a more sensitive and direct estimation.

3.6.2. *Direct spectroscopy*

For ligands which absorb at wavelengths above 250 nm it is possible to estimate the concentration of covalently attached ligand by direct spectroscopy of the gel itself. The gel is suspended in optically clear ethylene glycol, glycerol, concentrated sucrose solution or 1% aqueous polyethylene glycol (Polyox WSR 301) and read against a similar concentration of underivatised gel in a double beam spectrophotometer. It is important that the gels are thoroughly washed with these reagents prior to suspension to ensure uniformity of refractive index and thus reproducible results.

3.6.3. Solubilisation of gels

Several methods are available for solubilising derivatised agarose gels which permit quantitative spectrophotometry of the immobilised ligand. Derivatised agarose gels may be rendered soluble by warming at 75 °C with 0.1 M HCl, 0.1 N NaOH – 0.1% $NaBH_4$ or with 50% (v/v) acetic acid (Failla and Santi, 1973). Exact conditions to achieve optimal solubilisation vary from preparation to preparation and thus must be found by trial and error. Overtreatment of agarose with acid at 75 °C generates an absorption peak at 280 nm, probably due to acid-catalysed production of furfuraldehyde, whilst treatment with 0.1–1 N NaOH under similar conditions can lead to partial caramelisation of the beads with the appearance of amber-coloured solutions. The latter effect may be circumvented to some extent by incorporation of 0.1% $NaBH_4$ in the solubilisation solutions. The author routinely solubilises agarose-immobilised nucleotides in 7.5–10 N H_2SO_4 at temperatures up to 40 °C and records the optical densities directly in a spectrophotometer. Provided the gels are solubilised under the mildest conditions possible, the concentration of immobilised nucleotide determined by this method is in good agreement with that determined by more sophisticated methods.

3.6.4. Acid or enzymic hydrolysis

More vigorous treatment of immobilised ligands will hydrolyse the matrix–ligand bond and liberate either free ligand or a degradation product derived from it which may be assayed. Typically, complete hydrolysis of agarose-immobilised ligands may be achieved by heating for 1 h at 100 °C in 0.5 N HCl. Alternatively, hydrolysis in vacuo in 6 N HCl for 24 h at 110 °C is sufficient to yield products suitable for amino acid analysis and is particularly useful where the immobilised ligand is an amino acid or protein.

However, the estimation of the degree of substitution by acid hydrolysis is often complicated by the concomitant degradation of agarose to UV-absorbing furfuraldehyde derivatives (Failla and Santi, 1973). The problem may be circumvented in part by mild enzymic hydrolysis. For example, phosphate containing nucleotides may be estimated by the release of detectable fragments with a suitable diesterase or alkaline phosphatase. Enzymic hydrolysis however measures only 'accessible' ligand concentration and not total ligand concentration.

3.6.5. Elemental analysis

In cases where the immobilised ligand contains an unique element, elemental analysis can give unequivocal estimates of the ligand concentration. Thus, for example, phosphate analysis is particularly useful for immobilised nucleotides and nucleic acids and sulphur analysis has been used to estimate sulphanilamide coupled to CNBr-activated agarose (Falkbring et al., 1972). Elemental nitrogen or bromine analysis should be avoided in view of the fact that these elements are introduced during the cyanogen bromide activation step.

3.6.6. Radioactive methods

By far the most sensitive method for assaying the immobilised ligand concentration is to incorporate a radio-labelled ligand in the coupling step. The immobilised ligand concentration may be determined by difference analysis, by hydrolysis or by direct measurements on the gel. For example, aliquots of a gel containing [γ-^{32}P]GTP were resuspended in Bray's solution and counted directly in a liquid scintillation counter (Jackson et al., 1973). Likewise, ligands comprising immobilised ^{125}I-labelled proteins may be counted directly.

3.6.7. Other methods

Many methods are available for the estimation of specific immobilised ligands. For example, gel-bound sulphydryl groups may be assayed with Ellman's reagent, 5,5'-dithiobis-(2-nitrobenzoic acid), which liberates 1 mole of the strongly-coloured thionitro benzoate anion per mole of thiol. Alternatively, gel-bound thiols may be determined by their ability to react with [^{14}C]iodoacetamide when reacted with 0.01 M iodoacetamide in 0.1 M $NaHCO_3$ (pH 8.0) for 15 min at room temperature.

Subject index p. 519

TABLE 3.3
Colours produced in the sodium 2,4,6-trinitrobenzene sulphonate test

Derivative	Colour
Unsubstituted agarose or polyacrylamide	Pale yellow
Carboxyl and bromoacetyl	Yellow
Primary aliphatic amines	Orange
Primary aromatic amines	Red-orange
Unsubstituted hydrazides	Deep red
Thiols	Pink to red

The sodium 2,4,6-trinitrobenzene sulphonate (TNBS) test is a particularly convenient test for the presence of primary amino groups and others substituted on agarose or other matrices. A small quantity of the derivatised gel (0.2–0.5 g in distilled water) is added to a solution of saturated sodium borate (1 ml) or 10% $NaHCO_3$ and 1 ml of a 0.1% aqueous solution of TNBS added. The colour is allowed to develop for 2 h at 20 °C. Table 3.3 illustrates the range of colour products that may be formed with this reagent.

The test is a useful way of following the course of substitution of amino gels by carboxylic ligands and of hydrazide gels by amino ligands by assessing the relative colour intensity of the gels. The reaction can be quantified by exhaustive washing of the gels to remove unbound TNBS and solubilisation of the gels by warming in 50% acetic acid (Failla and Santi, 1973). The absorbance at 340 nm gives an estimate of the amino content of the gel.

Alkylamino substituted gels may be estimated by other methods. When alkylamines are coupled to CNBr-activated agarose, the amine moiety is converted into an isourea linkage. It is possible to determine the degree of substitution of the *N*-alkylisourea groups either by potentiometric titration (Jost et al., 1974), since isoureas retain their basicity, or by a method based on the nucleophilicity of the amine nitrogens. In the latter method, nucleophilic attack of the isoureas on the active chromogenic ester *p*-nitrophenyl acetate causes the release of stoichiometric amounts of *p*-nitrophenolate (Werber, 1976). The determination of the *p*-nitrophenolate anion at 400 nm constitutes the basis for a direct assay of alkylamino groups in substituted agaroses. Rosengren et al. (1975) determined the concentration of bound alkyl and phenyl groups in derivatised agarose by dissolution of the gel on refluxing in 85% formic acid and NMR analysis of the solubilised material.

A number of methods for the determination of bound ligands are based on the analysis of the ligand released on cleavage of the ligand–matrix bond. For example, reductive cleavage of azo- or disulphide bonds with sodium dithionite or thiols such as 2-mercaptoethanol or dithiothreitol respectively, may be used (Cuatrecasas, 1970; Brocklehurst et al., 1973). Furthermore, if a ligand is coupled to the matrix via an ester, thioester or amide bond, where the carbonyl-moiety is donated by the ligand, reaction with 1 M hydroxylamine yields soluble hydroxamic acid derivatives which are easily estimated colorimetrically at 500 nm on complexing with ferric chloride (Wolpert and Ernst-Fonberg, 1973). Tyrosine-containing proteins may be determined by applying the Lowry procedure to the intact affinity gel. Reaction of the Folin-Ciocalteau reagent with the protein generates the reduced phosphomolybdate complex which is released into the supernatant and thus determined spectrophotometrically (Koelsch et al., 1975).

3.7. *Other considerations in the preparation of affinity adsorbents*

The chemical conditions under which the ligand is coupled to the solid support must be sufficiently mild to be tolerated well by both components. These considerations could be important where the ligand being attached to the matrix is a highly labile compound such as a coenzyme or a protein. Furthermore, if the ligand is insoluble in aqueous media or does not contain a functional group which may be modified by conventional techniques, problems may be encountered in coupling the ligand to the matrix. Cross-linked agarose

Subject index p. 519

(Sepharose CL) is stable in a variety of organic solvents and thus may form a basis to prepare adsorbents with water-insoluble ligands. Pharmacia Fine Chemicals have developed a number of reactions for substituting Sepharose CL in several organic solvents and at temperatures up to 60 °C. Thus base-catalysed *O*-alkylations related to the Hakomori reaction (Hakomori, 1964) and using the methylsulphinyl carbanion as the base, provide a number of new routes for the preparation of adsorbents. Reactions for *O*-acylation in dichloroethane with pyridine as base have also been developed. These reactions are detailed in the Pharmacia literature on Sepharose CL.

Derivatised gels must be washed exhaustively to ensure not only complete removal of non-covalently bound ligand but also any products that may be formed during the coupling process. In some cases, particularly with aromatic or heterocyclic ligands, complete elimination of adsorbed material may require prolific washing over several days. In this context, organic solvents may prove useful.

The chromatographic techniques of affinity chromatography

Once the affinity adsorbent has been prepared by optimising the selection of matrix, ligand and chemistry, attention can be focussed on the conditions necessary for satisfactory adsorption and elution of the complementary macromolecule.

4.1. Considerations affecting the adsorption of complementary proteins

Affinity chromatography is normally effected by preparing a chromatographic column. The procedures and equipment necessary to establish and run chromatographic columns are discussed in detail by Peterson and by Fischer in this series of monographs. However, for routine testing of affinity adsorbents, the present author recommends the following procedure. The affinity gel (0.5 g) is weighed out and suspended in 1–2 ml of the selected equilibration buffer. A small column is prepared by cutting a Pasteur pipette to a suitable length, mounting it vertically and introducing a small wad of glass wool to act as a plug at the constricted end. Outlet tubing is connected, the column filled with equilibration buffer and the outflow controlled with a suitable clamp. The equilibration buffer is drained out until about 5 mm above the level of glass wool whence the affinity gel is introduced with a Pasteur pipette and the outlet opened. Once the bed starts to form, the gel suspension may be added continuously. The column may be packed at room temperature prior to equilibration at 4 °C or be packed directly at 4 °C. The column should be equilibrated

Subject index p. 519

by passing through a minimum of 10 column volumes of buffer at a flow rate of 8–10 ml/h controlled with a peristaltic pump. The column is now ready for sample application. Typically, the protein sample or extract will be dialysed overnight (16–24 h) against a 1000-fold excess volume of the equilibration buffer. A suitable aliquot of the dialysed sample (50–500 μl) is applied to the top of a moist bed of the affinity adsorbent, allowed to run in, a small volume of buffer added and the column connected to a reservoir of the equilibration buffer. The column is washed with equilibration buffer until protein, monitored at 280 nm, no longer appears in the eluant. The protein recovered in these void fractions is generally referred to as the *unretarded* or *inert protein* and is that which displays no affinity for the immobilised ligand. The column is now ready for development with a suitable eluant.

It is important to appreciate, however, that affinity purification need not be restricted to column procedures. Indeed in many cases it may be preferable to use a batchwise technique. Column procedures are often hampered by a deterioration in flow rates when crude samples are applied. When relatively small amounts of specific protein are to be extracted from a mixture containing a significant proportion of inert protein with an adsorbent of high affinity, the purification may be achieved more readily by adding a slurry of the specific adsorbent. The non-adsorbed proteins may be washed off either under batchwise conditions or by placing the adsorbent in a chromatographic column and proceeding as usual.

Batchwise adsorption may also be exploited as an invaluable aid to evaluate optimal conditions for adsorption and elution in affinity chromatography. Typically, a given weight of adsorbent is added to a tube or vial containing enzyme and kept in suspension by gentle agitation until binding between the enzyme and adsorbent reaches equilibrium. Brief centrifugation of the slurry then permits determination of the enzyme activity and protein concentration in the supernatant fractions. The percentage of enzyme bound to the gel may subsequently be related to the parameters being tested. A similar approach may also be utilised to assess elution conditions.

4.1.1. The choice of equilibration buffer

The buffer used to equilibrate the adsorbent should reflect the optimum pH, ionic strength, temperature and chemical composition (metal ions or other specific factors) necessary to achieve a strong interaction between the insolubilised ligand and the protein. In this context it is important to realise that conditions optimal for catalytic activity may not necessarily be optimal for binding of the affinity ligand. Data pertaining to these considerations may be obtained from the scientific literature relating to the enzyme or protein to be purified.

4.1.2. The sample volume, flow rate and equilibration time

Generally speaking, the volume of the sample applied to an affinity adsorbent is not critical if the substance of interest has a high affinity for the immobilised ligand. The substance will form a narrow zone at the top of the gel bed. On the other hand, weakly-bound proteins should be applied in a relatively small volume, to circumvent co-elution with the inert proteins in the void volume.

The adsorption equilibrium between the immobilised ligand and the macromolecule to be purified is often reached at a very slow rate. The sample should thus be applied to the column at the lowest flow rate acceptable from a practical point of view. If very high flow rates are used, especially when combined with high sample protein concentrations, small amounts of complementary enzyme may appear in the void volume along with the protein impurities (Cuatrecasas et al., 1968; Lowe et al., 1974). However, even with excessively high flow rates, the complementary enzymes could be retained if more dilute samples were applied.

The time dependence of the interaction between the enzyme and the immobilised ligand is also reflected in the effect of incubation time. For example, Fig. 4.1 shows that under batchwise conditions there is a rapid increase in the percentage of lactate dehydrogenase bound to N^6-(6-aminohexyl)-AMP–Sepharose during the initial time period, followed by a gradual progression to 100% binding after

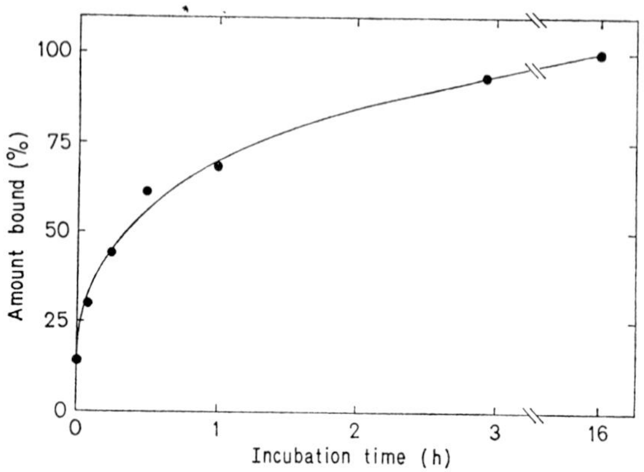

Fig. 4.1. The effect of incubation time on the capacity of N^6-(6-aminohexyl)-AMP–agarose for lactate dehydrogenase under batchwise conditions. The adsorbent (0.5 g moist weight containing 1.5 μmol AMP/ml) was suspended in 100 ml 10 mM KH_2PO_4–KOH buffer (pH 7.5) containing 10 IU pig heart lactate dehydrogenase for various times. The proportion of enzyme bound was deduced by assaying the supernatant for enzyme activity. Reproduced with permission from Lowe et al. (1974), Eur. J. Biochem., *41*, 341.

16 h (Lowe et al., 1974). Similar effects of incubation time may be demonstrated under conditions of column chromatography. For example, if glycerokinase and lactate dehydrogenase are applied to a column of N^6-(6-aminohexyl)-AMP–Sepharose, and allowed to incubate for times up to 3 days at 4 °C prior to elution, both the efficiency of the column and the strength of the interaction increase with time (Lowe et al., 1974). In the case of glycerokinase, the percentage of enzyme bound also increased as a function of time. Similar observations have been made for other enzymes and adsorbents (Lowe and Gore, 1977).

These effects of equilibrating the enzyme with the adsorbent for a period of time prior to elution are particularly relevant from a practical standpoint since not only is the strength of the interaction increased but often the gel bed shows enhanced resolution (Lowe et al., 1974). It is worthwhile, therefore, bearing in mind that if the interaction between the enzyme and immobilised ligand is weak some preincubation of the system prior to elution may pay dividends. However, it should be realised that non-specific adsorption may produce similar time dependent effects and result in a considerably reduced recovery of enzyme on subsequent elution. These effects are discussed in detail in § 4.4.

4.1.3. *The effect of protein concentration*

With interacting systems of average or high affinity there is no apparent effect of complementary enzyme concentration on the capacity of an affinity adsorbent under column conditions. The enzyme is bound to the top of the column and within practical limits is independent of the concentration initially applied. Enzyme will appear in the void volume eluate if high flow rates are used with highly loaded columns (Cuatrecasas et al., 1968; Lowe et al., 1974). Under batchwise conditions, however, an effect of enzyme concentration has been observed (Lowe et al., 1974). Thus the percentage of glycerokinase and lactate dehydrogenase bound to N^6-(6-aminohexyl)-AMP–Sepharose increased with enzyme concentration in a manner typical of an enzyme–ligand equilibrium.

The interaction between the complementary enzyme and the immobilised ligand appears to be almost entirely independent of the concentration of inert protein in the sample except at very high flow rates.

4.1.4. *The effect of temperature*

In general, the strength of adsorption to an affinity gel decreases with increasing temperature. Thus, for example, the effect of temperature on the binding of lactate dehydrogenase to immobilised-AMP is shown in Fig. 4.2. The concentration of NADH required to elute the enzyme decreased with increasing temperature and was particularly marked over the range 0–10 °C. The decreased binding in this range has particular significance since this range of temperature is that generally experienced in a typical laboratory cold room. Thus for reproducible purifications by affinity chromatography it is essential

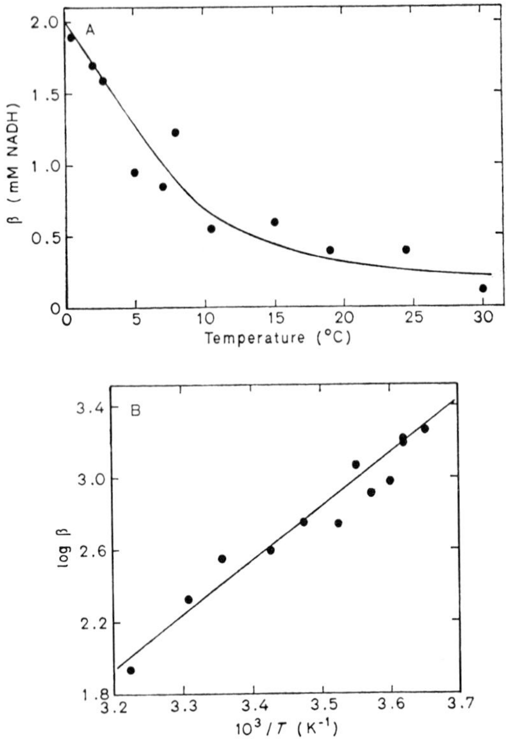

Fig. 4.2. The effect of temperature on the binding of lactate dehydrogenase to N^6-(6-aminohexyl)-AMP–agarose. The enzyme sample (5 IU) containing 1.5 mg bovine serum albumin (100 μl) was applied to a column (5 mm × 50 mm) containing 0.5 g immobilised-AMP (1.5 μmol AMP/ml) at several temperatures between 0.5 °C and 30 °C. The ordinate represents the concentration of NADH required to elute the peak of enzyme activity on a linear gradient of NADH (0–5 mM, 20 ml total volume). Reproduced with permission from Harvey et al. (1974), Eur. J. Biochem., *41*, 353.

that the temperature be carefully controlled, preferably by the use of jacketed columns. Furthermore, the use of different temperatures for adsorption and elution can have very beneficial effects on the subsequent purification. Thus, for example, tight binding may be effected at 4 °C and subsequently elution achieved under mild con-

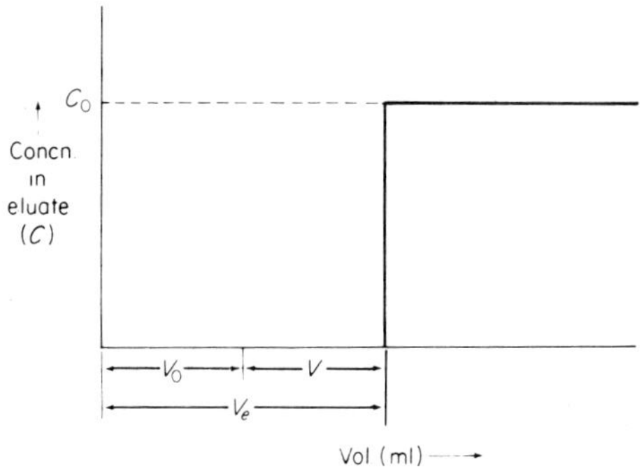

Fig. 4.3. Determination of the capacity of an affinity adsorbent by frontal analysis. Reproduced with permission from Lowe and Dean (1974) *Affinity Chromatography*, John Wiley and Sons Ltd., London.

ditions by raising the temperature to 25 °C or above (Harvey et al., 1974; Paulson et al., 1977).

4.2. The capacity of affinity adsorbents

The capacity of a selective adsorbent is determined principally by two interdependent sets of parameters: (1) the correct choice of matrix, spacer molecule and ligand to optimalise the enzyme–ligand interaction and (2) the way in which the capacity is determined by such dynamic factors as flow rate, equilibration time and adsorption technique. Assuming the design of the adsorbent has been optimised, the operational capacity of an affinity gel is best determined by frontal analysis. A given concentration of the complementary protein (C_0) is applied to the adsorbent continuously and its emergence monitored. As the bed becomes saturated with the adsorbate, the solution breaks through at the same concentration it had on entering the column (Fig. 4.3). The volume of eluant that appears up to the 'step', where the concentration of the complementary protein increases rapidly to C_0 over a small volume, is called the *retention volume* (V_e). It comprises the interstitial volume (V_0) and the volume of solution from which the adsorbate was removed by the adsorbent (V), i.e.,

$$V_e = V_0 + V$$

If m is the total weight of the affinity adsorbent in grams, the capacity of the gel, i.e., the amount of adsorbate specifically adsorbed per gram, q, is $q = (V/m)\, C_0$ and the total amount adsorbed by the bed is qm, i.e., $V\, C_0$.

It should be appreciated however that the capacity, consistent with the emergence point of the monitored species, is dependent on the rate of application of the original sample. At relatively high sample application rates, affinity equilibrium is not attained and premature emergence of adsorbate will be observed. This will lead to underestimation of the operational capacity of the adsorbent. The effective capacity of an adsorbent may also be deduced by incubating a known weight of adsorbent (m) with a given volume of solution of concentration C_0 and subsequently, after equilibrium has been established, measuring the new lower concentration C. The capacity, q, is then calculated from $q = (C_0 - C)/m$.

Little data is available to estimate the capacity theoretically from the known immobilised ligand concentration and other parameters of the system. It appears that the effective capacity of an adsorbent is considerably lower, in fact often $<1\%$, of the theoretical capacity based on the ligand concentration (Lowe et al., 1973; Harvey et al., 1974; Nishikawa et al., 1976). Presumably, the effective capacity of a specific adsorbent is determined by the concentration of immobilised ligand that is freely available for interaction with the complementary enzyme. Nevertheless, despite these difficulties, an estimate of the operational capacity of the adsorbent for the protein to be purified is useful to assess the loading required to achieve maximal efficiency of operation.

4.3. The elution of specifically adsorbed proteins

When a sample containing the protein to be purified is applied to a

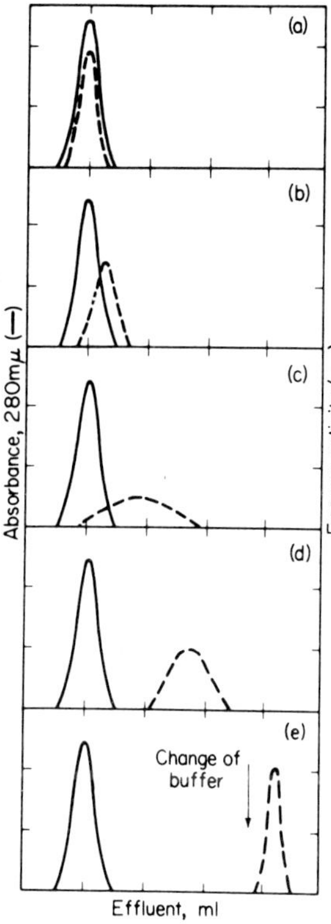

Fig. 4.4. Theoretical elution profiles for the purification of a specific enzyme (– – –) from a crude protein mixture by affinity chromatography. The specific enzyme is eluted in the void volume of the column of adsorbent (a) together with the inert non-adsorbed proteins (———), but is retarded to various extents by adsorbents (b) to (d). Successful application of affinity chromatography is depicted in (c) where an alteration in the buffer is required to effect elution of the specifically adsorbed protein. Reproduced with permission from Cuatrecasas (1972), Adv. Enzymol., *36*, 29.

column of the selective adsorbent and the column is washed with the equilibrating buffer, several elution profiles are possible depending on the effectiveness of the adsorbent under the experimental conditions chosen (Fig. 4.4). If the matrix is underivatised or if the adsorbent is wholly ineffective, the protein to be purified will emerge with the inert protein in the void volume (Fig. 4.4a). If the adsorbent displays relatively weak affinity for the desired protein, the latter may be retarded relative to the void volume by subsequent partition down the column and result in one of several elution patterns (Fig. 4.4b–d). If the adsorbent has been correctly designed and constructed the protein to be purified will be strongly adsorbed as a concentrated zone at the top of the column and will require a change in buffer composition to effect desorption. The change in buffer composition, i.e., the introduction of the eluant may be achieved in several ways.

The technique of 'stepwise' elution is commonly employed, particularly for routine and rapid separations. After inert proteins have been washed off the column the composition, pH, ionic strength or temperature of the buffer, is changed and elution effected by percolating the new buffer through the column. However, whilst this is a facile technique, the use of stepwise elution may generate spurious peaks, so-called band splitting, by collecting the tail end of a previous peak at the change in solvent and producing the illusion of an additional peak.

In 'pulse elution', the eluant is applied to the column in a small volume and washed through with the equilibrating buffer. The 'pulse' of eluant migrates through the chromatographic bed as a compact zone, carrying the eluted protein with it. Thus conversion to a new buffer or eluant medium can be accomplished within a fraction of the total bed volume. This elution technique is particularly applicable to situations where the cost of the eluant, such as a coenzyme, is a controlling factor.

The elution strength of the solvent is increased continuously in gradient elution. This generates a concentration gradient of eluant down the column and results in sharp elution of the adsorbate. Full details of the principles and equipment required to establish linear and non-linear gradients are given by Peterson in this series of monographs.

4.3.1. Non-specific elution techniques

The selection of the elution procedure will be determined to some extent by the cost of the eluant, the stability of the specific protein and adsorbent, the strength of the protein–ligand interaction and the specificity of the adsorbent. Non-specific elution techniques tend to be relatively inexpensive and require changing the pH, ionic strength, dielectric constant or temperature of the buffer. Ideally, the eluant buffer should sufficiently alter the conformation of the protein to reduce its affinity for the immobilised ligand without compromising the stability of the protein or adsorbent. In most cases, alteration of a single physical variable is sufficient although in some cases simultaneous alterations in two parameters may prove more effective than alterations in either alone.

In many cases, a change in pH is sufficient to elute adsorbed proteins. The pH shift required can often be evaluated from the known behaviour of the system in free solution, i.e., a change in the ability to form a binary complex or in enzyme activity as a function of pH. However, the stability of the substance of interest and of the matrix generally places a lower limit on the pH which may be used. For example, trypsin may be eluted from soybean trypsin inhibitor–Sepharose by a drop in pH from 7.8 to 3.0 and antigen–antibody complexes may be dissociated by exposure to glycine–HCl buffer (pH 2.5). Dissociation of proteins from adsorbents of very high affinity may require a combination of extremes of pH and protein denaturants such as guanidine–HCl or urea. For example, elution of hens egg-white avidin from biocytin–agarose requires a combination of 6 M guanidine–HCl and pH 1.5 (Cuatrecasas and Wilchek, 1968). In such cases, it is advisable to promptly restore the native protein structure by removal of the denaturant by neutralisation, dilution or dialysis. Protein denaturants such as urea and guanidine–HCl (Nishimura et al., 1976; Stassen, 1976), detergents and low concentrations of organic solvents such as dioxane, ethylene glycol and dimethylformamide (Lowe and Mosbach, 1975) are extremely effective eluants in some cases.

Alterations in the ionic strength of the eluting buffer to effect desorption is a particularly facile and easily monitored approach. Typically, the biospecifically adsorbed protein is eluted by the addition of 0.5 M or 1.0 M NaCl to the starting buffer solution

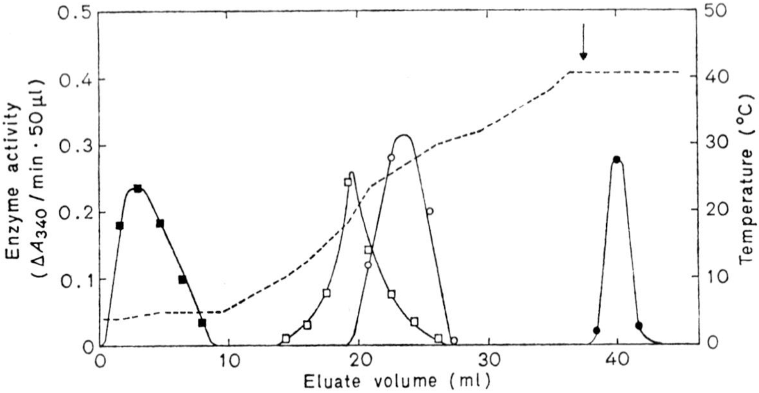

Fig. 4.5. The resolution of an enzyme mixture on N^6-(6-aminohexyl)-AMP–agarose by a temperature gradient. The enzyme sample (100 µl), containing 5 IU of each enzyme and 1.5 mg bovine serum albumin was applied to a column (5 mm × 50 mm) containing 0.5 g immobilised-AMP (1.5 µmol AMP/ml) at 4.7 °C. The column was equilibrated at each individual temperature for 5 min prior to elution with 1.6 ml equilibration buffer, 10 mM tricine–KOH (pH 7.5) containing 10 mM glycerol, 5 mM $MgCl_2$, 1 mM EDTA and 0.02% sodium azide. A 'pulse' (200 µl) of 5 mM NADH in the equilibration buffer was added at the arrow. Bovine serum albumin was located in the void volume (0–4) ml) and hexokinase (■), glycerokinase (□), yeast alcohol dehydrogenase (○) and pig heart lactate dehydrogenase (●) were assayed in the effluent. Reproduced with permission from Harvey et al. (1974), Eur. J. Biochem., *41*, 353.

although other agents such as 0.5 M NH_4Cl or 1 M Tris–HCl may also be used. In typical cases where the specific protein is tightly bound, chaotropic ions may prove useful. The elution of IgE from agarose-immobilised anti-IgE requires high concentrations of sodium thiocyanate and illustrates the use of chaotropes in the dissociation of antigen–antibody complexes (Bennich and Johansson, 1971).

The elution technique can often provide an additional means of enhancing the resolution and/or purification of the adsorbed proteins. Thus, the application of gradients of pH, ionic strength, dielectric constant or temperature may achieve a valuable secondary resolution by virtue of the different sensitivities of the adsorbed proteins, even

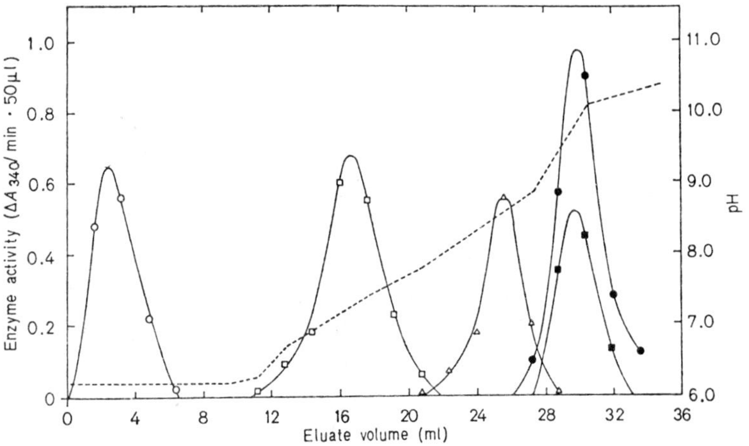

Fig. 4.6. The resolution of a mixture of dehydrogenases on N^6-(6-aminohexyl)-AMP–agarose by a pH gradient. The enzyme mixture (100 μl) containing bovine serum albumin (1.5 mg) and 5 IU of each enzyme, was applied to a column (5 mm × 50 mm) containing 0.5 g immobilised-AMP equilibrated with 10 mM KH_2PO_4–KOH (pH 6.0). The column was washed with equilibration buffer (pH 6.0) prior to development with a pH gradient (pH 6–10; 10 ml equilibration buffer against 10 ml 30 mM K_2HPO_4–KOH (pH 11.0) in a linear gradient apparatus). Bovine serum albumin (○), malate dehydrogenase (●), glucose-6-phosphate dehydrogenase (□), pig heart lactate dehydrogenase (■) and yeast alcohol dehydrogenase (△) were assayed in the effluent. Reproduced with permission from Lowe et al. (1974), Eur. J. Biochem., *41*, 347.

though their affinities may have been comparable under the original conditions of adsorption. Figures 4.5 and 4.6 illustrate the resolution of a mixture of several dehydrogenases and kinases on N^6-(6-amino-hexyl)-AMP–Sepharose with a temperature and a pH gradient, respectively. The elution of proteins with temperature gradients is a particularly valuable approach since cooling restores the eluant to the original buffer composition and thus permits further studies directly on the desorbed protein. Furthermore, the eluant may be re-applied directly to a second affinity adsorbent without the necessity of removing unwanted eluants by dialysis or gel filtration.

4.3.2. Special elution techniques

There are a number of elution techniques which are not directly related to the biological function of the complementary macromolecule, but which may be applicable in some circumstances. The special properties of borate buffer for example can be exploited to advantage. The galactosyltransferase of bovine milk lactose synthetase is readily adsorbed to UDP–Sepharose in the presence of manganous ions. Under these conditions the presence of high concentrations (0.5–1.0 M) NaCl were without effect. In contrast, at pH 8.5 borate buffers eluted the enzyme in good yield, presumably by forming a complex with the ribose moiety of UDP and thereby weakening the binding of the enzyme. Similarly, *E. coli* β-galactosidase is strongly adsorbed to *p*-aminophenyl-β-D-thiogalactopyranoside covalently attached to Sepharose (Steers et al., 1971). The enzyme could only be partially eluted with substrates but was effectively eluted with 0.1 M borate (pH 10.05). Subsequent investigations showed that 0.1 M borate (pH 8.0) was a satisfactory eluant for the enzyme, whilst Tris buffer was ineffective even at pH values up to 9.5. The use of salicylate to elute NAD^+-dependent dehydrogenases from immobilised-AMP is also an example of this type of approach (Ohlsson et al., 1972).

An alternative approach to elute specifically bound proteins is to remove the intact ligand–protein complex by selective cleavage of the matrix–ligand bond. This technique may be applied whenever ligands are attached to agarose via susceptible bonds such as azo, thiol ester, alcohol ester or disulphide (§ 3.5). The method is particularly suitable for high-affinity systems where the desired macromolecule would be irreversibly denatured by exposure to the extremes of pH, or protein denaturants, necessary to effect elution. Thus, for example, the serum oestradiol-binding protein displays high affinity for oestradiol ($K_d \sim 10^{-9}$ M) and is particularly susceptible to denaturation. The protein may be tightly adsorbed to oestradiol–agarose and subsequently removed in active form by reductive cleavage of the azo-linkage with dithionite (Cuatrecasas, 1970).

4.3.3. *Specific elution techniques*

In many cases, non-specific methods of elution are quite satisfactory to achieve the desired purification. However, there are a number of circumstances where additional selectivity in the elution procedures is desirable. For example, where the ligand is charged, non-specific proteins may be retained by ion exchange. The problem cannot be controlled by increasing the ionic strength in this case since this is accompanied by a reduction in the affinity of the complementary macromolecule for the immobilised ligand. The contaminants respond to increases in the ionic strength and are therefore co-eluted with the desired protein. This problem is well documented for acetyl-cholinesterase and has prompted the development of specific elution procedures. Co-elution with non-specific methods of proteins non-specifically bound to the matrix–spacer arm assembly must also be anticipated unless it can be demonstrated unequivocally that the adsorption and elution processes are specific. Furthermore, in cases where the ligand itself displays affinity for several macromolecules, such as in the case of immobilised 'general ligands' or 'group specific' adsorbents, some additional means of increasing selectivity is desirable. A gradient of a specific displacer should, in principle, effect the resolution of enzymes or isoenzymes with qualitatively similar but quantitatively different specificity.

Almost any free ligand which competes with the immobilised ligand for the enzyme is potentially able to effect elution of the bound enzyme. Thus, for example, high concentrations of the same ligand that is immobilised should elute the bound enzyme. In the case of ribonucleotide reductase it was found that concentrations of dATP, ATP and dAMP required to elute the enzyme from the equivalent immobilised analogues were roughly proportional to the concentrations required to stimulate the enzyme (Berglund and Eckstein, 1974). Similar correlations have been found for other systems, although no general rules for estimating the concentration of competing ligand necessary to elute a bound enzyme are available. In preliminary experiments the present author generally selects a concentration of eluant about 20 times greater than the K_m or K_i value in free solution. The conditions may subsequently be refined depending on the effectiveness of this initial trial elution. If the affinity of the enzyme for the immobilised ligand is particularly high a concentration of free competing ligand several orders of magnitude greater than the K_m or K_i value in free solution may be necessary to effect elution. Conversely, for ligand–enzyme systems of low affinity, comparatively low concentrations of eluant may be required.

As a general rule, it is preferable to use an eluant ligand other than that which is immobilised to the matrix, since the system may then exhibit dual specificity; firstly for the immobilised ligand and secondly for the eluant ligand. For example, cAMP-dependent protamine kinase may be eluted from 8-(6-aminohexyl)-cAMP–agarose by 5 mM AMP. Furthermore, greater selectivity of elution is achieved if the eluant ligand displays higher affinity for the enzyme to be purified than the ligand used for immobilisation. Thus, NAD^+-dependent dehydrogenases are quantitatively eluted from N^6-(6-aminohexyl)-AMP–agarose columns by low concentration NADH pulses or gradients with little concomitant non-specific elution of inert proteins. Low concentrations of allosteric effectors are equally effective in this approach. In general, therefore, the eluant ligand should be selected such that it displays high affinity for the enzyme to be purified and such that it is structurally distinct from the immobilised ligand. This ensures double specificity in adsorption and elution and minimises the release of non-specifically bound inert proteins along with the protein to be purified.

Dual specificity for more than one ligand also may be exploited in 'negative elution'. This extra degree of specificity may be introduced for bi- and multi-substrate enzymes which have compulsory ordered kinetic mechanisms (§ 2.4.1). For example, lactate dehydrogenase is strongly retained on an immobilised analogue of pyruvate in the presence of the leading substrate, 100 μM NADH, and is promptly eluted on its removal from the irrigating buffer (O'Carra and Barry, 1972). Likewise, the purification of lactose synthetase A protein from

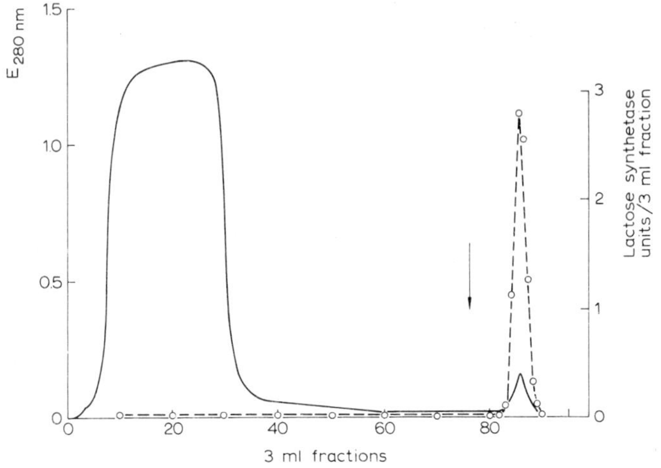

Fig. 4.7. Affinity chromatography of a solution (72 ml) of partially purified lactose synthetase A protein on a column (1.1 cm × 23 cm) of agarose–α-lactalbumin equilibrated with 0.01 M Tris–HCl (pH 7.5) containing 0.04 M KCl and 3 mM *N*-acetylglucosamine. Elution was continued with the same buffer after application of the sample until 76 fractions had been collected (arrow), whence elution was with the Tris buffer containing 0.04 M KCl only. A_{280}nm (———); lactose synthetase activity (O———O). Reproduced with permission from Andrews (1970), FEBS Lett., 9, 297.

human milk has been effected by affinity chromatography on α-lactalbumin–agarose in a buffer containing 3 mM *N*-acetyl-D-glucosamine (Andrews, 1970). Lactose synthetase A protein was strongly adsorbed to the column but could be released with a 40-fold enrichment in specific activity when *N*-acetyl-D-glucosamine was omitted from the eluant buffer (Fig. 4.7). It is clear therefore that a considerable degree of specificity may be achieved by 'negative elution', i.e., by discontinuation of the complementary ligand. However, the selectivity of 'positive' elution with a competitive counter ligand may also be enhanced by exploitation of ternary complex formation. Thus, for example, alcohol dehydrogenase may be specifically eluted from an immobilised-AMP column with a solution of 0.5 mM NAD^+ plus 3 mM hydroxylamine. Lactate dehydrogenase may be eluted subsequently with 0.5 mM NAD^+ plus

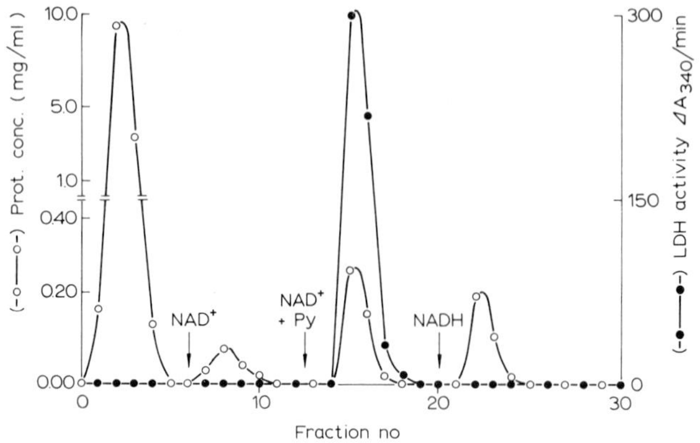

Fig. 4.8. Purification of lactate dehydrogenase from ox heart by affinity chromatography. Crude extract (1.0 ml) was applied to an N^6-(6-aminohexyl)-AMP–agarose column (15 mm × 40 mm, containing 1.6 g packed gel) equilibrated with 0.03 M potassium phosphate (pH 7.3) containing 1 mM cysteine. After washing to remove unbound proteins elution was effected with (1) 0.5 mM NAD^+, (2) 0.5 mM NAD^+ plus 5 mM pyruvate and (3) 0.5 mM NADH. To ensure adequate ternary complex formation at stage (2), the flow was stopped for 12 h following application of the NAD^+–pyruvate mixture, prior to elution. Reproduced with permission from Ohlsson et al. (1972), FEBS Lett., *25*, 234.

5 mM pyruvate, an abortive ternary complex (Ohlsson et al., 1972). The latter technique may be exploited to purify lactate dehydrogenase from a crude extract of ox heart. Figure 4.8 shows that whilst 0.5 mM NAD^+ alone is ineffective as an eluant, 0.5 mM NAD^+ plus 5 mM pyruvate readily elutes the enzyme from the column.

4.4. Non-specific adsorption

More often than not, despite optimalisation of the design of the affinity adsorbent, the adsorption and elution of the desired macromolecule does not acheive the anticipated enchancement in specific activity. It is now widely recognised that these operational shortcomings are attributable to non-specific adsorption of inert proteins to the affinity adsorbent (O'Carra et al., 1974; Nishikawa et al., 1976).

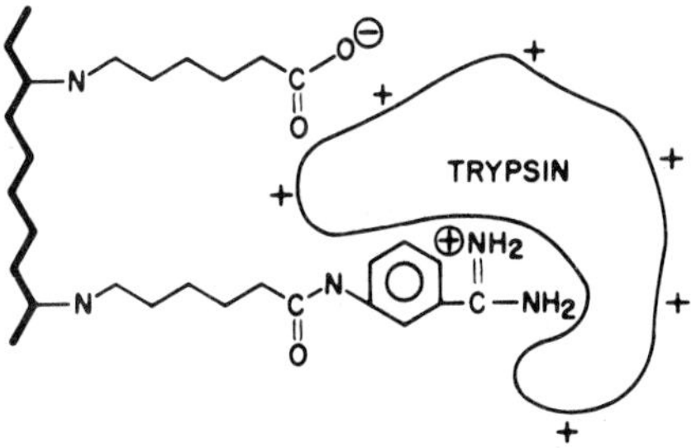

Fig. 4.9. The effect of extraneous ionic interactions in affinity chromatography. Reproduced with permission from Nishikawa et al. (1976), J. Macromol. Sci. Chem., *A10*, 149.

4.4.1. Ionic effects

The presence of ionic groups on any chromatographic adsorbent will affect the elution behaviour of polyelectrolytes such as proteins. These interactions, whilst central to the more established technique of ion-exchange chromatography, may generate problems of non-specific binding in affinity chromatography.

Incomplete attachment of ligands to preformed matrix–spacer arm assemblies can introduce extraneous ionic groups into the adsorbent. Such problems were encountered in the preparation of affinity adsorbents for the purification of trypsin and thrombin (Hixson and Nishikawa, 1973). Figure 4.9 demonstrates that the expected active site interaction of the benzamidine ligand with the trypsin is supplemented with an ionic interaction between the positively charged enzyme and the carboxylate anion of residual ε-amino caproate spacer molecules. Non-specific interactions with residual spacer molecules may be substantial even if 90% of the spacer molecules are substituted. Chymotrypsin may be used to test the functional specificity of the adsorbent. At pH 8, both trypsin and chymotrypsin are polycations but display affinities for basic and apolar substrates respectively, and thus if the adsorbent is quantitatively substituted with benzamidine only trypsin should be adsorbed. If chymotrypsin is adsorbed on a test run, the procedure for coupling the ligand to the matrix–spacer arm should be repeated until the resulting gel will not bind chymotrypsin. Similar tests for functional specificity may be desired for other systems and ensure minimisation of non-specific effects due to this cause. However, in the opinion of the author there is no substitute for prior synthesis and characterisation of the ligand–spacer arm assembly followed by introduction of the ensemble into the carrier gel. However, for many workers such an approach requires considerable expertise in conventional organic chemistry compared to the solid phase 'Aufbau' approach.

Unwanted charges may also be introduced into the adsorbent by linkages of the spacer molecule to the matrix backbone with the cyanogen bromide activation procedure (§ 3.1.1.1.7). The resulting isourea linkage exhibits a pK_a value of 10.4 and is thus protonated at physiological pH values. The use of acylhydrazides such as adipic dihydrazide circumvents this problem since the isourea linkage with acylhydrazides has a markedly lower pK_a value and is unprotonated at physiological pH values. Figure 4.10 illustrates the importance of these charge considerations in the binding of β-galactosidase to appropriate affinity and control adsorbents (Nishikawa et al., 1976). It is evident that the polyanionic character of this protein dominates and directs the binding to cationic charges inadvertently introduced into the gel. It is particularly interesting to note that where there is no proximal cationic group, as in the acylhydrazide linked adsorbent, the phenylthiogalactoside ligand displays the weak affinity expected from its inhibition constant of approximately 5 mM.

These problems are almost certainly present in many of the affinity purifications reported to date and may be relieved to some extent by the inclusion of NaCl in the buffer medium. A concentration of at least 0.15 M in the equilibrating buffer is recommended to optimalise specificity in binding. Alternatively, linkage via bis-epoxides may prove preferable.

4.4.2. *Ionic ligands*

Ionic ligands may also create troublesome ion-exchange effects in affinity adsorbents. For example, the purification of trypsin with an immobilised *m*-aminobenzamidine ligand (Hixson and Nishikawa,

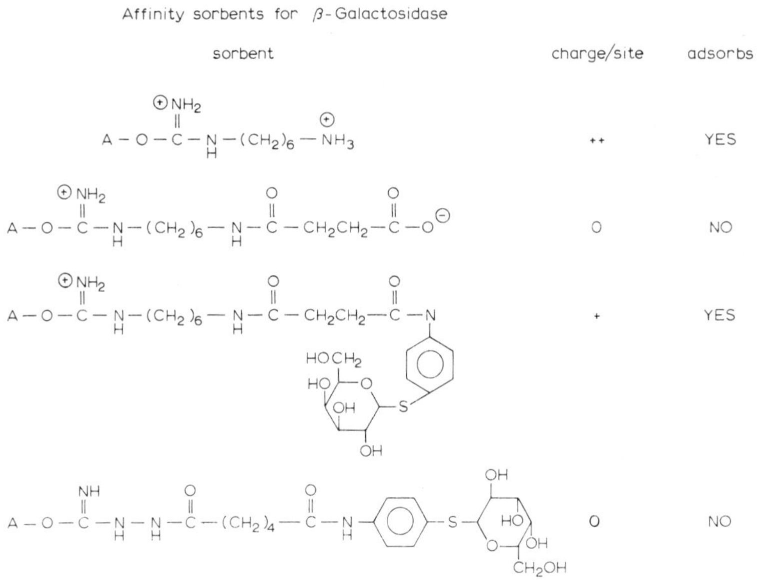

Fig. 4.10. The binding of β-galactosidase to several control and affinity adsorbents to illustrate the importance of ionic interactions in the adsorption process. A = agarose backbone. Reproduced with permission from Nishikawa et al. (1976), J. Macromol. Sci. Chem., *A10*, 149.

1973) and the adsorption of lactate dehydrogenase to immobilised oxamate (O'Carra and Barry, 1972) both required inclusion of 0.5 M NaCl or KCl to the irrigating buffer to enhance the specificity of binding. However, the concentration of salt required to minimise the non-specific effects should be carefully evaluated in the light of the effect of ionic strength on the binding affinity of the enzyme to the ligand. The effect of high ionic strengths on the binding of acetylcholinesterase to adsorbents comprising immobilised quaternary ammonium salts is well documented (Schmidt and Raftery, 1972). A careful balance must be achieved between the enhanced specificity in binding and the decreased affinity for the immobilised ligand occasioned by increasing the ionic strength of the equilibrating buffer.

4.4.3. Hydrophobic effects

Proteins are not only complex polyelectrolytes but also may have hydrophobic crevices or pockets at or near their otherwise hydrophilic surfaces. A number of recent reports have highlighted the effects of hydrophobic interactions in affinity chromatography (Yon, 1972; Lowe, 1977). It has been suggested that such effects emanate from the nature of the 'spacer arms' used to separate the immobilised ligand from the matrix backbone. The aliphatic spacer molecules commonly employed are believed to interact non-specifically with accessible hydrophobic patches on the enzyme surface. The indiscriminate use of long hydrophobic spacer molecules in the preparation of affinity adsorbents has been questioned by O'Carra and coworkers (O'Carra et al., 1974) in their studies with *E. coli* β-galactosidase. The presence of hydrophobic interactions in affinity chromatography has been demonstrated in the interaction between lactate dehydrogenase and immobilised-AMP (Lowe and Mosbach, 1975) and between 3α-hydroxy-steriod dehydrogenase and immobilised glycocholic acid (Aukrust et al., 1976). In the former case, supplementation of the irrigation buffers with low concentrations of ethylene glycol (0–20%), dioxane (0–10%) or urea (0–1 M) dramatically improved the recovery of lactate dehydrogenase from the immobilised-AMP adsorbent. Other organic solvents such as glycerol, butanol, ethanol or *N,N'*-dimethylformamide are equally applicable although the precise concentrations tolerated by the particular system under study should be ascertained by trial and error. The use of organic solvents in column irrigants is an extremely facile and effective way to improve performance in affinity purifications.

4.4.4. Hydrophobic ligands

The elimination of hydrophobic interactions introduced by virtue of the nature of the spacer molecule may also be achieved by the construction of adsorbents containing more hydrophilic arms (§ 3.2.2). However, where the interfering hydrophobic adsorption derives from the nature of the ligand itself, this approach is inapplicable and the use of organic solvents is recommended. Thus, the binding of *Pseudomonas testosteroni* 3α-hydroxysteroid dehydrogenase to glycocholic acid immobilised to Sepharose 4B is greatly enhanced in the presence of 1 mM NAD^+ (Aukrust et al., 1976), suggesting an ordered reaction sequence with NAD^+ as the leading substrate. A largely hydrophobic NAD^+-independent interaction between the enzyme and the immobilised ligand was also found and interfered with the biospecific NAD^+-dependent binding. The NAD^+-independent interaction with the adsorbent was extinguished completely by incorporating 10% *N,N'*-dimethylformamide in the adsorption buffer, and greatly reduced by 5% *n*-butanol or 20% glycerol.

4.4.5. Compound affinity

Affinity binding to an adsorbent may be considerably enhanced by the simultaneous expression of ionic and hydrophobic interactions, so-called compound affinity (O'Carra et al., 1974). More often than not these two weak types of interactions can mutually reinforce each other so that the resultant effect is much greater than the sum of the two individually. In systems displaying relatively weak biospecific interactions with the immobilised ligand, the reinforcing effect of compound affinity is a desirable feature without which no affinity and hence purification would be experienced. In such cases, the ionic strength should be manipulated by trial and error to achieve optimal separations. With high affinity systems, where non-specific interactions are a complicating feature, the use of both elevated concentrations of salt and organic solvents is recommended. The precise conditions of ligand concentration, pH, ionic strength and temperature to effect optimum purification must be found for each system under investigation.

4.5. Methods of regenerating 'used' adsorbents

Despite optimalisation of adsorption and elution conditions to reduce or eliminate non-biospecific adsorption, irreversible adsorption to affinity matrices may present a serious problem, particularly where enzymes are being isolated from crude extracts. This problem is manifested by a decreased effectiveness of the adsorbent to bind the enzyme after several passes through the same column. The physical appearance of the affinity matrix often suggests clumping of the gel particles after several operations and the gel may appear distinctly coloured when contrasted to the clean opalescent appearance of an unused matrix. These observations suggest an accumulation of denatured protein on the adsorbent and indicate that more vigorous washing procedures between chromatographic runs are required. Typically, the adsorbents should be routinely washed with 2 M KCl/6 M urea after each run. The incorporation of dioxane or dimethylformamide may prove beneficial in some cases. Furthermore, incubation of 'aged' matrices overnight with a non-specific protease such as pronase (Holroyde et al., 1976) restores the capacity of the columns almost completely (Fig. 4.11). Consequently, the working life of columns may be considerably prolonged by introducing the pronase treatment after every second use of the column followed by the 2 M KCl/6 M urea wash.

4.6. *Criteria for affinity chromatography*

The complications and restrictions imposed by the presence of non-specific and/or compound binding would suggest that some criteria to assess the extent of biospecificity of the system under study might prove useful. This would permit optimisation of the biospecific element and thus achieve more satisfactory purifications. Unfortunately, the multiplicity and diversity of biological interactions make generalisations difficult and the list of criteria below are intended as a guideline only.

(1) The enzyme or protein should not be bound to matrices to which an inactive substrate analogue has been attached by similar procedures nor to control matrices which bear no ligand (§ 2.2.3.2).

(2) The binding of the enzyme or protein to the affinity matrix should correlate well with the known properties of the system in free solution.

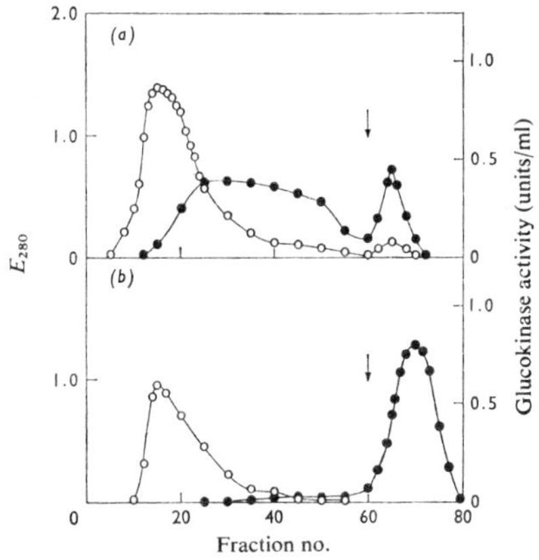

Fig. 4.11. The effect of pronase treatment on the regeneration of affinity adsorbents. The chromatography of partially purified rat hepatic glucokinase on agarose–*N*-(6-aminohexanoyl)-2-amino-2-deoxy-D-glucopyranose (a) that had been used in previous experiments with liver extracts at least 6 times. Between operations this matrix had been washed exhaustively with 2 M KCl–6 M urea. In (b) the same column was treated with pronase overnight. A_{280nm} (○); glucokinase activity (●).

Reproduced with permission from Holroyde et al. (1976), Biochem. J., *153*, 351.

(3) The retention of the enzyme in high ionic strength buffers is a useful criterion for affinity chromatography, although loss of binding capacity at high ionic strengths may reflect the electrostatic nature of the ligand–macromolecule interaction (§ 4.4.2).

(4) Strong evidence in favour of a specific interaction is suggested if complete inhibition of enzyme activity, such as can be achieved by reaction with active site-directed irreversible inhibitors, results in a loss of ability of the enzyme to be bound to the specific adsorbent. To take an interesting case in point, Stevenson and Landman (1971) selected 4-phenylbutylamine as a ligand to purify chymotrypsins from a variety of sources. As a control, to test the specificity of their gel, Stevenson and Landman inhibited chymotrypsin with the active site inhibitor tosylphenylalanyl chloroketone (TPCK) as shown in Fig. 4.12. Elution profiles A and B illustrate the behaviour of

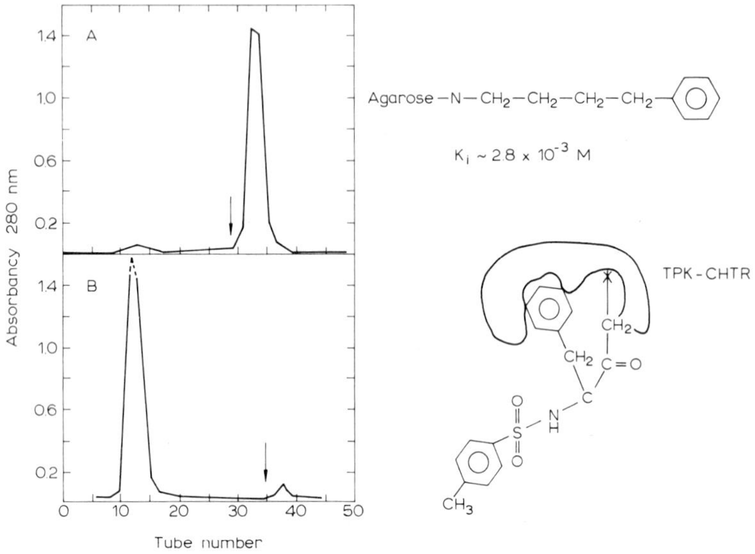

Fig. 4.12. The chromatography of native and active-site inhibited chymotrypsin on 4-phenylbutylamine–agarose. Reproduced with permission from Nishikawa et al. (1976), J. Macromol. Sci. Chem., *A10*, 149.

native chymotrypsin and TPCK-inhibited chymotrypsin on 4-phenylbutylamine–agarose on subsequent desorption with 0.1 M acetic acid. The lack of binding of the inhibited enzyme probably attests to a true biospecific interaction.

(5) The specific elution of an enzyme bound to an immobilised ligand with a suitable competing ligand or allosteric effector is indicative, though not conclusive proof, of affinity chromatography. Affinity elution of enzymes from non-specific adsorbents is well known (Scopes, 1977).

The prevention of adsorption by physiological concentrations of a competing ligand is a good criterion for affinity chromatography.

(6) The use of ternary complex formation to enhance binding to an immobilised ligand is a good indication of true biospecific behaviour (§ 2.4.1).

(7) Enzymic activity displayed by the insolubilised ligand is perhaps the best criterion of affinity chromatography since it implies an interaction at the active site of the enzyme. The enzymic reduction of immobilised pyridine nucleotide coenzymes has been demonstrated (Lowe and Mosbach, 1975).

It is thus essential to apply several independent criteria to establish the presence of biospecific affinity chromatography.

4.7. Large-scale affinity chromatography

Despite the impact of affinity chromatography on protein separation, only in relatively few instances has the technology been applied to the large-scale purification of enzymes to homogeneity (Robinson et al., 1972; Nicolas et al., 1972; Holroyde et al., 1976). In particular, the factors affecting the scale up of affinity chromatography of β-galactosidase on agarose columns substituted with *p*-aminophenyl-β-D-thiogalactoside have been investigated (Robinson et al., 1972). The continuous isolation of multigram quantities of pure enzyme per hour is envisaged. More recently, Pahud and Schwarz (1976) describe the automation of affinity chromatography using an ultrograd gradient maker as a process programmer.

Some applications of affinity chromatography in the purification and resolution of proteins and other macromolecules

Affinity chromatography has progressed rapidly over the past few years because of the conceptual attractiveness and apparent simplicity of the procedure. Indeed, the applications of affinity chromatography to biological separations are now too numerous to permit detailed discussion in a monograph of this type. In this chapter some examples of ligand immobilisation which represent the type of coupling reaction that have appeared to be successful for a number of investigators will be presented. When the ligands have some general interest, an outline of the preparative procedure will be given.

5.1. Immobilised specific adsorbents for protein purification

Over the last decade large numbers of procedures have been devised for the isolation of individual proteins based on their biospecific interaction with immobilised substrates, products, inhibitors or allosteric effectors. A number of these are given in the references associated with the previous chapters of this monograph and those that are not may be found in any of the recent reviews of affinity chromatography. In this section only one example will be given to illustrate the problems associated with a typical preparative procedure and the way in which they were circumvented.

In an attempt to isolate β-galactosidase from *Escherichia coli*, Steers et al. (1971) developed an adsorbent comprising a β-thiogalactoside ligand coupled to an agarose matrix via suitable spacer molecules. Subsequent work has shown, however, that the interaction between β-galactosidase and the adsorbent was not biospecific but largely ionic and/or hydrophobic in nature (O'Carra et al., 1974; Nishikawa et al., 1976). To overcome these difficulties, a new affinity method for the purification of β-galactosidase from wheat germ on lactose-coupled polyacrylamide gel has been suggested (Hamazaki and Hotta, 1977). The hydrazide derivative of polyacrylamide Bio-gel P-300 was prepared by the method of Inman and Dintzis (1969), suspended in 0.1 M lactose, and heated in a boiling water bath for 1 h. Free ligand was removed by exhaustive washing to yield a gel containing 27–30 μmol ligand/ml packed gel. A 0–60% ammonium sulphate fraction (270 ml) of crude wheat-germ extract was applied to the affinity column (2.5 cm × 16 cm) and washed with 300 ml column buffer consisting of 0.02 M sodium acetate buffer (pH 3.9) containing 0.1 M NaCl. All the glycosidases except β-galactosidase appeared almost completely in the void volume and the β-galactosidase adsorbed to the gel was effectively eluted with 200 ml of the same buffer containing 0.1 M galactosyl-β (1–4) glucitol, prepared by reduction of lactose with sodium borohydride, as shown in Fig. 5.1. The specific activity of β-galactosidase was increased 300-fold in this step and cross-contamination with other glycosidases removed almost completely. The column could be used repeatedly after washing with 0.1 M lactose and the acetate buffer.

Although the nature of the linkage between the lactose and the polyacrylamide–hydrazide is ill-defined, this example demonstrates how facile chemistry can yield an adsorbent capable of very useful purifications. Furthermore, this adsorbent should be useful for the isolation of β-galactosidase from other sources in view of its specificity and mildness of the elution system.

5.2. Immobilised 'general ligands'

Affinity chromatography using specific ligands requires not only that each new separation problem should have different and often elaborate syntheses, but that since the choice of the ideal ligand is still largely empirical, should be associated with extensive experimentation to achieve satisfactory separations. One way to circumvent

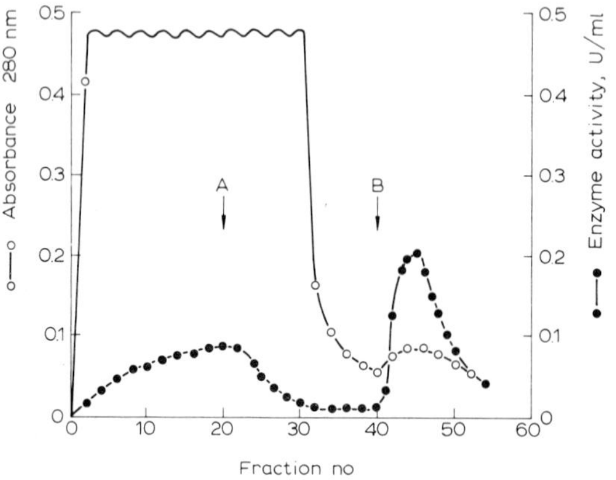

Fig. 5.1. The purification of wheat germ β-galactosidase on polyacrylamide derivatised with lactose. A partially purified fraction was applied to a column (2.5 cm × 16 cm) of immobilised lactose (27–30 μmol/ml gel) equilibrated with 0.02 M sodium acetate (pH 3.9) containing 0.1 M NaCl. The non-adsorbed proteins were washed off with 300 ml equilibrating buffer (A) whence β-galactosidase was eluted with 0.1 M galactosyl-(β,1-4)-glucitol in the same buffer (B). Fractions (15 ml) were collected at 50 ml/h. Adsorbance at 280 nm (O——O); Enzyme activity (●——●). Reproduced with permission from Hamazaki and Hotta (1977), FEBS Lett., *76*, 299.

these practical difficulties would be to prepare group specific or general adsorbents capable of interacting with a wide range of complementary proteins. This would eliminate the tedious process of screening all the ligands which interact with the macromolecule in order to find that which is best suited to the particular separation in mind. Furthermore, what selectivity is lost on adsorption is readily regained by appropriate choice of eluant conditions. It is therefore advantageous to use immobilised 'general ligands' that display affinity for a broad spectrum of enzymes.

5.2.1. Immobilised adenine nucleotide coenzymes

Nucleotide phosphates and many of their derivatives such as the vitamin-containing coenzymes and sugar nucleotides are substrates, inhibitors or cofactors for a wide variety of enzymes. Thus, nucleotide phosphates covalently attached to agarose should find wide application as biospecific adsorbents for the purification of a number of different enzymes including the kinases and dehydrogenases.

The simplest way of preparing a polymer-bound nucleotide would be to condense the parent nucleotide with a polymer matrix already functionalised with a suitable spacer molecule. Typically, ε-amino-caproic acid is attached to the CNBr-activated Sepharose 4B and the resulting gel suspended in 80% (v/v) aqueous pyridine (Larsson and Mosbach, 1971). The gel is transferred to a stoppered bottle and the nucleotide (NAD^+, $NADP^+$, FAD, ATP, AMP, coenzyme A, etc.) added, followed by dicyclohexyl–carbodiimide in aqueous pyridine. The suspension is shaken gently for 10–14 days at room temperature and the formed dicyclohexylurea removed by washing with water, ethanol, *n*-butanol, ethanol, water and finally 80% (v/v) aqueous pyridine. The washed gel is resubjected to nucleotide and carbodiimide for a further week and the gel exhaustively washed prior to use. The total amount of gel-bound nucleotide, typically about 2 μmol/g moist weight, is determined spectrophotometrically.

Adsorbents prepared by coupling pyridine nucleotides to ε-amino-hexanoyl–Sepharose with a carbodiimide-promoted reaction have proved extremely successful in the purification of dehydrogenases. Thus, for example, D-3-hydroxybutyrate dehydrogenase may be enriched 100-fold from a crude extract of *Rhodopseudomonas spheroides* by affinity chromatography on immobilised-NAD^+ (Lowe et al., 1973). In contrast, no increase in specific activity was obtained by chromatography of the crude bacterial extract on ε-aminohexanoyl-$NADP^+$–Sepharose since the enzyme passed through the column unretarded. Similarly very substantial purifications of L-threonine dehydrogenase from *Pseudomonas oxalaticus* may be obtained on immobilised-NAD^+ when linear gradients of NAD^+ are used in conjunction with the substrate, L-threonine, for elution (Lowe et al., 1973).

There are, however, a number of inherent disadvantages with this otherwise attractive procedure. The adsorbent will undoubtedly contain residual unreacted spacer molecules which can interact non-specifically with proteins and reduce the resolving power of the chromatographic material. Furthermore, most nucleotides have several potentially reactive groups which can react with a functionalised polymer to yield a heterogeneous adsorbent containing a number of chemically distinct ligands. Similar reservations may be expressed with immobilised dinucleotide adsorbents prepared by solid phase 'modular assembly' (Barry and O'Carra, 1973). For example, when polymer-bound AMP is condensed with free NMN^+, the resulting polymer-bound NAD^+ adsorbent will also contain varying amounts of bound AMP and NMN^+ which may obscure otherwise clean-cut separations.

In the authors' view, therefore, the recommended way to prepare a polymer bound nucleotide is to synthesise a suitable analogue carrying a spacer arm and to couple this directly to CNBr-activated agarose. Whilst this method is considerably more elaborate and time consuming, the resulting chemically defined adsorbents display a markedly improved performance and must be considered the only suitable ones when analytical work is to be undertaken. Such defined derivatives are now available commercially (Tables 5.1 and 5.2).

Figure 5.2 shows the structure of a typical mononucleotide, adenosine-5′-monophosphate, which will be used to illustrate the synthesis of analogues substituted with an alkyl or aryl-amino group at the purine base, phosphate or ribosyl ring. Adenine nucleotides substituted with an alkylamino group at the N^6 position of the purine base have proved particularly effective for the purification of NAD^+-dependent dehydrogenases (Guilford et al., 1972; Craven et al., 1974). The first analogue of this type suitable for attachment to polymers was N^6-(6-aminohexyl)-AMP prepared by nucleophilic displacement of either the 6-chloro (Guilford et al., 1972) or the 6-mercapto group (Craven et al., 1974) from appropriate purine riboside phosphates with 1,6-diaminohexane. Typically, 6-chloropurine riboside is phosphorylated for 10 h at 0°C by phosphoryl chloride in triethylphosphate and after appropriate purification steps, the crude 6-chloropurine riboside phosphate is treated with a 10-fold molar excess of 1,6-diaminohexane at 30°C for 3 h. The progress of the

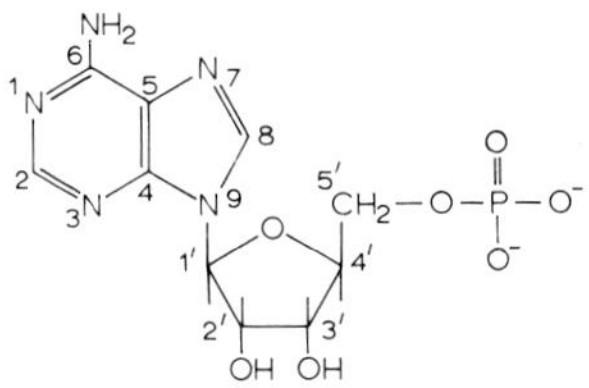

Fig. 5.2. The structure of adenosine 5′-monophosphate (AMP).

reaction may be followed by thin layer chromatography in suitable solvent systems or by the change in ultraviolet adsorption following substitution (Guilford et al., 1972). The yields in this and the synthesis involving displacement of the 6-mercapto group are approximately 75–80% overall. These procedures permit the synthesis of not only a homologous series of N^6-(ω-aminoalkyl)-AMP derivatives from the corresponding α,ω-diaminoalkanes (Hipwell et al., 1974), but also, the corresponding N^6-substituted ADP, ATP and NAD^+ analogues. Furthermore, the synthetic route commencing with 6-chloropurine riboside forms a suitable basis for the preparation of N^6-(6-amino-hexyl)-adenosine-2′,5′- and 3′,5′-bisphosphates (Brodelius et al., 1974). The latter derivatives have proved successful in the purification of $NADP^+$-dependent dehydrogenases (Brodelius et al., 1974) and coenzyme A-dependent enzymes (Barry et al., 1976) respectively.

An alternative procedure for introducing a spacer molecule at the N^6-amino group and generally applicable to adenine nucleotides involves alkylation at N^1 followed by alkaline Dimroth rearrangement (Mosbach et al., 1976) to yield the N^6-substituted derivatives. Typically, the adenine nucleotide is alkylated with a 3–10-fold excess of iodoacetic acid at pH 6.5 for 5 days in the dark at room temperature. The pH is periodically readjusted to 6.5 with 2 M LiOH and the progress of the reaction followed by thin layer chromatography. The resulting 1-carboxymethyl nucleotides are intrinsically unstable and rearrange in alkaline solution and at elevated temperatures into the corresponding N^6-substituted nucleotides. The 1-carboxymethyl derivatives of AMP, ADP and ATP are readily converted

TABLE 6.1
Commercially available adenine nucleotide adsorbents.

Ligand	Point of linkage* to nucleotide	Registered Trade Mark	Company	Typical Applications
Adenosine	C^8	AGADENOSINE	PL Biochemicals	Adenosine kinases and deaminases
Adenosine-3′-5′-cyclic monophosphate (cAMP)	N^6	Type 3	Sigma	cAMP-dependent protein kinases regulatory cAMP receptor subunit protamine kinase
Adenosine-2′,5′-diphosphate (2′,5′-ADP)	N^6	AG 2′,5′-ADP	PL Biochemicals	$NADP^+$-dependent dehydrogenases
	N^6	Type 2	Pharmacia	$NADP^+$-dependent malic enzyme
			Sigma	Transhydrogenase
Adenosine-3′,5′-diphosphate (3′,5′-ADP)	N^6	AG 3′,5′-ADP	PL Biochemicals	Coenzyme A-dependent enzymes
		Type 2	Sigma	
Adenosine-5′-diphosphate (ADP)	N^6	AGADP	PL Biochemicals	Myosin
		Type 2	Sigma	Hexokinase
	C^8	AGADP	PL Biochemicals	ATPase
		Type 3		Glucokinase
	Ribose hydroxyls	AGADP	PL Biochemicals	
		Type 4		

Adenosine-5′-monophosphate (AMP)	N^6	AGAMP Type 2	PL Biochemicals	NAD$^+$-dependent dehydrogenases kinases Phosporylase *b*
		5′-AMP Sepharose 4B	Pharmacia Sigma	Isoenzyme separations
	C^8	AGAMP Type 3	PL Biochemicals Sigma	NAD$^+$-dependent dehydrogenases Adenylate kinase DNAase and RNAase
	Ribose hydroxyls	AGAMP Type 4	PL Biochemicals	
Adenosine-5′-triphosphate (ATP)	N^6	AGATP Type 2	PL Biochemicals Sigma	ATP-dependent enzymes Kinases
	C^8	AGATP Type 3	PL Biochemicals Sigma	Hexokinase
	Ribose hydroxyls	AGATP Type 4	PL Biochemicals Sigma	Heavy meromyosin Citrate synthetase
Coenzyme A	Unknown	AGCoA Type 1	PL Biochemicals	Cl. Kluyveri phosphotransacetylase CoA transferase
	Thioester linkage	AGCoA Type 5	PL Biochemicals Sigma	Acyl-CoA enzymes
Nicotinamide adenine dinucleotide (NAD$^+$)	Unknown	AGNAD Type 1	PL Biochemicals	Apo-β-hydroxybutyrate dehydrogenase myoinositol-l-phosphate synthetase 5′-nucleotidase

Subject index p. 519

TABLE 5.1 *(continued)*

Ligand	Point of linkage* to nucleotide	Registered Trade Mark	Company	Typical Applications
				NAD$^+$-dependent dehydrogenases
	C^8	AGNAD Type 3	PL Biochemicals	
	Ribose hydroxyls	AGNAD Type 4	PL Biochemicals Sigma	
Nicotinamide adenine dinucleotide phosphate (NADP$^+$)	C^8	AGNADP Type 3	PL Biochemicals	NADP$^+$-dependent dehydrogenases
	Ribose hydroxyls	AGNADP Type 4	PL Biochemicals	Nucleases

* Agarose as matrix throughout. 5′-AMP-cellulose is available from PL Biochemicals.

TABLE 5.2
Commercially available nucleotide adsorbents.

Ligand	Point of linkage to nucleotide	Registered Trade Mark	Company	Typical applications
Cytidine-5′-diphosphate (CDP)	Ribose hydroxyls		Sigma	
Cytidine-5′-monophosphate (CMP)	Ribose hydroxyls		Sigma	
Cytidine-5′-triphosphate (CTP)	Ribose hydroxyls		Sigma	
Guanosine-5’-diphosphate (GDP)	Ribose hydroxyls		Sigma	
Guanosine-5′-monophosphate (GMP)*	Ribose hydroxyls		Sigma	
Guanosine-5’-triphosphate (GTP)	Ribose hydroxyls	AGGTP Type 4	PL Biochemicals	Dihydroneopterin triphosphate synthetase
			Sigma	Glutamate dehydrogenase GTP glycohydrolase
Uridine-5′-diphosphate (UDP)	Ribose hydroxyls	AGUDP Type 4	PL Biochemicals Sigma	
Uridine-5′-monophosphate (UMP)*	Ribose hydroxyls		Sigma	
Uridine-5′-triphosphate (UTP)	Ribose hydroxyls	AGUTP Type 4	PL Biochemicals Sigma	

* Also available from PL Biochemicals attached to cellulose.

to the N^6-derivatives simply by heating at 90 °C and pH 8.5 for 1.5 h. There appears to be no hydrolysis of the terminal phosphate under these conditions. On the other hand, the N^6-substituted derivatives of NAD^+ and $NADP^+$ can only satisfactorily be produced by reduction of the N^1-derivatives, either with dithionite or enzymically, prior to treatment at pH 11 and 70–75 °C for 1 h. The resulting N^6-carboxymethyl-NAD(P)H is then reoxidised enzymically and purified to homogeneity by ion-exchange chromatography (Mosbach et al., 1976). Condensation of the N^6-substituted nucleotides with a 10-fold excess of 1,6-diaminohexane in the presence of a water soluble carbodiimide generates N^6-(6-aminohexyl)-carbamoylmethyl nucleotides bearing a terminal amino group suitable for direct attachment to CNBr-activated supports. Figure 5.3 illustrates the synthesis of the N^6-substituted analogues of NAD^+, $NADP^+$ and ATP. The derivatives and intermediates are characterised by their spectral properties, their thin layer behaviour under several conditions and by their activity as coenzymes with several enzymes (Mosbach et al., 1976).

Whilst this procedure is straight forward and generally applicable to the adenine coenzymes, a simpler process involving the direct carbamoylation of the N^6-amino group has been developed (Yamazaki et al., 1977). Reaction of ADP with hexamethylene diisocyanate in hexamethylphosphoramide for 2 h at 75 °C followed by hydrolysis in acidic medium afforded the N^6-(N-(6-aminohexyl)-carbamoyl) derivatives of ADP, ATP and AMP in overall yields of 13%, 12% and 17%, respectively. The preparation of the N^6-carbamoyl analogues by this method is facile and may be completed within 24 h (Fig. 5.4). The progress of the carbamoylation is readily followed

Fig. 5.3. The synthesis of the 1-carboxymethyl-, N^6-carboxymethyl- and N^6-[(6-aminohexyl)carbamoyl-methyl]-derivatives of (a) NAD^+, (b) $NADP^+$ and (c) ATP. Figure 5.3 (c) illustrates also the solid phase enzymic interconversion of ATP and ADP. R = ribose, P = phosphate, ADH = alcohol dehydrogenase, Gl-6-P dehydrog. = glucose-6-phosphate dehydrogenase. Reproduced with permission from Mosbach et al. (1976), Methods Enzymol., *44*, 860.

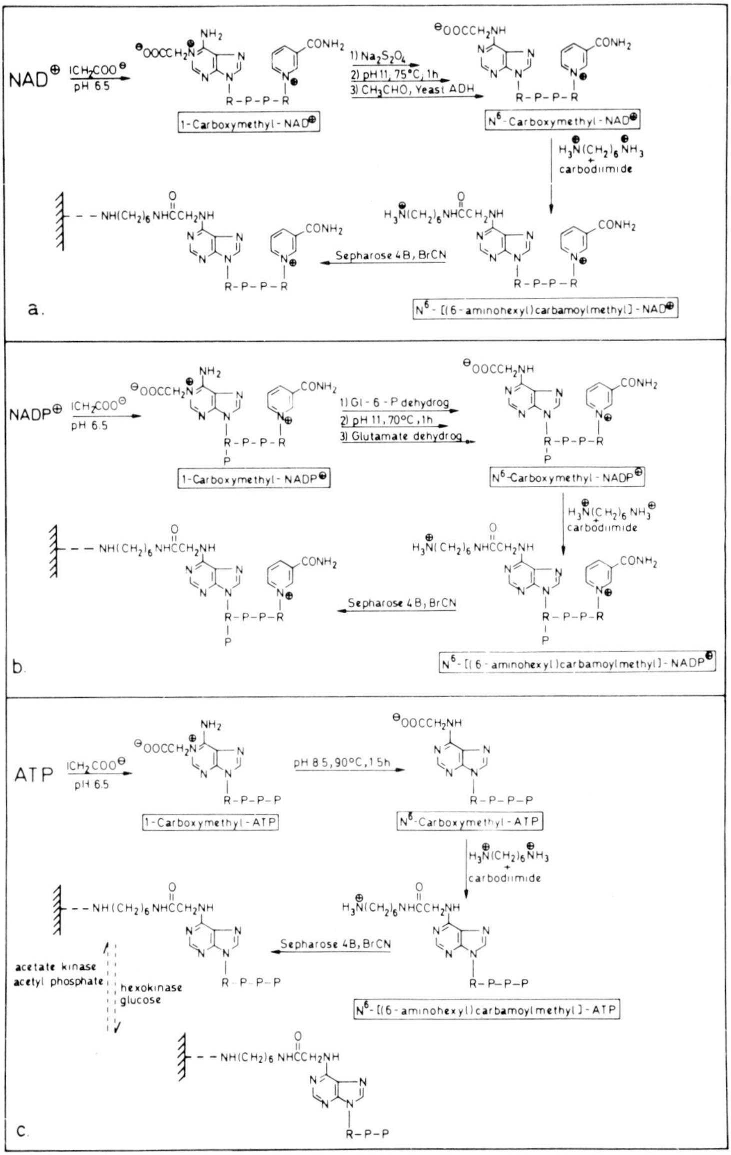
NAD⊕
ICH2COO⊖
pH 6.5
1-Carboxymethyl-NAD⊕
1) Na2S2O4
2) pH 11, 75°C, 1h
3) CH3CHO, Yeast ADH
N6-Carboxymethyl-NAD⊕
H3N(CH2)6NH3
carbodiimide
Sepharose 4B, BrCN
N6-[(6-aminohexyl)carbamoylmethyl]-NAD⊕
a.
NADP⊕
ICH2COO⊖
pH 6.5
1-Carboxymethyl-NADP⊕
1) Gl-6-P dehydrog
2) pH 11, 70°C, 1h
3) Glutamate dehydrog
N6-Carboxymethyl-NADP⊕
H3N(CH2)6NH3
carbodiimide
Sepharose 4B, BrCN
N6-[(6-aminohexyl)carbamoylmethyl]-NADP⊕
b.
ATP
ICH2COO⊖
pH 6.5
1-Carboxymethyl-ATP
pH 8.5, 90°C, 1.5h
N6-Carboxymethyl-ATP
H3N(CH2)6NH3
carbodiimide
Sepharose 4B, BrCN
N6-[(6-aminohexyl)carbamoylmethyl]-ATP
acetate kinase
acetyl phosphate
hexokinase
glucose
c.

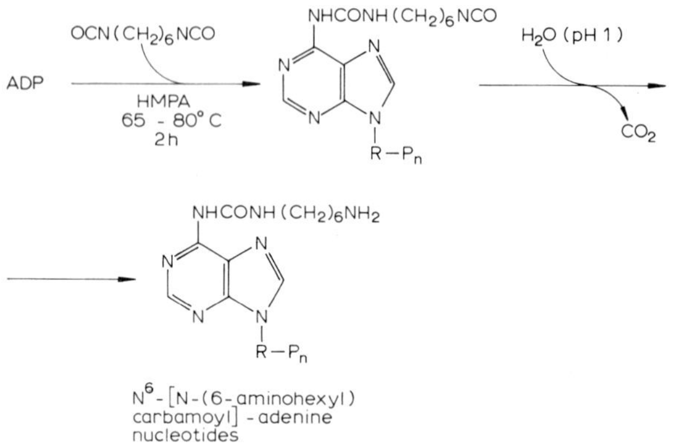

Fig. 5.4. Scheme for the synthesis of N^6-[*N*-(6-aminohexyl)carbamoyl]-adenine nucleotides for affinity chromatography. HMPA = hexamethylphosphoramide, R = ribose, Pn = monophosphate, diphosphate or triphosphate.

by a shift in the ultraviolet absorption maximum from 260 to 276 nm at acidic pH.

Whereas N^6-alkylated adenine nucleotides are prepared by somewhat tedious and indirect routes, N^6-acylated nucleotides may be obtained by quite simple procedures. For example, treatment of NAD^+ with succinic anhydride gives N^6-succinyl-NAD^+ which may subsequently be condensed with a suitable amino-containing polymer (Wykes et al., 1972). Similarly, the anhydride of carbobenzoxy-6-aminohexanoic acid was reacted with cAMP and following removal of the carbobenzoxy residue N^6-(6-aminohexanoyl)-cAMP was used as a ligand for affinity chromatography of protein kinases (Wilchek et al., 1971). These acylated derivatives are, however, particularly labile towards alkali and even in physiological media show a slow breakdown and are thus not attractive propositions for affinity chromatography. On the other hand, C^8-substituted nucleotides are ideal as ligands for affinity chromatography.

The synthesis of adenine nucleotides bearing a spacer molecule at the C^8 position of the purine is comparatively straightforward. The parent nucleotide is treated with bromine solution for 24 h in 1 M sodium acetate buffer (pH 4.0), whence the resulting 8-bromo-nucleotide is isolated in almost quantitative yield (Lowe, 1977). Attachment of the spacer molecule is then accomplished by heating the 8-bromo-nucleotide with a 10-fold excess of 1,6-diaminohexane. This method permits the synthesis of the 8-substituted derivatives of 5′-AMP, 2′-AMP, ADP, ATP, NAD^+ and $NADP^+$. In the case of the pyridine nucleotides however, the nucleophilic displacement of the bromine with 1,6-diaminohexane in hot alkaline solution can only be performed with the reduced form of the coenzyme to protect against degradation. The 8-(6-aminohexyl)-analogues of ADP and ATP may also be prepared by condensation of the corresponding AMP derivative with phosphate or pyrophosphate (Trayer et al., 1974).

The 8-position of adenine nucleotides (AMP, ATP, NAD^+, $NADP^+$) is also susceptible to electrophilic substitution with diazonium salts due to its high electron density. Diazo-linked NAD^+ has been exploited as an affinity adsorbent (Barry and O'Carra, 1973), although the rather bulky benzoyl group introduced in close proximity to the ligand might in some cases lead to spurious ligand–enzyme interactions. Nevertheless, diazo-linked nucleotides are simple to prepare and often yield quite respectable purifications.

5.2.2. *Other immobilised nucleotides*

The procedures described so far are almost exclusively applicable to the adenine nucleotides. A number of other modification procedures are available for the functionalisation of nucleoside phosphates in general. For example, phosphate-esterified nucleoside phosphates are useful affinity ligands for dehydrogenases, kinases and a number of other enzymes. Figure 5.5 illustrates a typical reaction sequence used to prepare the 6-amino-1-hexyl phosphoesters of all the common nucleosides. Essentially, 6-amino-1-hexanol is phosphorylated and the 6-amino group protected by trifluoroacetylation prior to coupling to the nucleoside phosphate via a 1,1′-carbonyl-diimidazole promoted reaction (Barker et al., 1974). Full practical details for this reaction sequence applicable to the synthesis of the 6-amino-1-hexyl esters of

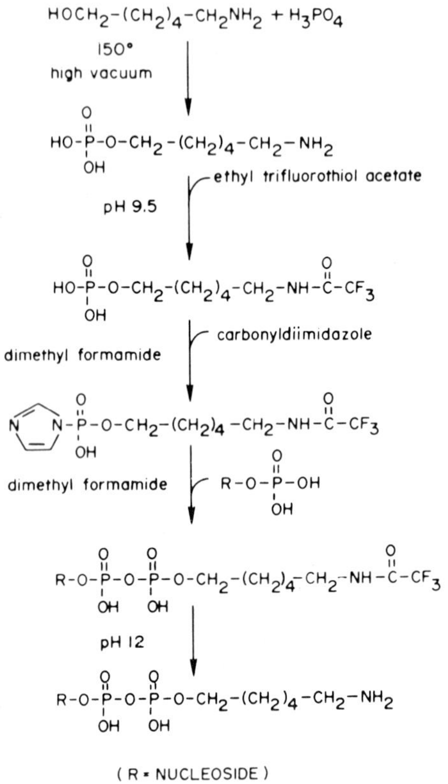

Fig. 5.5 The reaction sequence for the synthesis of 6-amino-1-hexyl nucleoside phosphomonoesters.

GDP, ADP, ATP, CDP, dGTP, UDP, galactosyl pyrophosphate and other sugar phosphates is given in Barker et al. (1974).

An alternative virtually universal procedure for immobilising ribonucleosides and ribonucleotides involves prior periodate oxidation followed by coupling to agarose hydrazides (Lamed et al., 1973). Adsorbents of this type containing up to 3–4 μmol bound nucleotide per gram of agarose have been prepared from AMP, ADP, ATP, UMP, CMP, CTP, UTP, IMP, GMP, GTP, NAD^+ and $NADP^+$. Typically, adipic acid dihydrazide (Barker et al., 1974) is coupled to

CNBr-activated agarose in 0.1 M sodium carbonate buffer (pH 9.5) for 12–18 h at 4 °C. An ice-cold solution of the nucleotide (20 μmol/ml) is adjusted to pH 8 and added to an equal volume of ice-cold 20 mM sodium periodate solution. After 1 h at 0 °C in the dark, the periodate oxidised nucleotide (4–5 μmol) is dissolved in 2.5 ml 0.1 M sodium acetate buffer (pH 5) and added to the agarose–hydrazide (1 g) in 1 ml of the sodium acetate buffer. The suspension is stirred gently for 3–4 h in the cold room and unbound nucleotide washed out with 2 M NaCl. The adsorbent when washed with equilibration buffer is ready for use. Many of these immobilised nucleotide derivatives are available commercially.

5.2.3. Other immobilised coenzymes

Detailed consideration of the synthesis and application of the variety of affinity adsorbents based on immobilised coenzymes other than the nucleotide coenzymes is beyond the scope of this monograph. Suffice it to say that detailed practical and theoretical discussions may be found in *Methods in Enzymology* (Volume 34) and Lowe and Dean (1974). In almost all cases the procedures employed are those to be found in the preceding chapters of this monograph. Table 5.3 lists some of the many applications of immobilised coenzymes in affinity chromatography.

5.2.4. Immobilised nucleic acids and polynucleotides

Immobilised-DNA has been used as a chromatographic adsorbent in nucleic acid research for almost a decade and has been applied to the purification of a number of DNA-specific proteins from several organisms. Traditionally, DNA has been immobilised by drying onto cellulose (Alberts et al., 1968). Acid washed cellulose and DNA are kneaded into a paste, air dried and suspended in absolute alcohol. The suspension is irradiated for 30 min at 15 cm from a low-pressure mercury lamp (100,000 erg/mm^2) with slow stirring. After filtration and washing to remove unbound nucleic acid the preparation is dried in air and may subsequently be stored at room temperature for several years, (Litman, 1968). A similar procedure may be used to prepare

TABLE 5.3
Immobilised coenzymes in affinity chromatography.

Immobilised coenzyme	Typical purification applications
Biotin	Avidin Acetyl-CoA carboxylase
Cobalamins	Vitamin B_{12} binding proteins Transcobalamin Ribonucleotide reductase Intrinsic factor
Flavins	Egg white ovoflavoprotein Flavokinase Glycollate apooxidase NADPH-cytochrome *c* aporeductase
Folate and folate analogues	Folate binding protein Dihydrofolate reductase from several sources
Haem	Albumin Haemopexin
Lipoic acid	Lipoamide dehydrogenase isoenzymes
Pyridoxal coenzymes	Aspartate aminotransaminase apoenzyme Tyrosine aminotransaminase and polysomes Glutamate-oxaloacetate transaminase
Thiamine pyrophosphate (TPP)	*E. coli* thiamine binding proteins *E. coli* pyruvate oxidase

RNA–cellulose. Unfortunately, it is difficult to immobilise low molecular weight DNA by this approach and the DNA that is imsolubilised slowly leaches out of the cellulose. Consequently, alternative procedures for immobilising DNA have been developed. For example, DNA and RNA can be immobilised to a variety of solid supports such as agarose (Schaller et al., 1972), polyacrylamide (Cavalieri and Carroll, 1970) and cellulose acetate in combination with agar (Bautz and Hall, 1962). While the cellulose and agarose derivatives exhibit a high capacity (0.5–5 mg DNA/g moist weight gel) their usefulness is limited to reduced temperatures and neutral pH, i.e., conditions under which desorption is minimised. The major disadvantage of polyacrylamide is its low capacity since only about 0.05 mg DNA/g moist gel can be entrapped.

To circumvent problems of leakage, Gilham (1968) introduced covalent coupling of nucleic acids to cellulose using water soluble carbodiimides. These adsorbents were able to selectively remove DNAs from mixtures. Furthermore, polymerised oligonucleotides when coupled to cellulose with dicyclohexylcarbodiimide were useful solid state primers and templates for polymerases. Poonian et al. (1971) used the CNBr-activation technique to covalently attach single-stranded RNA and DNA to agarose in 50 mM potassium phosphate buffer (pH 8.0) by incubation at 4 °C for 16–48 h. After exhaustive washing to remove unbound material, bound nucleic acid may be estimated by the difference between added and recovered radiolabelled material and/or absorbance at 260 nm. Under these conditions DNA does not undergo significant coupling to CNBr-activated agarose when present in the native double-stranded form and requires denaturation to the single-stranded form. For example, the percentage of HeLa DNA which can be attached to agarose increases as the amount of single-stranded ends increases (Weissbach and Poonian, 1974). There is however a limit to the capacity of agarose to bind DNA since the efficiency of coupling decreases at high DNA/agarose ratios. Thus Poonian et al. (1971) reported coupling of at most 0.045 mg DNA/g moist weight gel although with poly(A) they achieved 0.3 mg/g. Arndt-Jovin et al. (1975) have reappraised the conditions required to covalently couple DNA in high yield to CNBr-activated agarose. Activated agarose was prepared by reacting 1 g CNBr/100 mg dry weight Sepharose 2B or 4B and DNA (0.08–30 mM) coupled overnight to the activated gel (3–10 mg/ml) in 10 mM potassium phosphate buffer (pH 8.0). Under these conditions both single-stranded and double-stranded DNA may be coupled with yields up to 225 nmol/mg dry weight Sepharose. The resulting DNA–agarose is stable to high temperatures, prolonged storage and in the case of single-stranded DNA can be washed with NaOH to inactivate nucleases and to release any digested oligo- or mononucleotides.

DNA-agarose is suitable for the affinity chromatography of many enzymes and for nucleic acid hybridisation. Thus, agarose-bound DNA has proved to be particularly effective in the purification of HeLa cell DNA polymerase (Poonian et al., 1971) and deoxyribonuclease from bovine pancreas and hog spleen (Schabort, 1972). Furthermore, single-stranded DNA entrapped in agarose is effective in the purification of *E. coli* DNA-polymerases I and II, exonuclease III, RNA polymerase, T4 polynucleotide kinase (Schaller et al., 1972), ribonucleases and DNA unwinding enzyme II (Abdel-Monem et al., 1977).

Agarose activated with CNBr is also a suitable matrix for the attachment of single-stranded RNA and synthetic polynucleotides such as poly (I:C) (Wagner et al., 1971), poly(A), and poly(U) (Lindberg and Persson, 1972). Poly(A) and other polynucleotides are efficiently coupled to CNBr-activated agarose in 0.2 M 4-morpholinoethane sulphonic acid buffer (MES) (pH 6.0) under the conditions described by Wagner et al. (1971). Whilst the mechanism of this coupling procedure remains unclear it presumably involves multipoint attachment to the heterocyclic bases rather than a single point attachment of the terminal 5′-phosphate.

In general terms, immobilised polynucleotides are applicable to the purification of proteins with affinity for the respective polynucleotides, to the purification of a number of enzymes, to the detection and characterisation of polynucleotide sequences of RNA in mammalian viruses and oncornaviruses and for fractionating complementary oligonucleotides. These and other purifications of immobilised polynucleotides along with their commercial availability are given in Tabie 5.4. Two adsorbents are worthy of further consideration however. Immobilised poly(U) and oligo(dT) are widely applicable to the purification and resolution of mRNA. The biospecific interaction exploited is the hybridisation between a synthetic polynucleotide immobilised to a suitable matrix and the unique complementary run of polyriboadenylate (poly(A)), 50–150 base units in length, lying near the 3′-terminus of the mRNA. Poly(U)–Sepharose 4B is prepared by coupling a polyuridylate (poly(U)) chain

TABLE 5.4
Immobilised polynucleotides in affinity chromatography.

Immobilised polynucleotide	Commercial availability	Representative applications
Poly(A)–agarose*	Sigma PL Biochemicals Pharmacia	Myeloma RNA polymerase Hela cell cytoplasmic proteins σ-factor Poly(A) binding proteins
Poly(C)–agarose*	Sigma PL Biochemicals	RNA-dependent DNA polymerase
Poly(G)–agarose*	Sigma PL Biochemicals	*Citrobacter* ribonuclease Ribonuclease C
Poly(I)–agarose*	Sigma PL Biochemicals	Oligonucleotide fractionation Template/substrate in nucleotide Polymerising systems
Poly(I)(C)–agarose	Sigma PL Biochemicals	Interferon induction
Poly(U)–agarose*	Sigma PL Biochemicals Pharmacia	Poly(U)-binding proteins Reticulocyte ribosomes mRNA Insoluble template Polysomal RNA Cytoplasmic RNA from adenovirus Infected cells
DNA–agarose†	PL Biochemicals	Several DNA and RNA polymerases DNA binding proteins T4 polynucleotide kinase DNAses Exonuclease
DNA–cellulose†	PL Biochemicals	Several DNA and RNA polymerases DNA binding proteins DNAs Exonuclease
DNA–cellulose‡	PL Biochemicals	DNA polymerase Native DNA binding proteins Exonuclease
Oligo(dA)–cellulose	PL Biochemicals	Isolation of poly(U) sequences
Oligo(dC)–cellulose	PL Biochemicals	Poly(I) binding
Oligo(dT)–cellulose	PL Biochemicals	Isolation of RNA containing poly(A) mRNA DNA polymerases Polynucleotide phosphorylases

* Also applicable to the detection and characterisation of polynucleotide sequences of RNA in mammalian viruses and oncornaviruses.

† Denatured calf thymus DNA.

‡ Native calf thymus DNA.

approximately 100 nucleotides long to CNBr-activated Sepharose 4B. The long polynucleotide chain is multi-attached to the matrix backbone and thus generates an adsorbent that is both stable (§ 3.2.2) and encompasses its own 'spacer' to minimise steric hindrance. Furthermore, the long poly(U) chain ensures specific interaction with the complementary poly(A) segment of mRNA with little or no binding to oligo(A) sequences such as triplets found in other nucleic acids.

Poly(U)–Sepharose 4B is available from Pharmacia Fine Chemicals as a freeze-dried powder in packs of 5 g equivalent to approximately 25 g swollen gel containing 0.5 mg poly(U)/g. The gel is swollen in 1 M NaCl (pH 7.5), washed with a buffer comprising 10 mM EDTA, 0.2% lauroyl-sarcosine and 10 mM potassium phosphate (pH 7.5) in 90% formamide and finally equilibrated with 10 mM EDTA, 0.7 M NaCl, 50 mM Tris–HCl (pH 7.5) in 25% formamide. The polysomal samples are prepared in 0.03 M EDTA containing 1% lauroyl-sarcosine, diluted 5-fold with the buffer containing 0.7 M NaCl and applied to the column. The bed is washed with several volumes of the latter buffer and the mRNA eluted with high concentrations of formamide to dissociate the poly(U)–poly(A) complex, i.e., the buffer containing 90% formamide. The eluant is monitored by radioactivity if radio-labelled samples are used or by absorbance at 260 nm. The immobilised poly(U) adsorbent is remarkably stable. Several weeks storage in suspension at elevated temperatures results in a 1–2% loss of poly(U) but no loss of binding capacity for mRNA. The binding capacity for polysomal preparations from KB cells is approximately 150 μg mRNA/g moist gel with almost quantitative recoveries of the mRNA on subsequent elution.

Poly(U)–agarose is also available commercially from other sources (Table 5.4). However, immobilised poly(U) is nuclease and base sensitive and thus precautions must be taken to avoid degradation of the affinity adsorbent. Accordingly, Pharmacia recommend supplementing the column buffers with detergent or other ribonuclease inhibitors to prevent hydrolysis of the poly(U) chain.

Oligo(dT)–cellulose, on the other hand, is highly selective for its complementary poly(A) and is stable in alkaline solution, thus permitting repeated use of the same column. Oligo(dT)–cellulose is thus an ideal affinity adsorbent for the isolation of polyadenylate rich mRNA from a number of sources and is prepared according to the procedure of Gilham (1964). The adsorbent, containing about 40 mg oligo(dT)/g cellulose and comprising nucleotide chains up to 25 bases long, is available commercially from PL Biochemicals Inc. Typically, oligo(dT)–cellulose is used at room temperature with sterile glassware and reagents. A crude preparation of rabbit reticulocyte polysomal RNA is applied to a 2 ml (0.5 g dry weight) column of oligo(dT)–cellulose equilibrated with 10 mM Tris–chloride buffer (pH 7.5) containing 0.5 M KCl (Aviv and Leder, 1972). Under these conditions 1 g of adsorbent will adsorb 80–100 A_{260} units of poly(A). Non-adsorbed material amounting to approximately 95% of applied absorbance at 260 nm is washed through with the equilibrating buffer, whence bound material is eluted sequentially with 10 mM Tris–chloride (pH 7.5)–0.1 M KCl followed by 10 mM Tris-chloride (pH 7.5). Elution with the lowest ionic strength buffer releases UV-absorbing material which exhibits very high protein-synthesising activity in a cell free system.

Oligo(dT)–cellulose has also been applied to the purification of viral polymerases, to the fractionation of oligoadenylate nucleotides and as a template and primer for polymerases and transferases. Oligo(dA)–cellulose, which should prove useful for isolating poly-uridylate sequences such as those that occur in HeLa nuclear RNA, and oligo(dC)–cellulose are also available from PL Biochemicals Inc. The commercial availability of these and other polynucleotides are summarised in Table 5.4.

5.2.5. Immobilised lectins

Lectins, or phytohaemagglutinins, are sugar binding proteins isolated almost exclusively from plant sources that possess a number of remarkable properties. For example, they agglutinate erythrocytes and malignant cells, precipitate polysaccharides and glycoproteins and induce mitosis in lymphocytes. Not surprisingly, therefore, lectins have found widespread application, not only in the recognition of specific sugars or linkages, but also in the isolation and resolution of sugar-containing macromolecules. For example, lectins immobilised to agarose have been used for isolating glycoprotein hormones, enzymes such as glucose oxidase, horse radish peroxidase and neuraminidase, polysaccharides such as dextran, yeast mannan and glycogen, immunoglobulins and other human serum glycoproteins such as α_1-antitrypsin, membrane glycoproteins and whole cells (Kristiansen, 1974).

The literature on the purification of glycoproteins with immobilised lectins is prolific and burgeoning. The aim of this section is, therefore, to make generalisations about the selection of an appropriate lectin for a particular separation, its immobilisation and its consequent use as a chromatographic adsorbent. Identification of the terminal sugar(s) in the carbohydrate moiety of the glycoprotein or glycolipid permits selection of the appropriate lectin with the desired specificity. In this respect it is important to realise that lectins are generally not monospecific for a single sugar but may display broad specificity. Table 5.5 lists the sugar specificities, applications and commercial availability of a number of common lectins. Once the appropriate lectin has been selected, prepared or purchased, it may be covalently attached to agarose. Typically, the lectin is coupled overnight to CNBr-activated agarose in 0.5 M $NaHCO_3$ buffer (pH 8.5) to yield a gel containing 1–5 mg lectin/g moist weight gel. Untreated imido-carbonate is quenched by a similar treatment with saturated glycine for 24 h. In this case the potentially ionogenic α-carboxyl group of glycine counterbalances the basic imino group in the imidocarbonate. The resulting immobilised lectin adsorbent is then thoroughly washed and equilibrated with 20–50 mM phosphate buffer (pH 7.2) containing 0.15 M sodium chloride. The sample to be chromatographed is applied and the column washed with the equilibrating buffer until the absorbance at 280 nm returns to the baseline value. The bound material is subsequently eluted with the same buffer containing 0.1–0.3 M of the appropriate sugar given in the specificity column of Table 5.5. Greater selectivity in immobilised lectin chromatography may be achieved if the immobilised lectin chosen is inhibited to a

TABLE 5.5

Sugar specificities, commercial availabilities and applications of some common lectins.

Lectin	Specificity	Purification applications	Commercial availability
Canavalia ensiformis (jack bean)	α-D-Glucose α-D-Mannose	Lectins Glycopeptide hormones Glycoprotein enzymes Immunoglobulins Blood group glycoproteins α_1-Antitrypsin Pig serum lipoproteins	PL Biochemicals Pharmacia Sigma Miles Laboratories
Dolichos biflorus (horse gram)	*N*-Acetyl-D-galactosamine	Glycoproteins and glycolipids Blood group A substance	PL Biochemicals
Glycine max (soybean)	D-Galactose *N*-Acetyl-D-galactosamine	D-Galactose and acetyl-D-galactosamine containing polymers	PL Biochemicals Miles Laboratories
Lens culinaris (lentil)	D-Mannose D-Glucose	Virus envelope glycoproteins Pig lymphocyte membrane proteins Tissue culture cells	PL Biochemicals
Lotus tetragonolobus (peanut)	Galactose-β,1-3, *N*-Acetyl galactosamine	Desialyted glycoproteins Glycophorin Fetuin T-Antigen α-Acid glycoprotein	PL Biochemicals
Ricinus communis (castor bean)	D-Galactose	Plasma membrane glycoproteins Galactose-containing polymers	PL Biochemicals Miles Laboratories
Triticum vulgaris (wheat germ)	*N*-Acetyl-glucosamine *N*-Acetyl-neuraminic acid (NANA)	Membranes Lectin receptor from erythrocyte membranes Glycophorin A	PL Biochemicals Sigma Miles Laboratories
Ulex europaeus (gorse)	L-Fucose	Blood group O substance L-Fucose receptor sites on human group O enthrocytes	PL Biochemicals

greater extent by a sugar different from the one(s) in the glycoprotein and to use the latter sugar to effect biospecific desorption. Furthermore, preliminary data suggests that gradients of sugars achieve effective desorption of glycoproteins from immobilised lectin adsorbents.

Whilst the above considerations are generally applicable to most lectins, problems may be experienced with the lectin from jack bean, concanavalin A. Sepharose-bound concanavalin A (Con A–Sepharose) is a group specific adsorbent applicable to the purification of polysaccharides and glycoproteins containing α-D-mannosyl, α-D-glucosyl and sterically related residues. However, unlike most other phytohaemagglutinins, ConA is a metalloprotein requiring Mn^{2+} or a similar transition metal ion, and Ca^{2+}. Removal of the metal ions, for example, by dialysis against acetic acid (pH 3), destroys the carbohydrate binding capacity (Agrawal and Goldstein, 1968). Therefore, whilst the active ConA–metal ion complex is stable at pH 7 in the absence of added free metal ions, at low pH values the equilibration buffers should be supplemented with excess Mn^{2+} and Ca^{2+}. Concanavalin A–Sepharose is supplied by Pharmacia as a suspension of 100 ml sedimented gel in 0.1 M acetate buffer pH 6.0 containing 1 M NaCl and 1 mM each of $CaCl_2$, $MgCl_2$ and $MnCl_2$. In contrast, most other commercially available lectin adsorbents (Table 5.5) are supplied as suspensions in 1 M NaCl containing 0.02% sodium azide as a perservative.

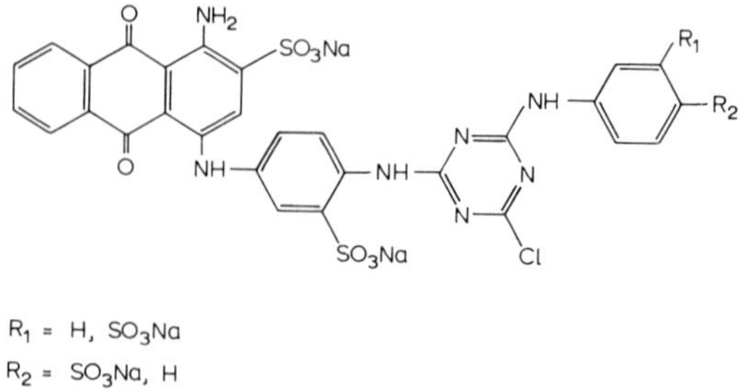

Fig. 5.6. The structure of Cibacron Blue F3G-A.

Elution of substances bound to immobilised-Con A may be effected with α-D-methylmannoside or α-D-methylglucoside either using a single increment in concentration or a gradient (Kristiansen, 1974). Alternatively, weakly interacting polysaccharides may be eluted in phosphate buffer whilst more tightly bound polysaccharides and glycoproteins may be eluted by borate buffer (Kennedy and Rosevear, 1973). Elution with borate buffer circumvents the problem of contamination of the eluted polysaccharide or glycoprotein by eluting the carbohydrate.

5.2.6. Immobilised dyestuffs

There are a number of examples of general ligand affinity chromatography based on the interaction of groups of proteins with entirely synthetic ligands. For example, the chromophore Cibacron Blue F3G-A (Fig. 5.6) when attached to agarose or other matrices is a group specific adsorbent which displays affinity for an exceptional variety of enzymes and proteins. Adsorbed proteins can subsequently be recovered by using specific elution procedures (Easterday and Easterday, 1974). Table 5.6 lists examples of the proteins and enzymes that have been purified by chromatography on immobilised Cibacron Blue F3G-A. The adsorbent appears to be particularly effective for binding kinases, dehydrogenases, glycolytic enzymes and a number of blood proteins. This observation suggests that specific affinity for this blue dye is shown by those proteins that possess a super-secondary structure termed the 'dinucleotide fold' which forms the binding site for a number of nucleotide substrates and effectors (Thompson et al., 1976; Thompson and Stellwagen, 1976). Inhibition studies suggest that the chromophore is a very strong competitive inhibitor of NADH and that it can assume a conformation which imitates the orientation of the principal functional groups of NAD^+ (Thompson et al., 1975). Immobilised Cibacron Blue F3G-A should thus find application not only as a diagnostic probe for the presence of the 'dinucleotide fold' in enzymes (Stellwagen et al., 1975) but also in their purification.

Cibacron Blue F3G-A may be coupled to polyacrylamide, dextran

TABLE 5.6

Some applications of Cibacron Blue F3G-A affinity chromatography.

Acetoacetate succinyl-CoA-transferase
Adenylate cyclase
Adenylate kinase
Albumin
Alcohol dehydrogenase
Aldolase
AMP deaminase
Amylopectin 1,6-glucosidase
Blood coagulation factors II (Prothrombin), IX and X
Carboxypeptidase G
Creatine kinase
Cyclic AMP phosphodiesterase
Cyclic GMP-dependent protein kinase
Cytochrome *c*
DNA polymerase
Enolase
Fructose diphosphatase
Glucose-6-phosphate dehyrogenase
Gluthathione reductase
Glyceraldehyde-3-phosphate dehydrogenase
Glycerol-3-phosphate dehydrogenase
Glyoxalase I
Hexokinase
3(17)α-hydroxysteroid dehydrogenase
Interferon
Isoleucyl-tRNA synthetase
Lactate dehydrogenase
Malate dehydrogenase
NAD-kinase
Nitrate reductase
Orotidylate decarboxylase
Phosphofructokinase
6-phosphogluconate dehydrogenase
Phosphoglycerate kinase
Phosphoglyceromutase
Phosphorylase a
Protein kinase
Pyruvate kinase
Ribonuclease A
RNA polymerase
Succinyl-CoA-transferase
Thymidine phosphotransferase

gels, dextran/agarose, CNBr-activated agarose, and agarose alone. The agarose-bound dye is available commercially from Pharmacia under the designation Blue Sepharose CL-6B and from Bio Rad Laboratories as Affi-Gel® Blue. Blue Sepharose CL-6B, prepared by coupling the chromophore to cross-linked agarose by the triazine coupling procedure (§ 3.1.1.2), is supplied as a lyophilised powder in packs of 25 g. The dry product should be stored below 8 °C. The swollen suspension, containing approximately 2 μmol/g moist weight, in neutral solution supplemented with a suitable bacteriostat should be stored in a fridge but without freezing. Affi-Gel® Blue is supplied ready to use as a hydrated slurry containing 0.02% sodium azide in two mesh sizes, 50–100 and 100–200.

Whilst there is no universal procedure for enzyme purification by Cibacron Blue–agarose chromatography, a suggested procedure is as follows. Prepare a column of Blue–agarose allowing 1 ml bed volume for every 4–5 mg of protein to be chromatographed and equilibrate with a starting buffer of low ionic strength, 0.05 M or less, and pH value 6.0–8.5. Apply the sample, which has been predialysed against the equilibrating buffer, wash through non-adsorbed proteins with 2–3 bed volumes buffer and apply the eluant solution. Elution may be effected by a change in pH or ionic strength or by supplementing the equilibrating buffers with low concentrations of nucleotides or effectors. For example, in general terms, the kinases, dehydrogenases and glycolytic enzymes may be eluted by low concentrations ($\leqslant$ 10 mM) of ADP/ATP, NAD^+/NADH or $NADP^+$/NADPH and their appropriate substrates respectively. Sequential application of eluants containing nucleotides may be used to fractionate enzymes present in a partially purified extract of rabbit muscle (Easterday and Easterday, 1974).

Immobilised Cibacron Blue F3G-A is also useful for selectively removing albumin either from crude serum or as a contaminant from other plasma protein preparations (Travis and Pannell, 1973). A column of Blue–agarose equivalent to a bed volume of 5 g moist gel/ml serum to be processed is equilibrated with either 0.02 M phosphate buffer (pH 7.0) or 0.05 M Tris–HCl buffer (pH 7.0) con-

Subject index p. 519

taining 0.1 M KCl. The predialysed serum protein sample is applied, non-adsorbed proteins washed through and albumin eluted with 1.4–1.5 M NaCl or KCl in the appropriate buffer. These procedures will remove over 95% of the albumin in serum with almost no non-specific adsorption of other serum proteins. For example, 90–99% of IgG applied to the column passes through unretarded. Affinity chromatography on immobilised Cibacron Blue F3G-A is thus recommended as an initial step in the purification of IgG from serum. However, immobilised Cibacron Blue F3G-A is only one of several adsorbents prepared by covalently attaching dyes to polysaccharide matrices.

Baird et al. (1976) have recently introduced polysaccharide–Procion dye conjugates for the purification of enzymes by affinity chromatography. In particular, it was found that immobilised Procion Red HE3B was effective in removing carboxypeptidase G from partially purified extract of *Pseudomonas* ATCC 25301. Biospecific elution was achieved with 2 mM *p*-aminobenzoyl glutamate (PABG). As a matrix, Sephadex G-100 was used in preference to Sepharose 4B in view of its markedly higher substitution capacity for the dye. The Dye–Sephadex conjugates were very stable, with no significant loss of enzyme binding capacity or leakage of the dye from the matrix detected after 6 months storage in buffer at 4 °C. Furthermore, the adsorbents could be used repeatedly without loss of capacity provided that they are regenerated after each use by washing with 1–2 M NaCl, 6 M guanidine hydrochloride or 8 M urea.

By analogy with Cibacron Blue F3G-A, structural similarity between nucleotide coenzymes and effectors and several Procion dyes can be revealed by constructing molecular models. Indeed, Procion–agarose columns have recently been found to be very effective in the purification of $NADP^+$-dependent dehydrogenases (Dean and Watson, 1978). The advantage of Procion and similar dyes is that they may be utilised for chemically facile synthesis of inexpensive, stable and reusable pseudo-affinity adsorbents. Similar arguments can be applied to phenol-containing adsorbents for the purification of some NAD^+-dependent dehydrogenases (Brenna et al., 1977).

5.2.7. *Thiol-specific adsorbents – organomercurials*

Agarose–mercurial adsorbents bind thiols and are thus capable of resolving thiol from non-thiol compounds:

$$\text{Matrix–HgCl} + \text{RSH} \rightleftharpoons \text{Matrix–HgSR} + \text{HCl}$$

Elution may be effected by a second thiol compound such as cysteine, β-mercaptoethanol or dithiothreitol:

$$\text{Matrix–HgSR} + \text{R'SH} \rightleftharpoons \text{Matrix–HgSR'} + \text{RSH}$$

The column may subsequently be regenerated by elution of the second thiol with mercuric chloride:

$$\text{Matrix–HgSR'} + \text{HgCl}_2 \rightleftharpoons \text{Matrix–HgCl} + \text{R'S HgCl}$$

Organomercurial derivatives of agarose may be prepared by coupling sodium *p*-chloromercuribenzoate to ω-aminoalkyl–agarose with water soluble carbodiimides (Cuatrecasas, 1970). Aminoethyl–Sepharose 4B is prepared by coupling 1,2–diaminoethane to CNBr-activated Sepharose 4B and washing extensively with distilled water to remove unbound diamine. The washed aminoethyl–Sepharose (25 g) is suspended in 20 ml 40% (w/v) *N*,*N'*-dimethylformamide and sodium *p*-chloromercuribenzoate (625 mg) added with stirring. The pH is adjusted to 4.8 with 6 N HCl and 1-ethyl-3-(dimethylamino-propyl)-carbodiimide (770 mg) added in small portions with gentle stirring. The pH of the reaction mixture is maintained at 4.8 for 1–2 h by periodic additions of 2 N HCl and the suspension gently stirred overnight at room temperature. The organomercurial–agarose is exhaustively washed with 40% (w/v) *N*,*N'*-dimethylformamide over a period of 48 h at room temperature to remove unbound mercuri-benzoate. The adsorbent should give a negative 2,4,6-trinitrobenzene sulphonate test (§ 3.6.7) after this treatment. If an orange colour is obtained, indicating residual free amino groups (Table 3.3), it is advisable to repeat the coupling step with the *p*-chloromercuri-benzoate.

The resulting organomercurial adsorbent may be stored at 4 °C as a suspension in water with a few drops of toluene as preservative for at least a year without noticeable deterioration in binding capacity. The binding capacity may be determined by reaction with [3–^{14}C]DL-cysteine in nitrogen-flushed 0.1 M sodium phosphate buffer (pH 6.0) and washing out unreacted thiol. The capacity may be calculated from the unbound radioactivity or by measuring the radioactivity bound to the gel in an automatic sample oxidiser (Ruizcarrillo and Allfrey, 1973).

Organomercurial–agarose may also be prepared by coupling *p*-aminophenylmercuric acetate directly to CNBr-activated agarose (Sluyterman and Wijdenes, 1974). The CNBr-activated agarose (40 g moist weight) is suspended in 80 ml 10% (v/v) dimethylsuphoxide (DMSO) at 0 °C and *p*-aminophenylmercuric acetate (600 mg) in 10 ml DMSO added. After 20 h at 0 °C the suspension is washed exhaustively with 20% (v/v) DMSO to remove unbound mercurial.

Organomercurial–agarose gels exhibit a high capacity for thiol-containing proteins by forming covalent mercaptide bonds in aqueous solution at pH 4–8. Typically, protein is adsorbed in 10–30 mM

TABLE 5.7
Some applications of organomercurial–agarose to the purification of thiol-proteins

Chinese gooseberry proteinase
Chymopapain
Creatine phosphokinase
Equine haemoglobin
Histones F3 and F2al
Lombricine kinase subunits
17β oestradiol dehydrogenase
Oestradiol receptor proteins
Papain
Platelet factor XIII
Thyroglobulin
Tryptic thiol containing peptides
Yeast phenylalanyl-tRNA synthetase subunits

sodium phosphate buffer (pH 6–7.2) or 50 mM sodium acetate buffer (pH 5.0) and non-adsorbed proteins washed through with the same buffers. Elution of the thiol proteins is effected with 10 mM–0.5 M cysteine, 10 mM β-mercaptoethanol, 1 mM dithiothreitol, 0.5 mM *p*-hydroxymercuribenzoate. The precise conditions of adsorption and elution will depend to some extent on the protein under consideration. Table 5.7 lists some proteins which have been purified to some extent on organomercurial adsorbents. Once used, the adsorbents may be regenerated by washing with $HgCl_2$–EDTA solutions (Barret, 1973). The organomercurial adsorbent is washed with 5 bed volumes of 50 mM sodium acetate buffer (pH 4.8) and then with 2 bed volumes of 10 mM $HgCl_2$–20 mM EDTA in the same buffer. Excess $HgCl_2$ is finally removed by extensive washing with 0.2 M NaCl–1 mM EDTA in 0.1 M sodium phosphate buffer (pH 6.0). Organomercurial-agarose is available commercially.

5.2.8. Immobilised amino acids

There are a number of cases where immobilised amino acids serve as useful 'general ligand' adsorbents for a variety of proteins. This may arise by virtue of their natural bioaffinity for these complementary proteins or their amphoteric or hydrophobic nature. In the former case, such as in the preparation of immobilised L-aspartic acid for the purification of asparaginase and several other aspartate-dependent enzymes, a hexamethylene-diamine spacer molecule is interposed between the matrix and ligand (Tosa et al., 1974). In cases where the amino acids are effective by virtue of their amphoteric or hydrophobic nature, as with immobilised lysine or phenylalanine, the ligand is often coupled directly to CNBr-activated agarose. Table 5.8 lists some immobilised amino acids together with their commercial availability and applications in affinity chromatography. The amphoteric derivative, lysine–agarose, is worthy of further consideration since it finds widespread application in the purification of ribosomal RNA (rRNA). L-Lysine is coupled to CNBr-activated agarose to yield an adsorbent containing 4–5 μmol/g moist weight gel. Crude rRNA may be applied to the adsorbent in 0.02 M Tris–HCl buffer

TABLE 5.8
Immobilised amino acids in affinity chromatography.

Immobilised amino acid	Purification application	Commercial availability
L-Arginine	Carboxypeptidase Prekallikrein	
L-Aspartic acid	Asparaginase Aspartate decarboxylase Aspartase	
L-Glutamic acid	Glutamate dehydrogenase	
L-Isoleucine	Threonine deaminase	
L-Leucine	α-Isopropylmalate synthetase	
L-Lysine	Plasminogen rRNA fractionation	PL Biochemicals Pharmacia Miles Laboratories Sigma
L-Methionine	Methionyl-tRNA synthetase	
L-Phenylalanine	Placental alkaline phosphatase γ-Globulin Thrombin Phenylalanine-tRNA ligase	Miles Laboratories PL Biochemicals
L-Tryptophan	Carboxypeptidase A Chymotrypsin Chorismate mutase Anthranilate synthetase complex	Miles Laboratories PL Biochemicals
L-Tyrosine	3-Deoxy-D-arabino-heptulosonate 7-phosphate synthetase Tyrosyl-tRNA synthetase	Miles Laboratories PL Biochemicals

(pH 7.5) containing 10 mM $MgCl_2$. Application of a linear gradient of 0.05–0.3 NaCl in the same buffer elutes 4 S tRNA, 5 S rRNA, 16 S rRNA and finally 23 S rRNA as distinct peaks. The affinity of rRNA for lysine–agarose is increased as the operating temperature is lowered. In the cold room salt concentrations in excess of 2 M are required to elute the larger rRNA species from the adsorbent. L-Lysine–Sepharose 4B is available commercially from Pharmacia.

5.3. Removal of trace contaminants from purified protein preparations

In principle, affinity chromatography is ideally suited to the removal of trace impurities from purified protein and enzyme preparations. The purified preparation would be passed through a column of a suitable affinity or immunoadsorbent to effect quantitative removal of the contaminant. Nowhere is this process more important than in the removal of protease and nuclease impurities from commercial enzyme preparations or in the preparation of pure protease devoid of contaminating zymogens or other functionally related enzymes. For example, many samples of cystalline trypsin contain significant proportions of chymotrypsin as impurity. Treatment of the trypsin sample with an irreversible chymotrypsin inhibitor, L-(1-tosylamido-2-phenyl)-ethylchloromethylketone (TPCK) reduces the chymotrypsin activity, but not to the point of extinction. These considerations underline the specific requirement for rapid and reliable techniques for the removal of proteases. A detailed discussion of the preparation and properties of affinity adsorbents for the purification of proteases is beyond the scope of this monograph; the reader is referred to Lowe and Dean (1974) for a general review of the area. Only the preparation of biospecific adsorbents for trypsin and chymotrypsin will be considered here.

Three general strategies for the preparation of protease-specific adsorbents can be recognised. The first employes high molecular weight protein inhibitors as ligands bound directly to CNBr-activated agarose. For example, Feinstein (1970) has attached chicken ovomucoid, a protein of egg white capable of inhibiting trypsin but not chymotrypsin, to CNBr-activated agarose. The adsorbent bound trypsin but not chymotrypsin at slightly alkaline pH and elution was effected by lowering the pH. Typically (Robinson et al., 1971), a crude bovine trypsin preparation containing 250 mg protein would be applied to a column (1.5 cm × 25 cm) of chicken ovomucoid-agarose equilibrated with 0.1 M Tris–HCl buffer (pH 7.5) containing 0.05 M $CaCl_2$ and 0.5 M KCl. The column is washed thoughly to remove unbound protein and active trypsin eluted with 0.1 M potassium formate containing 0.5 M KCl (pH 2.5). This adsorbent is particularly useful for removing contaminants from commercial crystalline trypsin preparations since it removes both chymotryptic impurities and autodigested fragments. Furthermore, the adsorbent will resolve α- and β-trypsin but will not bind either bovine trypsinogen or chymotrypsin. Alternatively, agarose-bound soybean trypsin inhibitor may be used to resolve chymotrypsin and trypsin (Porath and Sundberg, 1971). The adsorbent binds both enzymes at pH 7.5 with resolution being effected by a decreasing pH gradient or by specifically displacing chymotrypsin with tryptamine and trypsin with benzamidine (Porath and Sundberg, 1971). Immobilised Soybean Trypsin Inhibitor is available from Miles Laboratories in packs of 10 ml and 20 ml settled gel. The adsorbent is also available in smaller packs of 1 ml and 5 ml from P.L. Biochemicals and supplied as a suspension in 50% glycerol and 0.02% sodium azide. The gel, containing 2–6 mg soybean trypsin inhibitor per millilitre agarose, should be stored at −20 °C prior to use.

The heterogeneity of these adsorbents however, arising from the random coupling of the polypeptide inhibitors to the activated matrix, often results in poorly reproducible separations with immobilised ligands from different batches (Amneus et al., 1976). Consequently, adsorbents prepared by the immobilisation of oligopeptides with appropriate amino acids complementary to the specificity of the enzyme to be purified seems an attractive alternative. The tripeptide Gly–Ala–Arg when immobilised to agarose binds trypsin in the pH range 7.3–5.0 whence elution may be effected with 5 mM HCl or 10 mM benzamidine (Kasai et al., 1976). Agarose bound Gly–Gly–Tyr (*O*-benzoyl)-L-arginine specifically binds papain and is available commercially in 10 ml and 20 ml packs from Miles Laboratories.

A third approach to the preparation of trypsin-specific adsorbents involves the use of synthetic low molecular weight inhibitors such as *m*- and *p*-aminobenzamidine (Fig. 5.7). These derivatives may be immobilised via their free aromatic amino functions to polymethylene spacer molecules bound to agarose. Immobilised benzamidine binds trypsin at pH 8–5.5 whence elution may be effected by supplementing the starting buffer with 10 mM benzamidine or by reducing the pH to 2–2.5. These adsorbents also bind trypsin-like proteases such as acrosin, thrombin and kallikrein but do not retard the passage of chymotrypsin or chymotrypsinogen.

Chymotrypsin-free trypsin may also be prepared by using a matrix which specifically binds chymotrypsin but leaves trypsin unretarded. For example, agarose-bound ε-aminocaproyl-D-tryptophan methyl ester is almost completely specific for chymotryptic proteases (Cuatrecasas et al., 1968). The adsorbent strongly binds ε-chymotrypsin, weakly retains chymotrypsinogen A but does not bind trypsin, subtilisin or pancreatic ribonuclease. It is commercially available from Miles Laboratories. Stevenson and Landman (1971) have shown that 4-phenylbutylamine (Fig. 5.7) coupled directly to CNBr-activated agarose displays specificity for chymotrypsin but not trypsin or chymotrypsinogen. The sequential use of a chymotrypsin-specific adsorbent (4-phenylbutylamine–agarose) and a trypsin-specific adsorbent (*p*-aminobenzamidine–cellulose) has proved effective for the preparation of highly purified trypsin (Jany et al., 1976). Agarose-5′-

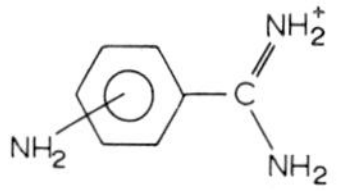

m, p - Aminobenzamidine

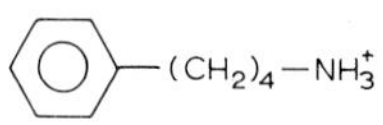

4 - Phenyl - butylamine

Fig. 5.7. The structures of synthetic protease inhibitors.

Subject index p. 519

(4-aminophenylphosphoryluridine-2′(3′)phosphate) is a potent inhibitor of ribonuclease and will free other nuclease preparations from this enzyme. The adsorbent is available in 10 ml and 20 ml packs from Miles Laboratories.

5.4. *Resolution of isoenzymes*

The resolution of isoenzymes has usually been effected by exploiting small differences in the physico-chemical properties of the individual forms. More recently, however, sulphanilamide–agarose has been used to resolve isoenzymes B and C of carbonic anhydrase (Falkbring et al., 1972) by affinity chromatography. Isoenzyme B was eluted

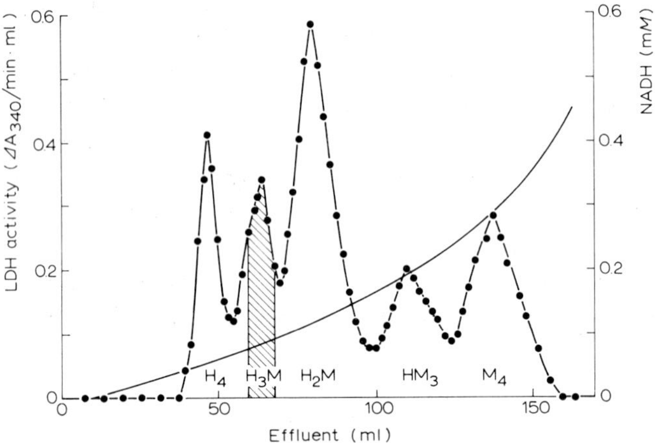

Fig. 5.8. The resolution of the five isoenzymes of lactate dehydrogenase by affinity chromatography on an immobilised-AMP adsorbent. Protein (0.2 mg) in 0.2 ml 0.1 M sodium phosphate buffer (pH 7.0), 1 mM 2-mercaptoethanol and 1 M NaCl was applied to an N^6-(6-aminohexyl)-AMP–agarose column (6 mm × 140 mm, containing 2.5 g wet gel) equilibrated with 0.1 M sodium phosphate buffer (10 ml) and the isoenzymes eluted with a concave gradient of NADH (0–0.5 mM) in the same buffer containing 1 mM 2-mercaptoethanol. Fractions (1 ml) were collected at 3.4 ml/h. Reproduced with permission from Brodelius and Mosbach (1973) FEBS Lett., *35*, 223.

from the immobilised sulphanilamide with 0.1 M NaI whilst isoenzyme C required 10 mM KCNO for quantitative release.

Several NAD^+-dependent dehydrogenases have also been resolved into isoenzymic forms by affinity chromatography on immobilised 'group specific' adsorbents. For example, all five isoenzymes of L-lactate dehydrogenase may be resolved on agarose-bound N^6-(6-aminohexyl)-AMP (Brodelius and Mosbach, 1973). The affinity adsorbent is available commercially (Table 5.1) or may be prepared by coupling N^6-(6-aminohexyl)-AMP to CNBr-activated agarose to yield a gel containing about 150 μmol nucleotide/g dry polymer. The isoenzyme sample is applied to a column of immobilised-AMP in 0.1 M sodium phosphate buffer (pH 7.5), non-adsorbed protein washed off with the same buffer and elution effected with a concave gradient of NADH (Fig. 5.8). The concave gradient, prepared by using two parallel-sided containers with a cross-sectional area ratio of 2 : 1 (mixing chamber: 0.5 mM NADH), elutes the H_4 isoenzyme first followed by H_3M, H_2M_2, HM_3, and M_4. The elution patterns obtained on several successive runs suggest excellent reproducibility with an overal recovery of enzyme activity of approximately 60%. The successful resolution of the isoenzymes of horse liver alcohol dehydrogenase and malate dehydrogenase suggests that these procedures may have wider application.

Applications of affinity chromatography to the purification of regulatory macromolecules and complex biological structures

A number of macromolecules associated with the recognition, transport, information transfer and regulatory aspects of cellular function exhibit high affinity and specificity for their complementary ligands. However, their presence in trace amounts makes their purification by conventional procedures almost impossible. Thus, selective adsorbents with biological specificity are ideally suited to the resolution and isolation of these macromolecules because they display specific binding functions with a high degree of affinity.

6.1. Antigens and antibodies

Immunoaffinity chromatography exploits the unique specificity and high affinity of antibodies for their respective antigens. In principle, the purification of antibodies from serum may be effected by binding to a column of their respective immobilised antigens. Inert proteins are washed through and the specific antibody released under conditions that disrupt the immune complex. Conversely, immobilised antibodies may be used to isolate their complementary antigens. These and a number of other applications are reviewed by Silman and Katchalski (1966), Robbins and Schneerson (1974) and Ruoslahti (1976).

Many procedures are available for insolubilising proteins to support matrices for the purification of monospecific antisera. Typically, the protein antigen is coupled to CNBr-activated agarose under mildly acid to weakly alkaline conditions to ensure retention of antibody activity. At more alkaline pH values excessive reaction of surface lysyl residues causes distortion of the immobilised protein and reduces the antigenic and antibody activity. Furthermore, over-immobilised antigen can also fail to release the complementary antibodies because the necessary conformational change cannot occur.

Sepharose 4B is activated with cyanogen bromide (33 mg/ml) at pH 10.8–11.0, thoroughly washed and added to a solution of the protein antigen. The protein (5–10 mg/ml) in a suitable buffer (pH 6–8), such as 0.1 M sodium citrate (pH 6.0) or 0.1 M sodium bicarbonate (pH 8.0), is incubated with the CNBr-activated agarose at 4 °C for 18 h and then thoroughly washed to remove unbound protein. Under these conditions, coupling efficiencies $\geqslant 90\%$ are often observed. The washed immunoadsorbent may be stored as a suspension in neutral buffer at 4 °C without loss of activity for at least 4–6 months.

The immunoadsorbent is poured into a column, equilibrated with a suitable buffer and antiserum applied. The unbound protein is washed through with starting buffer until the absorbance at 280 nm is negligible whence elution may commence. Elution of specific antibody from the immunoadsorbent may be effected, in principle at least, by establishing conditions that promote disruption of the immune complex. For example, extremes of pH and the supplementation of column irrigants with high concentrations of chaotropic ions, such as perchlorate, iodide and thiocyanate (Fig. 8.3), or protein denaturants, such as urea and guanidine hydrochloride, have proved effective. Table 6.1 lists some typical eluant conditions that have been employed in immunoaffinity chromatography. Unfortunately, the establishment of successful chromomatographic conditions is still largely an empirical process. By trial and error, one can establish elution conditions which commence with the minimum and approach the maximum tolerated disruption conditions. Clearly, the most desirable system for elution is that which is the least drastic means of releasing the maximal amount of active antibody. Elution with 0.1 M glycine–HCl buffer (pH 2.5) is highly reproducible and will therefore be described in more detail. The eluant buffer is applied to the top of the column and elution continued until protein is no longer detectable in the effluent fractions. The low pH of this buffer

TABLE 6.1
Typical eluants for immunoaffinity chromatography.

Acids
0.01–1 M HCl
0.1–1 M acetic acid (pH 2.0)
1 M propionic acid
20% formic acid
0.1 M glycine–HCl (pH 2.2–2.8)
0.015 M acetic acid/0.15 M NaCl
0.1 M glycine–H_2SO_4 (pH 2.8)
1% NaCl/HCl (pH 2.0)
Bases
0.01 M NH_4OH
1 M NH_4OH/0.3 M KCl
0.05 M NH_4OH/20% glycerol (v/v)/1% BSA
0.2 M NH_4OH/0.3 M KCl/1% BSA (pH 10.8)
0.3 M 1,4-diaminobutane
Salts
3 M potassium thiocyanate
5 M KI
2.5 M NaI (pH 7.5)
2.8 M $MgCl_2$
Protein denaturants
4–8 M urea
6 M guanidine–HCl (pH 3.1)

will eventually denature the antibody, so the eluted fractions must be titrated to pH 8–8.5 with solid Tris. A second batch of antibody may be recovered by elution with 0.1 M glycine–HCl buffer (pH 2.5) containing 10% (v/v) dioxane and is neutralised in the same way. The neutralised samples are concentrated by vacuum dialysis or by dialysis against sucrose or polyethylene glycol. Under these conditions, at least 90% of the recovered protein should be antibody with demonstrable antigen-binding activity.

Immunoadsorbents based on matrices other than agarose, such as glass, Sephadex, cellulose, metal oxides and synthetic polymers and involving a variety of chemical coupling methods are reviewed by Robbins and Schneerson (1974) and by Line and Becker (1975). Interestingly, cellulose-based antigens appear to exhibit less non-specific interaction with some proteins than the corresponding derivatised agaroses. Protein antigens may be coupled to bromo-acetyl– or carboxymethyl–cellulose and the resultant immunoadsorbents used for the purification of complementary antibodies. A typical procedure involving the use of bromoacetyl–cellulose (BAC) derivatives is as follows (Robbins and Schneerson, 1974). Cellulose is washed with acetone and anhydrous dioxane and dried over P_2O_5 to constant weight. To 10 g of the cellulose is added bromoacetic acid (100 g) in 30 ml anhydrous dioxane and the mixture stirred for 20 h at room temperature in a tightly stoppered flask. Bromoacetyl bromide (75 ml) is added and stirring continued for a further 20 h at room temperature. The HBr produced in the reaction is removed with a NaOH trap. The final solution is carefully poured into 6 litres of well stirred ice-cold water and the precipitated cellulose washed exhaustively with 0.1 M $NaHCO_3$ and water. The BAC can be stored at 4 °C in the moist state prior to coupling to the protein antigens. As a preliminary to covalent coupling of protein antigens, it is necessary to determine the optimal pH for physical adsorption to the BAC. For this, 10 mg of moist BAC is added to 10 ml antigen solution (1.0 mg/ml) in 0.15 M citrate–phosphate buffer (pH 2–7), homogenised for 5 min and the unbound protein determined. The antigen (300–500 mg) is subsequently coupled to 1 g BAC by gentle agitation for 24 h at room temperature in the citrate–phosphate buffer of optimal pH. The suspension is centrifuged and suspended in 30 ml 0.1 M $NaHCO_3$ (pH 8.9) for 24 h with gentle agitation. The antigen–cellulose is then exhaustively washed with 0.1 M $NaHCO_3$ (pH 8.9) containing 50 mM 2-aminoethanol to block residual activated groups, 8 M urea to remove non-covalently attached antigen and finally with saline. The immunoadsorbent may be stored at 4 °C in saline – 0.01% sodium azide (pH 7.4).

As an alternative to immunoadsorbents prepared by covalently coupling the antigen to an insoluble support matrix, adsorbents comprising cross-linked protein antigens offer some advantages. This method of insolubilisation is facile and the product is stable; it is frequently used for the preparation of serum proteins present in low concentrations. Typically, IgG (150 mg) is dissolved in 0.1 M potassium phosphate buffer (pH 7.2) and 0.1 ml ethyl chloroformate added with gentle stirring (Avrameas and Ternyck, 1967). After about 1 h the turbid suspension is centrifuged and the precipitate washed with water, homogenised with an all glass homogeniser and washed with 1 M glycine–HCl (pH 2.4) until protein no longer leaches into the supernatant. Appropriate volumes of antiserum are incubated for 2 h at room temperature with protein immunoadsorbent in neutral buffer and the precipitate centrifuged and washed. The washed adsorbent is suspended in 0.1 M glycine (pH 2.2), 5.0 M KI or any of the eluants listed in Table 6.1 for 2 h at room temperature. The adsorbent is removed by centrifugation and the eluate passed through a 0.4 μm Millipore filter and equilibrated with neutral buffer. The yield of active antibody is often high (30–60%).

Immunoadsorbents have also been effective in the purification of antihapten antibodies especially when the hapten is attached to the matrix backbone by a 'spacer molecule'. It is usually possible to dissociate the hapten–antibody complex at pH 7 with a 0.1–0.5 M solution of the appropriate hapten.

6.2. Binding and transport proteins

Binding proteins are characterised by a high affinity for their complementary vitamin or hormone with dissociation constants in the range 10^{-7}–10^{-16} M and by their presence in trace amounts. For example, 1000 litres of human plasma contain only 20 mg of the the vitamin B_{12} binding protein, transcobalamin II. Affinity chromatography is thus an attractive method in these cases since purification to this extent is beyond the scope of classical techniques.

The hormone or vitamin is coupled to agarose via an intermediate 'spacer' by standard organic chemistry. Thus pre-assembled ligand-spacer arms may be coupled directly to CNBr-activated agarose, or a suitable ligand analogue attached to a preformed matrix-spacer arm for assembly. For example, corticosteroid-binding globulin (CBG) may be extracted from serum by an adsorbent prepared by coupling cortisol hemisuccinate to 3,3′-diaminodipropylamino–agarose with dicyclohexylcarbodiimide in dioxane (§ 3.4.1.1) (Rosner and Bradlow, 1971). Similarly, oestradiol-binding proteins may be effectively bound by an adsorbent comprising 3-*O*-succinyl-[^{3}H]-oestradiol coupled to 2-aminoethyl–agarose with a carbodiimide promoted reaction. Alternatively, 17-β-oestradiol hemisuccinate may be coupled to agarose derivatives containing diaminodipropylamine, serum albumin, poly-L-lysine or poly(L-lysyl)-alanine as spacer molecules. The major problem, however, associated with affinity adsorbents for steroid binding proteins is leakage of the free steroid from the matrix. Steroids and other aromatic compounds are strongly adsorbed to some solid supports and require exhaustive washing with organic solvents for prolonged periods of time to ensure complete removal of unbound steroid. If these precautions are not taken, the free steroid binds tightly to the binding protein and inhibits the binding of the radio-labelled hormone used in the assay. This disappearance of binding capacity has often erroneously been interpreted in terms of binding to the affinity adsorbent. These problems may be circumvented by using a stable linkage between the matrix and spacer–ligand assembly and exhaustive washing conditions to remove unbound ligand.

The second major problem with the purification of binding proteins relates to their elution from the affinity adsorbent. Adsorption may be effected under batchwise conditions. For example, corticosteroid binding globulin may be adsorbed by adding cortisol–agarose batchwise to a 2-litre serum sample, gently stirring for 5 h at 4 °C and thoroughly washing off unbound protein on a sintered funnel. Batchwise adsorption obviates the need for passing large volumes of serum through a relatively small column of adsorbent. The elution regime is to some extent dictated by the affinity of the ligand for its complementary binding protein. Where the dissociation constant of the ligand-binding protein complex is $>10^{-8}$ M, the irrigant buffer is supplemented with the appropriate ligand and the temperature is often raised from 4 °C to 25–30 °C. Under these conditions yields are rarely quantitative but purifications amounting to several thousand-fold are commonplace. Where the dissociation constant of the complex is 10^{-8}–10^{-16} M, conditions for elution are more drastic and range from 1 M guanidine–HCl (pH 2.1) through 7.5 M guanidine–HCl (pH 7.5) to 6 M guanidine–HCl (pH 1.5) depending on the system. Alternative eluants include 8 M urea in 0.1 M sodium acetate buffer (pH 5.0) and 0.002 M KOH (pH 9.3).

6.3. Receptor proteins

The interaction of hormones with their complementary membrane bound receptor proteins is both specific and of high affinity, with dissociation constants ranging from 10^{-6}–10^{-12} M. The problems associated with the purification of receptor proteins have been due in part to the chemical difficulties in the development of suitable receptor assay systems, in part to solubilisation problems and in part to the exceedingly low concentration of receptors present on the membrane fractions. Unfortunately, detergent solubilised membrane proteins have similar physical properties and this precludes the use of conventional protein fractionation methods. Consequently, affinity chromatography is becoming a valuable technique for the isolation of membrane-bound receptor proteins.

In most cases the receptor protein can be solubilised by mild detergents without impairing its ability to bind the appropriate ligand. For example, the cardiac β-adrenergic receptor may be effectively solubilised with the non-ionic detergent, Lubrol-PX, or the ionic detergent, sodium deoxycholate; Triton X-100 proved less effective. Table 6.2 lists some typical receptor proteins that have been solubilised and purified by affinity chromatography. Typically, solubilisation may be effected with 1–2% (v/v) Triton X-100 or 2% Lubrol-PX. The detergent concentration may subsequently be reduced to 0.05–

TABLE 6.2
Some receptor proteins that have been purified by affinity chromatography

Acetylcholine
ACTH
Cardiac β-adrenergic receptor
Concanavalin A
Cyclic AMP
Glucagon
Growth hormone
Histamine
IgE
Insulin
Morphine
Norepinephrine
Oestradiol
Penicillin
Progesterone
Prolactin
Thyrotropin

0.1% by dialysis or dilution without fear of accompanying precipitation on storage. Dialysis of the extracts against detergent-free neutral buffer often results in gradual and virtually quantitative precipitation of the receptor proteins. Therefore, low concentrations of detergent, 0.1–0.2% (v/v) Triton X-100 or 0.02% Lubrol-PX are present in all column equilibrating, washing and eluting buffers.

The affinity adsorbents may be prepared by a variety of procedures. For example, the affinity gel for the cytosolic uterine oestradiol receptor may be prepared by using 17β-oestradiol-17-hemisuccinate as ligand and 3,3′-diaminodipropylamine as spacer. Similar adsorbents for penicillin, morphine, histamine and norepinephrine have been synthesised (Lowe and Dean, 1974). Alternatively, adsorbents for the receptors of protein hormones such as insulin may be prepared by the techniques given in Chapter 3. For example, insulin may be attached directly to CNBr-activated agarose, to the *N*-hydroxysuccinimide ester of 3,3′-diaminodipropylaminosuccinyl–agarose or to

Subject index p. 519

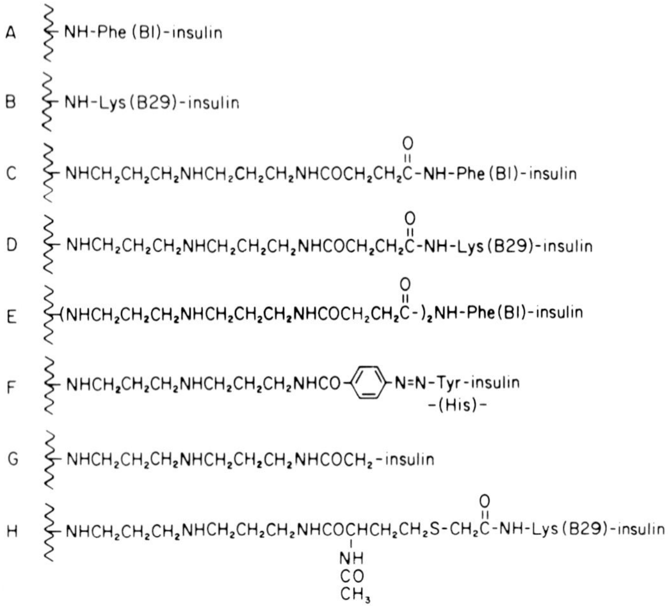

Fig. 6.1. The structures of several insulin–agarose derivatives for the affinity chromatography of soluble insulin receptors from liver membranes. Reproduced with permission from Cuatrecasas and Parikh (1974), Methals Enzymol., *34*, 657.

the bromoacetyl-derivative of agarose (Cuatrecasas and Parikh, 1974). Figure 6.1 illustrates the structures of several insulin–agarose derivatives for the affinity chromatography of solubilised liver insulin receptors. Full details of the preparation of these adsorbents are given in Cuatrecasas and Parikh (1974). Similar procedures are applicable to the preparation of adsorbents for the isolation of receptors for ACTH, glucagon, prolactin, and other lactogenic hormones. For example, a receptor for prolactin, solubilised by 1% (v/v) Triton X-100 from a crude membrane of pregnant rabbit mammary glands, was purified 1,500–fold by affinity chromatography on human growth hormone (hGH) coupled directly to the *N*-hydroxysuccinimide ester of succinylated aminoalkyl–agarose (Shiu and Friesen, 1974). Elution

TABLE 6.3
Typical eluants for receptor purification.

Non-specific eluants
0.1 M glycine–HCl (pH 2.2–2.5)
0.1 M sodium citrate (pH 2.2)
0.02 M ammonium acetate (pH 6.4) 0.5 M NaCl
5 M $MgCl_2$
Linear salt gradient containing 0.1% Triton X-100
4.5 M urea/0.05 sodium acetate pH 6.0/0.1% Triton X-100
5 M urea (pH 6.0)
6 M guanidine–HCl
Specific eluants
0.1 M epinephrine/0.15 M NaH_2PO_4 (pH 3.8)
0.001 M insulin

of the receptor from the hGH-agarose adsorbent was effected with 5 M magnesium chloride. Generally speaking, elution of bound receptor proteins from affinity adsorbents presents similar problems to the elution of antibodies from immunoadsorbents (§ 6.1). Table 6.3 lists some typical elution conditions that have been employed. When optimal elution conditions are employed, recoveries in the range 40–100%, typically 50–80%, and several hundred to many thousand-fold purifications may be expected from these procedures.

Immobilised hormones may also be applied to the investigation of hormone–receptor interactions and receptor topography with intact cells. Several studies with insulin, glucagon, ACTH and norepinephrine have been reported. For example, the surface localisation of insulin receptors has been investigated by Cuatrecasas (1969) with insulin bound to CNBr-activated agarose. However, the 'insulin-like' biological activity, in terms of its antilipolytic action and stimulation of glucose oxidation, which was originally ascribed to the immobilised hormone, is probably a result of leakage of hormone from the matrix. Similar leakage of prolactin and placental lactogen from CNBr-activated agarose has been observed (Topper et al.,

Subject index p. 519

1975). It is suggested therefore, that careful consideration should be given to the problem of leakage of hormone from the matrix (§ 3.1.1.1).

6.4. Affinity chromatography of cells

The resolution and purification of functionally distinct cells by affinity chromatography offers unique opportunities for achieving separations which are unobtainable by conventional procedures. This is because the physical characteristics of cells in a tissue overlap and represent a wide spectrum of maturation and differentiation.

A number of special considerations are applicable to the affinity chromatography of cells. Many of these revolve around the choice of the most appropriate insoluble matrix material. A major obstacle to the application of matrices to cell separations is that these materials should not encourage non-specific adhesion of cells. The characteristic 'stickiness' of cells leads to non-specific adsorption of large numbers of cells to glass and other charged or hydrophobic matrices. Nevertheless, polyacrylamide, nylon and agarose supports have been used successfully to separate cells from viruses. However, even in these cases, recoveries of viable cells was often low with some physical entrapment of cells being experienced. CNBr-activated Sepharose 6MB is a product from Pharmacia Fine Chemicals which is specially designed for the affinity chromatography of cells. It comprises large beads of Sepharose 6B (200–300 μm) to minimise physical entrapment of cells and combines excellent flow properties with quick and efficient coupling of sensitive bio-molecules. The product is supplied in airtight packs of 15 g freeze-dried material and is reconstituted in 1 mM HCl (200 ml/g) prior to the coupling of the biomolecule. Full details of handling and coupling procedures for Sepharose 6MB are given in the manufacturers technical literature.

The selection of the biospecific ligand to be immobilised is determined by the nature of the surface receptors on the cells to be purified. Immuno-specific procedures are particularly appropriate since it is known that cells exhibit different antigenic properties and that some surface antigens appear early in cellular maturation. For example, the resolution of lymphocyte populations by virtue of their affinity for antigen or antibody-containing immunoadsorbents permits quantitative isolation of immunoadsorbent cells. Thus, human peripheral blood lymphocytes could be resolved into two distinct populations by chromatography on an adsorbent comprising immobilised human immunoglobulin (Chess et al., 1974). Almost all of the cells which appeared in the void volume were devoid of immunoglobulin on the cell surface, whilst the retained population had immunoglobulin receptors and were elutable from human immunoglobulin. Furthermore, the specific resolution of antihapten antibody-producing cells from the spleens of mice immunised against an azo-linked phenyl-β-lactoside (lac) hapten has been achieved (Truffa-Bachi and Wofsy, 1970). A large number of cells (10^6–10^8) were applied to an adsorbent comprising *p*-aminophenyl-β-lactoside attached to polyacrylamide P6 beads in phosphate buffered saline (pH 7.2). Most of the non-specific cells were removed in the void volume whence the anti-*lac* cells were detached when 1×10^{-5} M *p*-nitrophenyl-β-lactoside was included in the irrigating buffer.

One problem associatied with the use of immunoadsorbent procedures for cell separations is the failure to effect quantitative recovery of the bound cells in some cases. Procedures have been developed whereby immunoglobulin-bearing β-lymphocytes bound to immunoadsorbents may be eluted by mild enzymic treatment, either by digestion of the immunoadsorbent in toto (Schlossman and Hudson, 1973) or by collagenase-digestion of a gelatin bridge between ligand and matrix (Thomas and Phillips, 1973). Furthermore, as an alternative to immunoadsorbents, specific lectin adsorbents are particularly suitable for cell separations. The Pharmacia technical brochure *CNBr-activated Sepharose 6MB for Affinity Chromatography of Cells* describes a procedure for lymphocyte fractionation on immobilised *Helix pomatia* Lectin. The lectin was coupled to CNBr-activated Sepharose 6MB (0.4–0.5 mg/g swollen gel) and a 3 ml column equilibrated with phosphate buffered saline containing 0.4% human serum albumin and 0.02% sodium azide

(PBS). Human peripheral blood lymphocytes ($\sim 10^8$) were treated with neuraminidase to expose receptors for the *Helix pomatia* lectin and incubated on the column for 15 min at room temperature. Unbound cells were washed from the column with the PBS buffer (60 ml) whence the adsorbed cells were eluted with two sequential steps (60 ml) of *N*-acetyl-D-galactosamine, 0.1 mg/ml and 1 mg/ml. The cell population eluted in the void volume was highly enriched in cells bearing the surface immunoglobulin and C3 receptors (B lymphocytes) whilst the cell fraction eluted specifically with *N*-acetyl-D-galactosamine comprised the T lymphocytes. A small fraction of cells bearing receptors for complement and surface bound immunoglobulin were also retained. The average total cell recovery from the adsorbent was approximately 70%.

Other lectins, such as wheat germ lectin, appear to be equally effective in resolving lymphocyte populations into subpopulations of cells with distinct biological properties. Immobilised lectins are also effective in the purification of erythrocytes (Edelman et al., 1971). Thus, erythrocytes bind to agarose-immobilised concanavalin A but not to ConA–agarose pretreated with α-methyl-D-mannoside, D-glucose or EDTA. The EDTA-treated columns could have their binding capacity restored by washing with $NiCl_2$ or $CaCl_2$, thus confirming the specific interaction between the lectin and receptor sites on the cell surface. For all these separations, however, the columns used for the resolution of cells must be fitted with an appropriate bed support net to permit passage of the cells. Nylon nets of 80 μm mesh size are available for Pharmacia columns.

6.5. *Applications to cellular biology*

The mechanism by which cells are stimulated by extracellular effectors such as hormones and mitogens remains a central problem in cell biology. The use of insoluble derivatives which can bind to the cell surface without fear of endocytosis is potentially able to resolve some outstanding problems. Thus, Cuatrecasas (1969) reported investigations into the localisation of insulin receptors on the fat cells with insulin covalently attached to CNBr-activated agarose. The immobilised insulin displayed apparent insulin-like activity in terms of its stimulation of glucose oxidation and antilipolytic activity. More recent investigations, however, have observed a continual leakage of insulin from supports prepared from CNBr-activated agarose (Kolb et al., 1975) and that the 'leaking' insulin may represent the total biological activity of the gel. Furthermore, detachment of so-called 'super' insulin by the action of albumin on immobilised insulin prepared from CNBr-activated agarose has been reported (Oka and Topper, 1974) and a mechanism for its formation proposed by Wilchek et al. (1975). Analogous leakages of insulin, prolactin and placental lactogen immobilised to CNBr-activated agarose have also been observed. Careful control experiments are clearly necessary before speculations about the effect of insoluble derivatives on cells can be made. Alternatively, derivatives containing the more stable ether linkage, such as those prepared by activating agarose with epichlorohydrin or bis-oxiranes (§ 3.1.1.4) may prove superior.

6.6. Clinical applications

Safe and effective techniques for the removal of protein-based drugs and toxic metabolites from the body would be particularly appropriate to the treatment of a number of clinical situations including drug intoxication and hepatic failure. Extracorporeal haemoperfusion through various specific adsorbents has been applied to the treatment of glutethimide and phenobarbital poisoning and the in vivo clearance from plasma of bromosulphophthalein, chenodeoxycholate and conjugated bilirubin. In vitro studies indicate that bilirubin and other albumin-bound substances can be effectively removed from plasma by filtration through an adsorbent comprising human serum albumin (HSA) covalently attached to agarose (Scharschmidt et al., 1974). The conjugates contained 30–50 mg HSA per gram wet weight gel and almost quantitatively bound plasma bilirubrin. The adsorbents were regenerated by elution of the bound bilirubin with 50% (v/v) ethanol.

Some analytical applications of affinity chromatography

The potential of affinity chromatography has been most fruitfully exploited in the isolation and purification of proteins from crude sources. Nevertheless, there has been general appreciation of the value of the technique for the resolution of chemically modified, affinity labelled, synthetic or mutant enzymes from their native counterparts. Furthermore, affinity chromatography has been used in the exploration of enzyme mechanisms and the estimation of dissociation and equilibrium constants.

7.1. The resolution of chemically modified and native proteins

In principle, affinity chromatography is ideally suited to the resolution of native and chemically or otherwise modified enzymes. Thus, the preparation of fully active papain provides a good example of the separation of active and inactive enzyme species by affinity chromatography. Crystalline papain prepared by the method of Kimmel and Smith (1954) contains three species of protein; active papain in which the active site cysteine is present as a free thiol and two inactive papain species, one of which is reactivatable with thiols or KCN and the other of which is not. Chromatography of the papain preparation on Sephadex G-25 yields a mixture containing approximately equal proportions of active and non-reactivatable protein. Three procedures for the resolution of these active and inactive components have been proposed. Thus, mercaptopapain may be resolved from non-mercaptopapain by affinity chromatography on agarose bound mercurials (§ 5.2.7) and subsequent elution of the active enzyme with 0.5 mM $HgCl_2$ or 0.5 mM 2-mercaptoethanol. Alternatively, Blumberg et al. (1969) have attached the inhibitory peptide glycyl-glycyl-(*O*-benzoyl)-L-tyrosyl-L-arginine to agarose and shown that active enzyme binds in the presence of 20 mM EDTA and can be eluted with buffers of low ionic strength. The purified enzyme had a thiol content approaching 1 mol thiol/mol protein. Likewise, Brockelhurst et al. (1973) prepared papain with 1 intact catalytic site thiol per mole protein from dried papaya latex and from commercial twice crystallised partially active papain by covalent chromatography on an agarose-(glutathione-2-pyridyl-disulphide) adsorbent (§ 8.1).

Affinity chromatography may also be used to determine whether the residual activity of a modified enzyme preparation reflects the presence of a proportion of unmodified native protein or a quantitatively modified protein with a diminished catalytic activity. Thus, an active site tyrosyl residue of staphylococcal nuclease may be reacted stoichiometrically with the diazonium salt of deoxythymidine-3′-(*p*-aminophenylphosphate)-5′-phosphate with a concomitant loss of about 83% of the catalytic activity. Affinity chromatography of the modified nuclease on agarose-bound deoxythymidine-3′-(*p*-aminophenylphosphate)-5′-phosphate resolved the protein into two equally yellow fractions; one appeared in the void volume and was devoid of activity whilst the other was strongly adsorbed and enzymically active. The residual activity of the nuclease thus represented approximately 20% of protein labelled at a site other than the catalytic site (Cuatrecasas, 1970). Similarly, DFP-inactivated trypsin and chymotrypsin were not adsorbed to agarose coupled with ε-amino-caproyl-D-tryptophan methyl ester (Cuatrecasas et al., 1968) and functional and non-functional xanthine oxidase were resolved on a pyrazolo-pyrimidine adsorbent (Massey et al., 1970).

7.2. The purification of affinity-labelled active-site peptides

Site-specific reagents are used extensively for affinity labelling of

Subject index p. 519

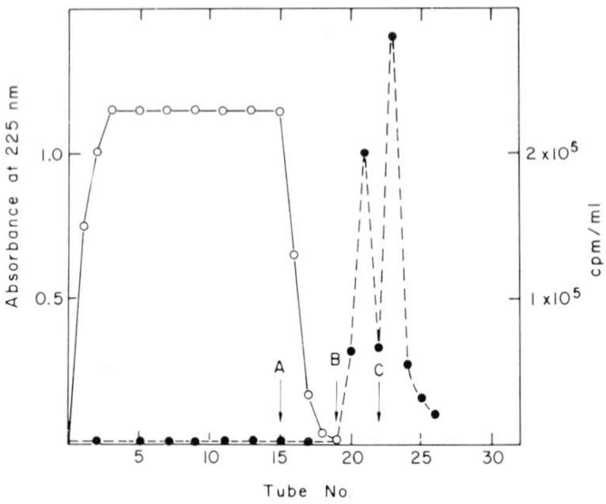

Fig. 7.1. The isolation of affinity labelled peptides: the purification of ^{14}C-labelled DNP-peptide on anti-DNP–agarose. A Sephadex G-50 fraction (100 ml) of tryptic digest of light chain derived from goat anti-DNP antibodies that were affinity labelled with [1-^{14}C]bromoacetyl-1′-DNP–ethylenediamine (BADE) was run through a 1 cm × 2.5 cm column of anti-DNP–agarose (25 mg antibody). The column was eluted as follows: (A) wash with water; (B) 10% formic acid; (C) 20% formic acid. Absorbance (○) at 225 nm and cpm (●) were monitored in the effluent. Reproduced with permission from Givol et al. (1970) Biochem. Biophys. Res. Commun., *38*, 825.

the active sites of enzymes or antibodies. The subsequent isolation of the labelled peptide from the protein digest is usually performed by the conventional techniques of peptide fractionation and can present problems with proteins whose sequence has not been fully elucidated. Givol et al. (1970) have reported a general method for the isolation of such peptides by single-step affinity chromatography. The technique utilises the affinity of the native enzyme or antibody for the ligand used to label the same protein. The labelled peptide is the only one in the digest that displays affinity for the immobilised protein and can be released with eluants that dissociate the protein–ligand complex. For example, goat anti-dinitrophenyl (DNP) antibodies were prepared from goat serum by affinity chromatography on DNP-ovalbumin–agarose and labelled with [1-^{14}C]bromoacetyl, 1′-DNP ethylenediamine (BADE). The BADE-labelled antibodies were reduced and alkylated and the heavy and light chains were resolved on a Sephadex G-100 column equilibrated with 8 M urea–1 M propionic acid. The DNP-peptides derived from a tryptic digest of the light chain of affinity-labelled goat anti-DNP were passed through a Sephadex G-50 column equilibrated with 0.05 M NH_4OH, pooled, adjusted to pH 8.0 and applied at room temperature to an anti-DNP–agarose column containing about 10 mg antibody/g gel. The DNP-peptide appeared as a yellow band on the column. The effluent had an absorbance at 220 nm very similar to that of the applied mixture but contained no radioactivity. The column was washed with water and the DNP-peptide eluted with 10–20% formic acid in 2–3 ml with a recovery of 75% of the applied radioactivity (Fig. 7.1). The eluted yellow DNP-peptide was lyophilised and subsequently purified by paper electrophoresis at pH 3.5 to yield a dodecapeptide containing lysine and the DNP-labelled tyrosine. The authors suggest that the final electrophoretic purification procedure is necessary in all cases in view of the fact that some agarose-bound protein is released under the conditions of elution (Givol et al., 1970). Elution with DNP-glycine represents a possible alternative.

For a general review of the methodology involved in the isolation of specific and modified peptides derived from proteins, the reader is referred to Wilchek (1974).

7.3. The purification of synthetic peptides and proteins

The purification of synthetic proteins and peptides, especially those prepared by the solid phase Merrifield procedure, by conventional techniques has often proved inadequate. This is because the solid phase approach generates unwanted failure sequences which differ from the desired polypeptide by subtle variations in amino acid sequence. Affinity chromatography is capable of purifying those polypeptides with the functionally 'correct' sequence which are thus able to bind to an appropriate adsorbent. The purification of synthetic lysozyme by affinity chromatography is a case in point (Cornelius et al., 1974). The ligand, a β-1,4-linked trisaccharide of *N*-acetyl-

Subject index p. 519

glucosamine was prepared by reductive-hydrogenolysis of the phenylhydrazone of $(GlcNAC)_4$ to yield an amino group suitable for coupling to CNBr-activated agarose. The tri-(*N*-acetylglucosamine)–agarose adsorbent quantitatively bound native lysozyme in 0.1 M ammonium acetate, 0.1 M acetic acid, 0.1 M sodium phosphate, 0.8 M Tris–HCl (pH 8) or 1 M NaCl. Quantitative elution was effected with 1.0 M NaCl, 0.1 M acetic acid (pH 2.5) containing $\geqslant$0.1 mM $(GlcNAC)_3$.

Polypeptide material with the enzymatic activity of lysozyme has been synthesised by the Merrifield procedures and purified from the crude mixture of products by affinity chromatography. The crude product was applied to a 0.2 cm × 15 cm affinity adsorbent and a single enzymically active species was eluted by $(GlcNAC)_3$. This species had a 30-fold improved specific activity compared to the crude product but only 20% of the activity of native lysozyme. Subsequent fractionation of the affinity product on Bio-Rex 70 improved the specific activity to 70% of that of native lysozyme by removing impurities eluted from the affinity column (Barstow et al., 1972). It appears from this and other preparations of synthetic material that impurities in the synthetic mixture affect elution behaviour.

Affinity chromatography has also been exploited to purify semisynthetic staphylococcal nuclease-T′, semisynthetic bovine pancreatic ribonuclease-S′ and synthetic ribonuclease A (Chaiken, 1974).

7.4. The estimation of dissociation and equilibrium constants

A number of theoretical models for affinity chromatography have been developed and based on the combination of enzyme kinetics with gel filtration theory (Dunn and Chaiken, 1974). Under the latter authors' approach, equilibrium conditions, the interaction of an enzyme (E) with an immobilised ligand (LM) is determined by the distribution of the enzyme between the bulk and gel phases and by the dissociation constant for the enzyme–ligand complex, i.e.

$$E \overset{K_d}{\rightleftharpoons} \underline{E} + \underline{LM} \overset{K_{LM}}{\rightleftharpoons} \underline{E - LM}$$

where E = free enzyme in void volume

$\underline{E}$ = free enzyme included in matrix beads

$\underline{LM}$ = ligand attached to matrix

$\underline{E - LM}$ = enzyme–immobilised ligand complex

$$K_d = \frac{[\underline{E}]}{[E]},$$

the chromatographic distribution coefficient (§ 1.3)
and

$$K_{\underline{LM}} = \frac{[\underline{E}][\underline{LM}]}{[\underline{E - LM}]} \tag{7.1}$$

For normal gel filtration chromatography

$$V_e = V_o + K_d \cdot V_i \tag{7.2}$$

where V_e, V_o and V_i have their usual significance (§ 1.3). The presence of the immobilised ligand in affinity chromatography induces an additional interaction between the enzyme and gel beads:

$$V_e' = V_o + K_d' \cdot V_i \tag{7.3}$$

where $K_d' = \dfrac{[\underline{E - LM}] + [\underline{E}]}{[E]}$

whence since $K_d = \dfrac{[\underline{E}]}{[E]}$ it follows that,

$$K_d' = K_d \frac{[\underline{E - LM}]}{[\underline{E}]} + K_d \tag{7.4}$$

and combining eqs. 7.1, 7.2, 7.3 and 7.4 gives,

$$V_e' = V_e + (V_e - V_o) \frac{[\underline{LM}]}{K_{\underline{LM}}} \tag{7.5}$$

or

Subject index p. 519

$$\frac{V_e'}{V_e} - 1 = \frac{V_e - V_o}{V_e} \cdot \frac{[\underline{\mathrm{LM}}]}{K_{\underline{\mathrm{LM}}}} \tag{7.6}$$

Equations 7.5 and 7.6 show that the increase in elution volume occasioned by the interaction of an enzyme with an immobilised ligand is proportional to the concentration of bound ligand ($\underline{\mathrm{LM}}$) and inversely proportional to the dissocation constant of the enzyme–immobilised ligand complex ($K_{\underline{\mathrm{LM}}}$). Furthermore, for an unsubstituted matrix ($[\underline{\mathrm{LM}}] = 0$), $V_e' = V_o$ and gel filtration effects only operate.

If $V_e \gg V_o$, eq. 7.6 simplifies to

$$\left(\frac{V_e'}{V_e} - 1\right) = \frac{[\underline{\mathrm{LM}}]}{K_{\underline{\mathrm{LM}}}} \tag{7.7}$$

However, whilst these equations represent a conceptually simple approach to the determination of $K_{\underline{\mathrm{LM}}}$ from the observed elution volume (V_e') of the enzyme expressed in column units (V_e), in practice, most adsorbents bind the enzyme so tightly that measurement of V_e'/V_e is impractical. This problem can be circumvented by incorporating a soluble ligand or competitive inhibitor (I) in the eluant solutions to reduce the apparent affinity of the enzyme for the immobilised ligand (Dunn and Chaiken, 1974):

$$\begin{array}{ccccc}
\mathrm{E} & \overset{K_d}{\rightleftharpoons} & \underline{\mathrm{E}} + \underline{\mathrm{LM}} & \overset{K_{\mathrm{LM}}}{\rightleftharpoons} & \underline{\mathrm{E-LM}} \\
+ & & + & & \\
\mathrm{I} & & \underline{\mathrm{I}} & & \\
\downarrow\uparrow K_{\mathrm{I}} & & \downarrow\uparrow K_{\underline{\mathrm{I}}} & & \\
\mathrm{EI} & & \underline{\mathrm{EI}} & &
\end{array}$$

where $K_{\mathrm{I}} = [\mathrm{E}][\mathrm{I}]/[\mathrm{EI}]$ and $K_{\underline{\mathrm{I}}} = [\underline{\mathrm{E}}][\underline{\mathrm{I}}]/[\underline{\mathrm{EI}}]$ and EI and $\underline{\mathrm{EI}}$ represent the enzyme–inhibitor complex outside and inside the beads respectively. Expressions analogous to 7.5 and 7.6 may be derived:

$$V_e' = V_e + (V_e - V_o) \cdot \frac{[\underline{\mathrm{LM}}]}{K_{\underline{\mathrm{LM}}}} \cdot \frac{K_{\mathrm{I}}}{K_{\mathrm{I}} + [\mathrm{I}]} \tag{7.8}$$

or

$$\left(\frac{V_e'}{V_e} - 1\right) = \left(\frac{V_e - V_o}{V_e}\right) \cdot \frac{[\underline{LM}]}{K_{\underline{LM}}} \cdot \frac{K_I}{K_I + [I]} \tag{7.9}$$

whence the retardation of an enzyme by an immobilised ligand is reduced in the presence of a soluble competitive ligand, to an extent depending on the relative concentrations and dissociation constants of the soluble and immobilised ligands.

Equation 7.8 may be rearranged by taking reciprocals to:

$$\left(\frac{1}{V_e' - V_e}\right) = \underbrace{\left(\frac{1}{V_e - V_o}\right) \frac{K_{\underline{LM}}}{[\underline{LM}]}}_{C} + \underbrace{\left(\frac{1}{V_e - V_o}\right) \frac{K_{\underline{LM}}}{[\underline{LM}]} \cdot \frac{1}{K_I}}_{M} \cdot [I] \tag{7.10}$$

whence a plot of $(1/V_e' - V_e)$ as a function of $[I]$should be linear with an intercept C and a slope M. The dissociation constant for the free ligand, K_I, may be deduced from C/M and the dissociation constant for the immobilised ligand from C. Equation 7.10 can thus be readily interpreted experimentally in terms of the known parameters of the system. Generally speaking, the chromatographically derived values of K_I are in good agreement with those determined in free solution by other methods. However, uncertainties in the value of $[\underline{LM}]$ due to steric occlusion, make estimates of $K_{\underline{LM}}$ for the immobilised ligand somewhat dubious at best. More rigorous theoretical treatments of affinity chromatography propounded by several authors incorporate parameters such as diffusion rates, reaction rate, flow rate and steric hindrance by enzyme molecules already bound. The reader is referred to Nichol et al. (1974) as an example of a more rigorous treatment.

A more qualitative approach to the determination of dissociation constants by affinity chromatography has been reported by Brodelius and Mosbach (1976). A number of lactate dehydrogenases from different sources whose dissociation constant for NADH (K_{NADH}) was known from literature data were chromatographed on N^6-(6-aminohexyl)-AMP–agarose (§ 5.2.1). The known K_{NADH} values were plotted as a function of the concentration of NADH required to effect elution to yield a linear calibration curve. Lactate dehydrogenases with unknown K_{NADH} values could be interpolated from their observed elution behaviour, as could the dissociation constants for other nucleotides or analogues if used as eluants. One considerable advantage of this approach is that data on dissociation constants may be deduced even from crude supernatants although it is doubtful if the technique is generally applicable.

7.5. Exploration of enzyme mechanisms

The effectiveness of affinity chromatography in enzyme purification may be significantly enhanced by exploiting characteristics of the enzyme mechanism. Conversely, affinity chromatography is capable of yielding unequivocal information about the mechanism and nature of multi-component reaction mechanisms. For example, the elegant work of O'Carra and Barry (1972) has confirmed the compulsory order of substrate binding to lactate dehydrogenase, with NADH as leading substrate and pyruvate as trailing substrate. The enzyme displayed no affinity for an immobilised-oxamate adsorbent in the absence of NADH but was strongly retained in its presence. Oxamate is an analogue of pyruvate and the adsorbent was prepared by the carbodiimide-promoted condensation of oxalic acid to 6-aminohexyl–agarose. Subsequent removal of NADH from the irrigant buffers caused prompt elution of the enzyme. Furthermore, the presence of an 'abortive' ternary complex of enzyme–NAD^+ pyruvate may be inferred from the retardation of lactate dehydrogenase on the immobilised oxamate adsorbent in the presence of saturating levels of NAD^+.

Similar studies with other enzymes and proteins have confirmed the presence of hitherto unsuspected complexes and the formation of these could be monitored under a variety of conditions.

Some special techniques of affinity chromatography

The basic principle of affinity chromatography, i.e., the specific interaction between the ligand and its complementary macromolecule may be exploited in a number of related affinity techniques.

8.1. Covalent chromatography

This procedure differs from conventional affinity chromatography in that instead of a reversible interaction between the macromolecule and the immobilised ligand, the adsorbent is designed to covalently trap the protein to be purified. Release is subsequently effected by suitable chemical treatments. For example, cholinesterases are covalently modified by organic phosphates such as diisopropylfluorophosphate (DFP) by phosphorylation of the active serine residue. Ashani and Wilson (1972) showed that acetylcholinesterase was covalently bound to agarose-bound *p*-nitrophenylmethyl phosphonate and could subsequently be released by reaction with oximes such as 2-(hydroxyiminomethyl)-1-methyl pyridinium iodide (2 PAM) or non-specific nucleophiles such as fluoride or hydroxylamine. Likewise, Blumberg and Strominger (1972) reported the purification of penicillin-binding proteins from solubilised membrances of *Bacillus subtilis* on a covalent affinity adsorbent comprising 6-aminopenicillanic acid coupled by a carbodiimide-promoted reaction to succinylated diaminodipropylamino–agarose. *B. subtilis* membranes were solubilised with 2% Nonidet P-40 and applied batchwise to the 6-aminopenicillanic acid–agarose adsorbent at 25 °C. The slurry was poured into a squat column at room temperature, washed

Subject index p. 519

throughly with 0.05 M potassium phosphate (pH 7.0) containing 1 M NaCl and 0.1% Nonidet P-40 to remove unbound proteins and incubated with one column volume of elution buffer, 0.05 M potassium phosphate (pH 7.0) containing 0.5 M NaCl, 1% Nonidet P-40, 1 mM 2-mercaptoethanol and 0.8 M neutral hydroxylamine for 30 min. The eluted proteins were finally washed off and the process repeated several times. This batchwise adsorption-column elution technique could be scaled up to yield 50–100 mg product with an overall recovery of 50%. The binding proteins are covalently bound by penicilloylation and released by neutral hydroxylamine.

Whilst there are a number of other examples where covalent chromatography is applicable, for example, in the isolation of tryptophan containing peptides (Rubinstein et al., 1976) or in a novel purification of N^5-methyltetrahydrofolate homocysteine methyltransferase on a cobalamin–agarose adsorbent (Sato et al., 1978) covalent chromatography is most frequently applied to the purification of thiol-containing proteins. The principle was developed from the observation that 2,2′-dipyridyldisulphide reacts with the active site thiol of papain at pH 3.5–4.0 at least 100-times faster than with the same thiol in denatured papain or with low molecular weight thiols at equivalent concentrations (Brocklehurst et al., 1973). In practice, an agarose–(glutathione-2-pyridyldisulphide) conjugate is prepared (Fig. 8.1a) by activating 50 g agarose with cyanogen bromide in the usual manner (§ 3.1.1.1) and coupling glutathione (1 g) in 0.1 M $NaHCO_3$–Na_2CO_3 buffer (pH 8.5) for 24 h at 4 °C. The resulting agarose–glutathione gel is thoroughly washed with 0.1 M $NaHCO_3$–Na_2CO_3 buffer (pH 8.5) on a sintered-glass funnel and then with 0.1 M Tris–HCl buffer (pH 8.0) containing 0.3 M KCl and 1 mM EDTA. The gel is suspended in 100 mM 2-mercaptoethanol in the same buffer for 30 min at room temperature whence the reduced gel is washed in turn with 1 M NaCl and 0.1 M Tris–HCl (pH 8.0) containing 0.3 M KCl and 1 mM EDTA. The washed gel is allowed to react with 1.5 mM 2,2′-dipyridyldisulphide in the Tris buffer for 30 min at room temperature and then exhaustively washed to remove all unbound disulphide.

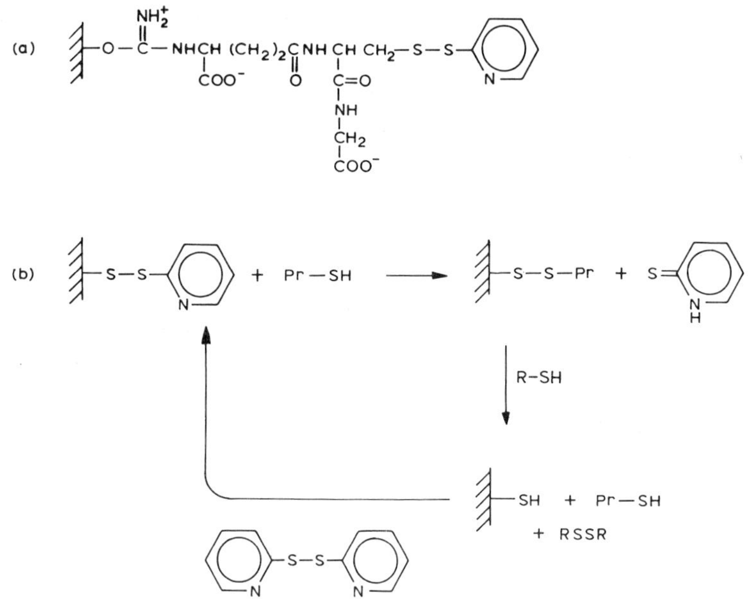

Fig. 8.1. Thiol–disulphide interchange chromatography. (a) The agarose–(glutathione-2-pyridyldisulphide) conjugate. (b) The principle of thiol–disulphide interchange chromatography. Pr–SH, thiol protein; R–SH, low molecular weight thiol such as 2-mercaptoethanol, L-cysteine, dithiothreitol.

Papain is applied to a column of the adsorbent at pH 4.0, non-adsorbed protein washed off with the same buffer whence the covalently bound papain is subsequently detached by elution with 50 mM L-cysteine (pH 8.0). Regeneration of the adsorbent is effected by reduction with dithiothreitol, followed by oxidation with 2,2′-dipyridyldisulphide (Fig. 8.1b). This type of covalent chromatography, thiol–disulphide interchange chromatography, is applicable to the purification of a number of other thiol-containing proteins, including urease, bovine mercaptalbumin, creatine phosphokinase, phosphofructokinase and collagen. Furthermore, the technique is also applicable to the specific isolation of the thiol containing peptides

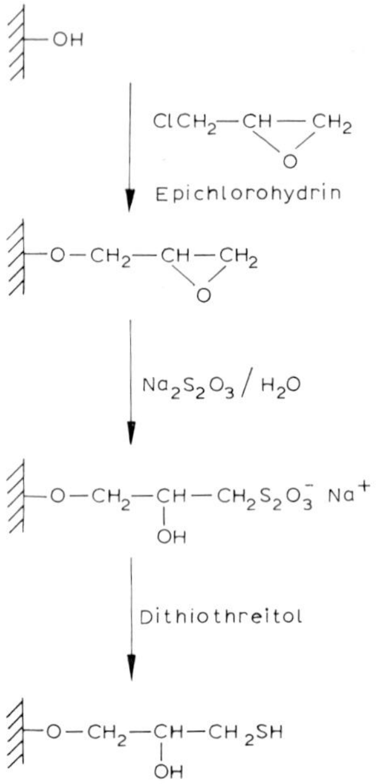

Fig. 8.2. The preparation of thiopropyl–agarose.

(Egorov et al., 1975) and the reversible immobilisation of enzymes (Lowe, 1977).

A number of variations on this original approach are reported in the literature. For example, thiol–agarose may be prepared by the reaction of an ω-aminoalkyl–agarose with *N*-acetylhomocysteine thiolactone in the presence of imidazole as catalyst (Lowe, 1977) or by treatment of epichlorohydrin or bis-oxirane activated agarose with sodium thiosulphate followed by reduction with dithiothreitol (Axén et al., 1975). The latter procedure generates thiopropyl–agarose from epichlorohydrin-activated agarose and is shown in Fig. 8.2. Alternatively, a covalent adsorbent may be prepared by coupling 5,5′-dithiobis-(2-nitrobenzoic acid) (DTNB) to 6-aminohexyl–agarose with a carbodiimide-promoted reaction (Lin and Foster, 1975). Several of these adsorbents are available commercially (Table 8.1).

TABLE 8.1

Commercially available materials for covalent chromatography of thiol proteins.

Product designation	Chemical structure	Supplier
Activated Thiol Sepharose 4B	Agarose–(glutathione-2-pyridyldisulphide)	Pharmacia
Thiopropyl–Sepharose 6B		
Agarose–cysteamine		PL Biochemicals
Agarose–(glutathione-2-pyridyl disulphide)		
Servachrom A-SH	*N*-Acetylhomocysteine derivative of 3,3′-diaminodipropylamino–agarose	SERVA
Affi-Gel 401	*N*-Acetylhomocysteinyl–agarose	Bio-Rad
Cysteamine–agarose		Sigma
GPC/Thiol	*N*-Acetylhomocysteine derivative of 3-aminopropyl-controlled pore glass	Corning Pierce

8.2. Hydrophobic chromatography

The principle of exploiting the interactions between an immobilised hydrophobic ligand and hydrophobic surface sites on proteins for chromatographic separations has been known for a number of years. For example, Er-El et al. (1972) used a homologous series of hydrocarbon coated agaroses prepared by coupling different alkylamines ($CH_3(CH_2)_nNH_2$; $n = 1$–6, 8) to CNBr-activated agarose in order to purify glycogen phosphorylase *b*. The enzyme was not retained by methyl–agarose ($n = 1$), was retarded on ethyl– ($n = 2$) and propyl–agaroses ($n = 3$) and was adsorbed on butyl– ($n = 4$), pentyl– ($n = 5$) and hexyl– ($n = 6$) agaroses. Elution was effected with a deforming buffer comprising 0.4 M imidazole, 0.05 M 2-mercaptoethanol adjusted to pH 7.0 with citric acid. In the case of hexyl–agarose ($n = 6$), adsorption of the enzyme was so strong that recovery was only possible in the denatured form, with 0.2 M acetic acid. The separation potential of these adsorbents is greatly amplified by the introduction of other homologous series of hydrocarbon-coated agaroses whose polymethylene chains terminate in a functional group such as a primary amino, carboxyl, hydroxyl etc. For example, ω-aminoalkyl–agaroses, prepared by coupling a homologous series of α,ω-diaminoalkanes ($NH_2(CH_2)_nNH_2$; $n = 1$–10) to CNBr-activated agarose, have been used in the purification of glycogen synthetase (Shaltiel and Er-El, 1973). The enzyme was adsorbed to 4-aminobutyl–agarose and subsequently eluted with a linear gradient of NaCl. Interestingly, with this type of hydrophobic adsorbent, glycogen phosphorylase required at least a pentamethylene chain for effective retention. Similar adsorbents have been used extensively in the purification of a wide range of proteins, of which some examples are summarised in Table 8.2, and in the fractionation of erythrocytes (Halperin and Shaltiel, 1976). It has been suggested that the discriminating power of these adsorbents may be ascribed to differences in the size and distribution of hydrophobic crevices or enclaves in the specific proteins to be purified. Adsorption is achieved through hydrophobic, ionic and possibly other types of interactions

TABLE 8.2
Some proteins purified by hydrophobic chromatography.

Alkaline phosphatase
Aspartate transcarbamoylase
β-Amylase
γ-Chymotrypsin
β-Galactosidase
γ-Globulin
Glutamate dehydrogenase
Glutamine synthetase
Glycogen synthetase
Histidine binding protein J
Histidine decarboxylase
Immunoglobulin A
Lipoamide dehydrogenase
N. crassa mitochondrial membrane proteins
Ovalbumin
Phosphorylase *b*
Phycoerythrin
Serum albumin
Serum proteins
Sialoglycoprotein from erythrocyte membrane

and the contribution of the hydrophobic interactions may be systematically increased with the the homologous series to effect optimal resolution of the desired protein.

Two hydrophobic chromatography kits, based on the above adsorbents, are available from Miles Laboratories. Hydrophobic chromatography kit I, comprises the alkyl–agarose series (agarose–C_n; $n = 0,2,4,6,8,10$) in the form of 6 small plastic columns ready filled with 1.0 ml of each of the alkyl–agaroses. Hydrophobic chromatography kit II comprises the ω-aminoalkyl–agaroses (agarose–C_n–NH_2; $n = 0,2,4,6,8,10$) in the same format. Both kits are extremely useful for determining the most appropriate alkyl or ω-aminoalkyl–agarose for the purification of a particular protein. Small samples (50–200 μl) of the protein mixture are applied to each of the alkyl– or aminoalkyl–agarose columns equilibrated with an appropriate

Subject index p. 519

buffer, washed with 2.0 ml of the same buffer to remove unbound protein, and finally eluted with 2.0 ml of an appropriate eluant. The % activity in the void effluent is then plotted against the number of carbon atoms (n) in the alkyl chains of the adsorbent. The adsorbent with the smallest n which is capable of retaining the protein to be purified is the column choice for the purification. The procedure may now be scaled up to the preparative scale with the appropriate hydrophobic adsorbent. The individual alkyl– and aminoalkyl–agaroses are available separately from Miles Laboratories in 20 ml and 50 ml package sizes for this purpose. The gels are supplied as 50% gel suspension in water preserved with 0.2% toluene and contain 10–25 μmol alkyl groups/ml swollen gel. The capacity of these adsorbents depends to some extent on the protein being purified but may be as much as 10 mg protein/ml swollen gel. The adsorbents may be used repeatedly provided they are washed exhaustively with 1 M NaCl, water and the appropriate buffers between applications. The gels should be stored at +4 °C.

A number of these hydrophobic media are also available from other firms, often under the designation of 'spacers' or 'control' adsorbents, and designed to be used in control experiments where the specific ligand is omitted.

Jennissen and Heilmeyer (1975) have considered a number of parameters involved in the hydrophobic chromatography of several enzymes, especially phosphorylase *b* and glycogen synthetase, on alkyl-amino–agaroses. They noted that an increased hydrophobicity of the gel necessitated a corresponding increase in NaCl concentration for subsequent desorption. Furthermore, not only was the ionic strength important, but also the nature of the salt employed. They found that the elution efficiency was in accord with the Hofmeister series: $KSCN > CaCl_2 > NH_4Cl > NaCl > Na_2SO_4$. Figure 8.3 lists the various salts (or ions) according to their relative chaotropic potency. It is known that ions of electrolytes such as LiCl or $MgCl_2$ have high charge density and therefore increase the structure of water ('structure forming') whilst ions such as Cs^+, K^+ or Br^- have low charge density and thus decrease the ordered structure of

$PO_4^{3-} < SO_4^{2-} <$ Acetate $< Cl^- < Br^- < NO_3^- < ClO_4^- < I^- < SCN^- <$ Guanidinium

$(CH_3)_4N^+ < NH_4^+ < Rb^+, K^+, Na^+, Cs^+ < Li^+ < Mg^{2+} < Ca^{2+} < Ba^{2+}$

Na_3citrate $< Li_2SO_4 < Na_2SO_4 < KH_2PO_4 < Na_2HPO_4 < (NH_4)_2SO_4 < MgSO_4 <$ KOAc $<$ NaOAc $<$ NaCl $< NaNO_3$

Increasing chaotropic potency ; decreasing salting out effectiveness →

Fig. 8.3. The 'lyotropic' or 'Hofmeister' series.

water. The former salts enhance the association of hydrophobic moieties whilst the latter, structure breaking or chaotropic ions, weaken hydrophobic interactions. Jennissen and Heilmeyer (1975) also observed that a 'critical hydrophobicity' was required to effect positive binding of each individual protein. The hydrophobicity may be increased either by increasing the degree of substitution or by elongation of the alkyl or alkylamine chain at a given degree of substitution. The degree of substitution of alkylamine–agaroses is readily determined by saturating the column with the dyestuff Ponceau S, followed by exhaustive washing and subsequent elution with either 0.3 M NaCl/50% (v/v) ethylene glycol or 1 M *n*-octylamine–HCl/50% (v/v) ethylene glycol. The released dyestuff is monitored by adsorption at 525 nm. The binding of Ponceau S to adsorbents is directly proportional to the acid-base titre of the substituted matrix and to the binding of ovalbumin (Hofstee, 1974).

It is evident from this discussion that many of the properties ascribed to hydrophobic chromatography are, in part at least, due to the co-introduction of charged moieties along with the hydrophobic ligands when CNBr-activated agarose is used. The introduction of charged isouronium functions in CNBr-activated agarose may be circumvented to some extent by using alkylhydrazides instead of alkylamines. For example, Nishikawa and Bailon (1975) studied the effects of salts from the Hofmeister series on the adsorption of several proteins to a gel comprising caprylic hydrazide coupled to CNBr-

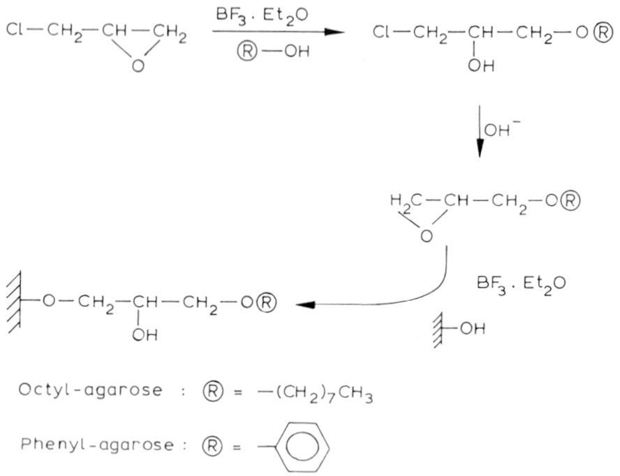

Fig. 8.4. The synthesis of non-charged hydrophobic adsorbents.

activated agarose. The data were predictable from the accepted order of the series.

Charge-free hydrophobic alkyl– and aryl–agarose adsorbents have been reported by Hjerten et al. (1974). Octyl and phenyl groups were introduced into the gel by reacting the corresponding glycidyl ethers prepared by the reaction of the alcohols with epichlorohydrin in boron trifluoride diethyletherate ($BF_3.Et_2O$) as shown in Fig. 8.4. The degree of substitution was measured by NMR spectroscopy following hydrolysis of the gel with formic acid (Rosengren et al., 1975). Both phenyl– and octyl–agarose are available commercially from Pharmacia under the designations Octyl–Sepharose CL-4B and Phenyl–Sepharose CL-4B. Both gels contain about 40 μmol coupled ligand/ml settled gel and bind 3–5 mg *β*-lactoglobulin or 15–20 mg human serum albumin/ml gel in 0.01 M sodium phosphate buffer (pH 6.8) containing 1 M ammonium sulphate. Elution may be effected by altering the eluant in one or more of the following ways:

(a) changing to an ionic composition with an increased chaotropic potency (Fig. 8.3),

(b) lowering the ionic strength,
(c) including a detergent or polarity reducing agent (ethylene glycol) in the eluant,
(d) altering the pH or temperature.

For example, a typical elution regime may involve a decreasing ammonium sulphate gradient co-running with an increasing gradient of ethylene glycol, to yield final concentrations of 0% and 50% respectively. Alternatively, a 0–0.5% increasing gradient of the non-ionic detergent Berol 072 may be substituted for the ethylene glycol. It is recommended that several elution regimes employing a variety of the above approaches should be applied in each situation to assess the optimal purification obtainable. It is important to thoroughly wash these adsorbents after experiments using detergents in order to remove the detergent and any strongly adsorbed proteins. The capacity of the adsorbents for subsequent steps can be seriously affected if this is not done. The Pharmacia technical brochure recommends the following washing procedure for the gel in the column: Wash with 1 bed volume of distilled water, followed successively with 1 bed volume ethanol, 2 bed volumes *n*-butanol, 1 bed volume ethanol, 1 bed volume water and finally the buffer required for re-equilibration.

8.3. Charge transfer and metal chelate affinity chromatography

The isolation of ethynyl steroids from biological fluids has been effected with an adsorbent comprising the silver salt of sulphoethyl–cellulose (Pellizari et al., 1973). Presumably this separation operates by an interaction of the π-electrons of the olefinic parts of these steroids with hybrid orbitals of the metal.

Metal chelate affinity chromatography exploits the specific complexing of zinc or copper ions by cysteine or histidine in neutral solution. Immobilised chelates of zinc or copper were prepared by passage of solutions of $ZnCl_2$ or $CuSO_4$ through agarose-bound iminodiacetic acid and could be used to fractionate human serum

(Porath et al., 1975). Small amounts of protein were adsorbed even in the presence of 1 M NaCl and could subsequently be eluted with EDTA.

Charge transfer affinity chromatography has been used to resolve tryptophan and *N*-acetyltryptophan on an adsorbent bearing 1,3-dinitrophenyl moieties attached to the terminus of a long hydrophilic spacer molecule (Porath, 1976).

8.4. Affinity density perturbation

Wallach et al. (1972) developed a general procedure for the fractionation of membrane fragments bearing specific receptors for drugs, toxins and hormones. For example, concanavalin A (ConA) was converted into a density perturbant by coupling to purified coliphage K29 with glutaraldehyde. Membrane fragments bearing a large number of ConA receptors were prepared from pig lymphocyte plasma membrane. The reversible interaction between the receptor-bearing membrane fragments and the ConA-coliphage perturbant increased the buoyant density in a CsCl gradient from about 1.18 for unmodified fragments to a broad band at 1.30–1.40. Incubation with α,α'-trehalose permits dissociation of the complex and shifts the membrane distribution to lower densities. In principle, therefore, density perturbates may be attached to hormones, drugs, specific antigens or antibodies or transmitters and be used to isolate and map receptor sites on membranes.

8.5. Affinity electrophoretic techniques

The electrophoretic mobility of phosphorylase on polyacrylamide gel electrophoresis was retarded when glycogen was added to the buffer system (Takeo and Nakamura, 1972). The retardation was shown to be substrate specific. In a similar application of affinity electrophoresis, Hořejší and Kocourek (1974) resolved several phytohaemagglutins on *O*-glycosyl polyacrylamide gels prepared by co-polymerisation of acrylamide, *N,N'*-methylenebisacrylamide and

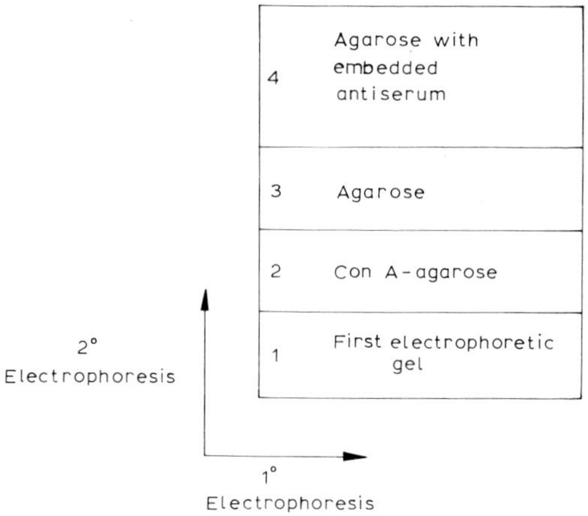

Fig. 8.5. The principle of crossed immuno-affinoelectrophoresis. The technique offers a rapid approach for the prediction of the eventual outcome of an affinity chromatography experiment. For example, human serum is electrophoresed in the first dimension; other gels are arranged as above and electrophoresis performed in the second dimension. The glycoproteins in the sample are removed by the conA–agarose, whereas the remaining proteins pass through the agarose gel (gel 3) and are precipitated by the reference gel containing the antiserum (gel 4).

synthetic allyl-glycosides of appropriate specificity.

A very convenient procedure for assessing the relative efficiences of affinity chromatography has been developed by Bøg-Hansen (1973) and subsequently refined. The technique, crossed immuno-affino-electrophoresis, uses four gels arranged as in Fig. 8.5. Gel 1 comprises the first dimension gel in which the electrophoresis of say human serum has already been performed. The second dimension electrophoresis is then performed through gel 2 (concanavalin A–agarose), gel 3 (agarose) and gel 4 (agarose gel containing antibodies) whence glycoproteins are abstracted at the ConA-gel and all other proteins precipitated in the antibody containing gel.

8.6. *Affinity partitioning*

The original two phase partition approach developed by Albertsson (1971) is based on the fact that when aqueous solutions of dextran (mol.wt. 500,000) and polyethylene glycol (mol.wt. 6,000) are mixed, two phases separate, the upper phase rich in polyethylene glycol and the lower rich in dextran. Components of a protein mixture distribute themselves between the two phases and may be separated by several extractions or by counter current distribution.

In a subsequent refinement, Flanagan et al. (1975) attached affinity ligands to one of the polymers. Thus, cholinergic ligands attached to polyethylene glycol could be used to purify membrane fractions bearing acetylcholine receptors. Furthermore, polyethylene glycol derivatised with palmitoic acid was effective in isolating human serum albumin (Shanbhag and Johansson, 1974). In this context, Pharmacia market a dextran of mol.wt. 70,000 that is activated with 1-hydroxy-2,3-dibromopropane and thus suitable for immobilising molecules containing nucleophiles such as primary amines, hydroxyl and thiol groups. The product, BHP-Activated Dextran 70 is available in 5 g packs and is ideally suited for direct application in affinity partitioning in aqueous 2-phase systems.

8.7. *Affinity histochemistry*

Several techniques based on the biospecificity of ligand–macromolecule interactions are applicable to the visualisation of cell surface and intracellular carbohydrates. Since glycosylated derivatives of horseradish peroxidase and ferritin react with many phytohaemagglutinins, they have been used as cytochemical markers (Kieda et al., 1977). Figure 8.6 illustrates the principle of this approach. The effectiveness is based on the fact that lectins have at least two sugar-binding sites and that a lectin bound to a cellular sugar will still be able to bind the glycosylated cytochemical marker. A typical procedure will involve the sequential incubation of cells with native concanavalin A followed by the marker, ferritin covalently modified

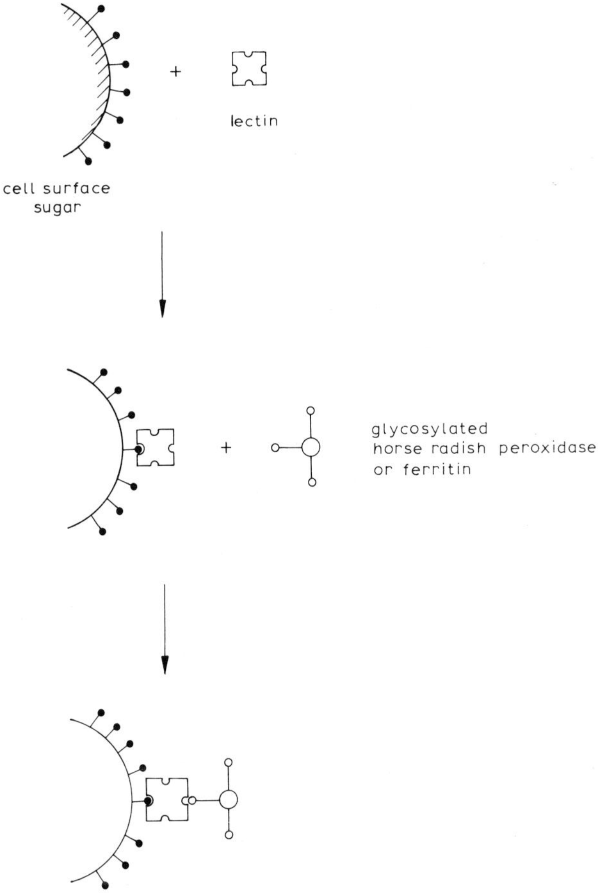

Fig. 8.6. The principle of affinity histochemistry. 1

with mannosyl residues. The cell surface carbohydrates are then visualised by electron microscopy. Alternatively, glycosylated peroxidase is visualised by light microscopy. These cytochemical markers are available commercially from L'Industrie Biologique Française.

Appendix

Manufacturers and suppliers of affinity adsorbents and related materials

Supplier	Products
Amicon Corporation 21 Hartwell Avenue Lexington Massachusetts, 02173 U.S.A.	Nylon based affinity matrices Matrex 102 (*N*-hydroxysuccinimide ester of carboxylated nylon)
57 Queens Road High Wycombe Bucks. England	
Bio-Rad Laboratories 32nd and Griffin Avenue Richmond California, 94804 U.S.A.	Affinity adsorbents comprising spacer molecules terminating in amino, carboxyl, thiol, phenylmercurial and activated carboxyl; stable ether linkages Agarose gels (Bio-Gel A) Chromatography equipment and columns CM-Biogel A Polyacrylamide gels (Bio-gel P)
27 Homesdale Road Bromley Kent England	
Limestrasse 71 8 Munich 66 G.F.R.	

Via Gran Sasso 8
20131 Milan
Italy

Corning/Pierce Corning Glass Works Corning New York, 14830 U.S.A. **Pierce Chemical Company** P.O. Box 117 Rockford Illinois, 61105 U.S.A. **Pierce & Warriner (U.K.) Ltd.** 44 Upper Northgate Street Chester Cheshire CH14 EF England **Pierce Eurochemie B.V.** P.O. Box 1151 Rotterdam Holland	Corning Controlled Pore Glass (CPG) CPG/Dextran CPG/long chain alkylamine CPG/Phenylhydrazine–HCl CPG/*N*-hydroxysuccinimide ester CPG/p-nitrophenyl ester CPG/carboxyl CPG/Thiol CPG/Stable Diazonium salt CPG/8-Hydroxy Quinoline (Chelating support) CPG/lipoamide (Reducing support)
Electro-Nucleonics Inc. 368 Passaic Avenue Fairfield New Jersey, 07006 U.S.A.	Controlled Pore Glass (CPG) Glyceryl-CPG Aminopropyl-CPG Carboxyl-CPG Polyethylene Glycol Dimethacrylate Gel (EM Gel® PGM 2000)
ICN Life Sciences Group 26201 Miles Road Cleveland Ohio, 44128 U.S.A.	Agarose coupled with: Albumin 1,2-Diaminoethane 1,6-Diaminohexane 3,3′-Diaminodipropylamine ε-Aminocaproic acid Adipic acid dihyrazide

Subject index p. 519

L'Industrie Biologique Française
Departement Reactifs
16 Boulevard du General Leclerc
92115 Clichy
France

Ultrogel AcA 34
1,6-Diaminohexane
(HMD-Ultrogel)
ε-Aminocaproic acid
(AC-Ultrogel)
Magnogel 44
(magnetic support)
Reagents for affinity histochemistry

Koch-Light Laboratories Ltd.
Colnbrook, SL3 OBZ
Bucks.
England

Spheron ArA 1000
Polyacrylamide based adsorbents:
Enzacryl AA (aminoaryl)
Enzacryl AH (hydrazido)
Enazcryl Polythiolactone
Enzacryl Polyacetal

LKB–Producter AB
S-161 25 Bromma
Sweden

Ultrogel AcA 22, 34, 44, 54
Chromatographic equipment

E. Merck
Darmstadt
G.F.R.

Succinylated aminododecyl–cellulose
Succinylated aminohexyl–cellulose
Aminohexyl–cellulose
Carboxymethyl–cellulose hydrazide
4-Aminobenzamidine bound to succinylated aminododecyl–cellulose
3-Aminobenzene boronic acid bound to succinylated aminododecyl–cellulose
Trypsin inhibitor bound to succinylated aminohexyl–cellulose
Trypsin (bovine) bound to CM–cellulose

Miles Laboratories Inc.
Elkart
Indiana 46514
U.S.A.

P.O. Box 37
Stoke Court
Stoke Poges
Slough SL2 4LY
England

Hydrophobic Kits I and II
Hydrophobic-agaroses
Agarose bound:
ε-Aminocaproyl-D-tryptophan methyl ester
Soybean Trypsin Inhibitor
Gly-Gly L-Tyr (*O*-Bzl)L-Arg
L-Lysine
L-Thyroxine
L-3,5,3-Tri-iodothyronine
ε-Aminocaproylfucosamine

Lyoner Strasse, 32
6 Frankfurt/Main
G.F.R.

75755 Paris Cedex 15
France

via F.L. Miles 10
20040 Cavenago Brianza
Milano
Italy

p-Aminophenyl mercuric acetate
5′-(4-Aminophenyl-phosphoryl-uridine-2′(3′)Phosphate)
Casein
L-Tryptophan
L-Tyrosine
L-Phenylalanine
Concanavalin A
Fucose Binding Protein
Wheat Germ Lectin
Soybean Agglutinin
Lens Culinaris Haemagglutinin A
Lens Culinaris Haemagglutinin B
Ricinus Communis Agglutinin
Immobilised Immunochemicals
Poly(L-lysine)
Succinyl-Poly(L-lysine)
Succinyl-Poly(DL-Alanine)Poly(L-lysine)
Adipic Hydrazide
Succinyl adipic Hydrazide
Polyacrylic hydrazide
Succinyl polyacrylic hydrazide
Bromoacetyl–cellulose
m-Aminobenzyloxymethyl–cellulose
Immobilised Enzymes

Marine Colloids Inc.
P.O. Box 308
Rockland
Maine 04841
U.S.A.

Agarose beads
(SeaSep AC Beads)

P. L. Biochemicals Inc.
1037 West McKinley Avenue
Milwaukee
Wisconsin 53205
U.S.A.

P–L Agarose 2%
4%
6%
8%
10%
Agarose bound:
1,6-Diaminohexane
Concanavalin A
Cysteamine
Haemoglobin

Subject index p. 519

Hexane
Adenosine
2′,5′-ADP
3′,5′-ADP
5′-ADP
AMP
ATP
Coenzyme A
NAD^+
$NADP^+$
GTP
UDP
UTP
Thiol
Thiol coupler
Hexanoic acid
N-Hydroxysuccinimide ester
Lima Bean Trypsin inhibitor
Poly(A)
Poly(C)
Poly(G)
Poly(I)
Poly(I) Poly (C)
Poly(U)
Soybean Trypsin Inhibitor
Cellulose
Cellulose bound:
5′-AMP
DNA (Denatured and Native)
5′-GMP
Oligo-(dT)
5′-UMP

Pharmacia Fine Chemicals AB
Box 175
S-751 04 Uppsala 1
Sweden

Sepharose 2B, 4B, 6B
Sepharose CL
Sepharose Ion Exchangers
Sephadex G-10–G-200
Sephadex LH-20
Sephacryl S-200
Sephadex Ion Exchangers
2′,5′-ADP–Sepharose 4B
Activated Thiol–Sepharose 4B

Activated CH–Sepharose 4B
AH–Sepharose 4B
5′-AMP–Sepharose 4B
Blue Dextran
Blue Sepharose CL-6B
BHP-activated Dextran 70
CH-Sepharose 4B
CNBr-activated Sepharose 4B
CNBr-activated Sepharose 6MB
Epoxy-activated Sepharose 6B
Lysine–Sepharose 4B
Octyl–Sepharose CL-4B
Phenyl-Sepharose CL-4B
Protein A–Sepharose CL-4B
Poly(A)–Sepharose 4B
Poly(U)–Sepharose 4B
Wheat Germ Lectin Sepharose 6MB
Chromatographic columns, pumps and equipment

SERVA Feinbiochemica GmbH & Co.
Postfach 105260
Karl-Benz-Strasse 7
D-6900 Heidelberg 1
G.F.R.

Agarose bound:
1,2-Diaminoethane (Servachrom AA-1)
3,3′-Diaminodipropylamine (Servachrom AA-2)
Succinylated analogues (Servachrom AC-1, AC-2)
Servachrom G-550 Aminopropyl
p-Aminobenzyl-cellulose (Servacel PAB 23)
p-Aminobenzoyl-3,3′-diaminodipropylamine–agarose (Servachrom A-PAB)
Servachrom A-SH
Servachrom A-Hg

Sigma Chemical Company
P.O. Box 14508
St. Louis
Missouri 63178
U.S.A.

Agarose bound:
3′,5′-cyclic-AMP
2′,5′-ADP
3′,5′-ADP
5′-ADP
5′-AMP
5′-ATP
Adipic dihydrazide

Subject index p. 519

ε-Aminocaproic acid
N-(*p*-aminophenyl)-oxamic acid
Coenzyme A
Concanavalin A
Cysteamine
5′-CDP
5′-CMP
5′-CTP
Diaminohexane
Fetuin
5′-GDP
5′-GMP
5′-GTP
Haemoglobin
α-Lactalbumin
NAD^+
$NADP^+$
Poly(A)
Poly(C)
Poly(G)
Poly(I)
Poly(I) poly (C)
Poly(U)
Protein A–Sepharose 4B and other
Pharmacia products
Trypsin Inhibitor
5′-UDP
5′-UMP
5′-UTP

This list is by no means comprehensive: full details of the above and other products are available from the respective firm.

References

ABDEL-MONEM, M., M.C. CHANAL and H. HOFFMANN-BERLING (1977) Eur. J. Biochem. *79*, 33.

ACKERS, G.K. (1964) Biochemistry *93*, 723.

AGRAWAL, B.B.L. and I.J. GOLDSTEIN (1968) Can. J. Biochem. *46*, 1147.

ALBERTS, B.M., F.J. AMODIO, M. JENKINS, E.D. GUTMANN and F.L. FERRIS (1968) Cold Spring Harbor Symp. Quant. Biol. *33*, 289.

ALBERTSSON, P.Å. (1971) *In:* Partition of Cell Particles and Macromolecules (Almqvist and Wiksell, Stockholm; Wiley, New York).

AMNEUS, H., D. GABEL and V. KASCHE (1976) J. Chromatogr. *120*, 391.

ANDREWS, P. (1964) Biochem. J. *91*, 222.

ANDREWS, P. (1970) FEBS Lett. *9*, 297.

ARAKI, C. (1937) J. Chem. Soc. Jn. *58*, 1338.

ARAKI, C. (1956) Bull. Chem. Soc. Jn. *29*, 543.

ARNDT-JOVIN, D.J., T.M. JOVIN, W. BÄHR, A.M. FRISCHAUF and M. MARQUARDT (1975) Eur. J. Biochem. *54*, 411.

ARNOTT, S., A. FULMER, W.E. SCOTT, I.C.M. DEA, R. MOORHOUSE and D.A. REES (1974) J. Mol. Biol. *90*, 269.

ARSENIS, C. and D.B. MCCORMICK (1964) J. Biol. Chem. *239*, 3093.

ASHANI, Y. and I.B. WILSON (1972) Biochim. Biophys. Acta *276*, 317.

AVKRUST, L.E., K.R. NORUM and B.A. SKÅLHEGG (1976) Biochim. Biophys. Acta *438*, 13.

AVIV, H. and P. LEDER (1972) Proc. Natl. Acad. Sci. U.S.A. *69*, 1408.

AURAMEAS, S. and T. TERNYCK (1967) J. Biol. Chem. *242*, 1651.

AXÉN, R. and S. ERNBACK (1971) Eur. J. Biochem. *18*, 351.

AXÉN, R., J. PORATH and S. ERNBACK (1967) Nature (London) *214*. 1302.

AXÉN, R., P. VRETBLAD and J. PORATH (1971) Acta Chem. Scand. *25*, 1129.

BARKER, R., I.P. TRAYER and R.L. HILL (1974) Methods Enzymol. *34*, 479.

BARRET, A.J. (1973) Biochem. J. *131*, 809.

BARRY, S., P. BRODELIUS and K. MOSBACH (1976) FEBS Lett. *70*, 261.

BARRY, S. and P. O'CARRA (1973) Biochem. J. *135*, 595.

BARRY, S. and P. O'CARRA (1973) FEBS Lett. *37*, 134.

Subject index p. 519

BARSTOW, L.E., D.A. CORNELIUS, V.J. HRUBY, T. SHIMODA, J.A. RUPLEY, J.J. SHARP, A.B. ROBINSON and M.D. KAMEN (1972) *In:* Chemistry and Biology of Peptides (Ann Arbor Science Publishers, Ann Arbor, Michigan, U.S.A.) p. 331.

BAUTZ, E.K.F. and B.D. HALL (1962) Proc. Natl. Acad. Sci. U.S.A. *48*, 400.

BENGTSSON, S. and L. PHILIPSON (1964) Biochim. Biophys. Acta *79*, 399.

BENNICH, H. and S.G.O. JOHANSSON (1971) Adv. Immunol. *13*, 1.

BERGLUND, O. and F. ECKSTEIN (1974) Methods Enzymol. *34*, 253.

BLUMBERG, S., I. SCHECHTER and A. BERGER (1969) Isr. J. Chem. *7*, 125.

BLUMBERG, P.M. and J.L. STROMINGER (1972) Proc. Natl. Acad. Sci. U.S.A. *69*, 3751.

BØG-HANSEN, T.C. (1973) Anal. Biochem. *56*, 480.

BORCH, R.F., M.D. BERNSTEIN and H.D. DURST (1971) J. Am. Chem. Soc. *93*, 2897.

BRANDT, J., L.O. ANDERSSON and J. PORATH (1975) Biochim. Biophys. Acta *386*, 196.

BRENNA, O., B. ROUX, P.G. PIETTA and M. PACE (1977) FEBS Lett. *76*, 246.

BROCKLEHURST, K., J. CARLSSON, M.P.J. KIERSTAN and E.M. CROOK (1973) Biochem. J. *133*, 573.

BRODELIUS, P., P.O. LARSSON and K. MOSBACH (1974) Eur. J. Biochem. *47*, 81.

BRODELIUS, P. and K. MOSBACH (1973) FEBS Lett. *35*, 223.

BRODELIUS, P. and K. MOSBACH (1976) Anal. Biochem. *72*, 629.

BRÜMMER, W. (1974) Kontakte (Merck) 1/74, p. 23; 2/74, p. 3.

CAVALIERI, L.F. and E. CARROLL (1970) Proc. Natl. Acad. Sci. U.S.A. *67*, 807.

CHAIKEN, I.M. (1974) Methods Enzymol. *34*, 631.

CHANDRASEKHAR, K., A. MCPHERSON, M.J. ADAMS and M.G. ROSSMANN (1973) J. Mol. Biol. *76*, 503.

CHESS, L., R.P. MACDERMOTT and S.F. SCHLOSSMAN (1974) J. Immunol. *113*, 1113.

CORNELIUS, D.A., W.H. BROWN, A.F. SHRAKE and J.A. RUPLEY (1974) Methods Enzymol. *34*, 639.

CRANE, L.J., G.E. BETTINGER and J.O. LAMPEN (1973) Biochem. Biophys. Res. Commun. *50*, 220.

CRAVEN, D.B., M.J. HARVEY, C.R. LOWE and P.D.G. DEAN (1974) Eur. J. Biochem. *41*, 329.

CRUFT, H.J. (1961) Biochim. Biophys. Acta *54*, 609.

CUATRECASAS, P. (1969) Proc. Natl. Acad. Sci. U.S.A. *63*, 450.

CUATRECASAS, P. (1970) J. Biol. Chem. *245*, 574.

CUATRECASAS, P. (1970) J. Biol. Chem. *245*, 3059.

CUATRECASAS, P. (1971) J. Agric. Food Chem. *19*, 600.

CUATRECASAS, P. and C.B. ANFINSEN (1971) Annu. Rev. Biochem. *40*, 259.

CUATRECASAS, P. and G. ILLIANO (1971) Biochem. Biophys. Res. Commun. *44*, 178.

CUATRECASAS, P. and C.B. ANFINSEN (1971) Methods Enzymol. *22*, 345.

CUATRECASAS, P. and I. PARIKH (1974) Methods Enzymol. *34*, 653.

CUATRECASAS, P. and I. PARIKH (1972) Biochemistry *11*, 2291.

CUATRECASAS, P., M. WILCHEK and C.B. ANFINSEN (1968) Proc. Natl. Acad. Sci. U.S.A. 61, *636*.

CUATRECASAS, P. and M. WILCHEK (1968) Biochem. Biophys. Res. Commun. *33*, 235.
CUATRECASAS, P., H. TANIUCHI and C.B. ANFINSEN (1968) Brookhaven Symp. Biol. *21*, 172.
DEAN, P.D.G., P.H. ROWE and D. EXLEY (1972) Steroids Lipids Res. *3*, 82.
DEAN, P.D.G. and D.H. WATSON (1978) *In:* 1st International Conference on Affinity Chromatography (Pergamon Press, New York) In press.
DOLEY, S.G., M.J. HARVEY and P.D.G. DEAN (1976) FEBS Lett. *65*, 87.
DUNN, B.M. and I.M. CHAIKEN (1974) Proc. Natl. Acad. Sci. U.S.A. *71*, 2382.
EASTERDAY, R.L. and I. EASTERDAY (1974) *In:* Immobilised Biochemicals and Affinity Chromatography (Plenum Press, New York) p. 123.
EDELMAN, G.M., U. RUTISHAUSER and C.F. MILLETTE (1971) Proc. Natl. Acad. Sci. U.S.A. *68*, 2153.
EGOROV, T.A., A. SVENSSON, L. RYDEN and J. CARLSSON (1975) Proc. Natl. Acad. Sci. U.S.A. *72*, 3029.
ER-EL, Z., Y. ZAIDENZAIG and S. SHALTIEL (1972) Biochem. Biophys. Res. Commun. *49*, 383.
FAILLA, D. and D.V. SANTI (1973) Anal. Biochem. *52*, 363.
FALKBRING, S.O., P.O. GÖTHE, P.O. NYMAN, L. SUNDBERG and J. PORATH (1972) FEBS Lett. *24*, 229.
FEINSTEIN, G. (1970) FEBS Lett. *7*, 353.
FLANAGAN, S.D., P. TAYLOR and S.H. BARONDES (1975) Croat. Chem. Acta *47*, 449.
FLODIN, P. (1961) J. Chromatogr. *5*, 103.
FLODIN, P. (1962) *In:* Dextran gels and their applications in gel filtration (Uppsala, Sweden).
GILHAM, P.T. (1964) J. Am. Chem. Soc. *86*, 4982.
GILHAM, P.T. (1968) Biochemistry *7*, 2809.
GIVOL, D., Y. WEINSTEIN, M. GORECKI and M. WILCHEK (1970) Biochem. Biophys. Res. Commun. *38*, 825.
GLENNER, G.G., M. HARADA and C. ISERSKY (1972) Prep. Biochem. *2*, 39.
GUILFORD, H., P.O. LARSSON and K. MOSBACH (1972) Chem. Scripta *2*, 165.
HAKOMORI, S.-I. (1964) J. Biochem. (Tokyo) *55*, 205.
HALPERIN, G. and S. SHALTIEL (1976) Biochem. Biophys. Res. Commun. *72*, 1497.
HAMAZAKI, H. and K. HOTTA (1977) FEBS Lett. *76*, 299.
HARVEY, M.J., C.R. LOWE, D.B. CRAVEN and P.D.G. DEAN (1974) Eur. J. Biochem. *41*, 335.
HARVEY, M.J., C.R. LOWE and P.D.G. DEAN (1974) Eur. J. Biochem. *41*, 353.
HIPWELL, M.C., M.J. HARVEY and P.D.G. DEAN (1974) FEBS Lett. *42*, 355.
HIXSON, H.F. and A.H. NISHIKAWA (1973) Arch. Biochem. Biophys. *154*, 501.
HJERTÉN, S. (1961) Biochim. Biophys. Acta *53*, 514.
HJERTÉN, S. (1964) Biochim. Biophys. Acta *79*, 393.
HJERTÉN, S. and R. MOSBACH (1962) Anal. Biochem. *3*, 109.
HJERTÉN, S., J. ROSENGREN and S. PÅHLMANN (1974) J. Chromatogr. *101*, 281.

Subject index p. 519

HOFSTEE, B.H.J. (1973) Anal. Biochem. *52*, 430.

HOFSTEE, B.H.J. (1974) *In:* Immobilised Biochemicals and Affinity Chromatography (Plenum Press, New York) p. 43.

HOLROYDE, M.J., J.M.E. CHESHER, I.P. TRAYER and D.G. WALKER (1976) Biochem. J. *153*, 351.

HOŘEJŠÍ, V. and J. KOCOUREK (1974) Biochim. Biophys. Acta *336*, 338.

HORNBY, W.E. and L. GOLDSTEIN (1976) Methods Enzymol. *44*, 118.

INMAN, J.K. and H.M. DINTZIS (1969) Biochemistry *8*, 4074.

JACKSON, R.J., R.M. WOLCOTT and T. SHIOTA (1973) Biochem. Biophys. Res. Commun. *51*, 428.

JAGENDORF, A., A. PATCHORNIK and M. SELA (1963) Biochim. Biophys. Acta *78*, 516.

JANY, K.D., W. KEIL, H. MEYER and H.H. KILTZ (1976) Biochim. Biophys. Acta *453*, 62.

JENNISSEN, H.P. and L.M.G. HEILMEYER (1975) Biochemistry *14*, 754.

JOST, R., T. MIRON and M. WILCHEK (1974) Biochim. Biophys. Acta *362*, 75.

JOUSTRA, M.K. (1969) *In:* Modern Separation Methods of Macromolecules and Particles, Vol. 2 (John Wiley and Sons, New York).

JOVIN, T.M. and A. KORNBERG (1968) J. Biol. Chem. *243*, 250.

KASAI, K., M. NISHIKATA, T. OSHIZAWA and S. ISHII (1976) Abstracts, Tenth International Congress of Biochemistry, Hamburg, p. 193.

KAY, G. and E.M. CROOK (1967) Nature (London) *216*, 514.

KENNEDY, J.F. and A. ROSEVEAR (1973) J. Chem. Soc. Perkin Trans. 2041.

KERN, V.W., T. HUCKE, R. HOLLANDER and R. SCHNEIDER (1957) Makromol. Chem. *22*, 31.

KIEDA, C., F. DELMOTTE and M. MONSIGNY (1977) FEBS Lett. *76*, 257.

KIMMEL, J.R. and E.L. SMITH (1954) J. Biol. Chem. *207*, 515.

KOELSCH, R., J. LASCH, I. MARQUARDT and H. HANSON (1975) Anal. Biochem. *66*, 556.

KOLB, H.J., R. RENNER, K.D. HEPP, L. WEISS and O.H. WIELAND (1975) Proc. Natl. Acad. Sci. U.S.A. *72*, 248.

KRISTIANSEN, T. (1974) Methods Enzymol. *34*, 331.

LAMED, R., Y. LEVIN and M. WILCHEK (1973) Biochim. Biophys. Acta *304*, 231.

LARSSON, P.O. and K. MOSBACH (1971) Biotechnol. Bioeng. *13*, 393.

LASCH, J., M. IWIG, R. KOELSCH, H. DAVID and I. MARK (1975) Eur. J. Biochem. *60*, 163.

LASCH, J., M. IWIG and H. HANSON (1972) Eur. J. Biochem. *27*, 431.

LAURENT, T.C. and J. KILLANDER (1964) J. Chromatogr. *14*, 317.

LERMAN, L.S. (1953) Proc. Natl. Acad. Sci. U.S.A. *39*, 232.

LEVIN, Y., M. PECHT, L. GOLDSTEIN and E. KATCHALSKI (1964) Biochemistry *3*, 1905.

LIN, L.J. and J.F. FOSTER (1975) Anal. Biochem. *63*, 485.

LINDBERG, U. and T. PERSSON (1972) Eur. J. Biochem. *31*, 246.

LINE, W.F. and M.J. BECKER (1975) *In:* Immobilised Enzymes, Antigens, Antibodies and Peptides (Marcel Dekker Inc., New York) p. 495.

LITMAN, R. (1968) J. Biol. Chem. *243*, 6222.

LOWE, C.R. (1977) Eur. J. Biochem. *73*, 265.

LOWE, C.R. (1977) Eur. J. Biochem. *76*, 391.

LOWE, C.R. and P.D.G. DEAN (1974) *In:* Affinity Chromatography (John Wiley and Sons Ltd., London).

LOWE, C.R. and M.G. GORE (1977) FEBS Lett. *77*, 247.

LOWE, C.R., M.J. HARVEY, D.B. CRAVEN and P.D.G. DEAN (1973) Biochem. J. *133*, 499.

LOWE, C.R., M.J. HARVEY, D.B. CRAVEN, M.A. KERFOOT, M.E. HOLLOWS and P.D.G. DEAN (1973) Biochem. J. *133*, 507.

LOWE, C.R., M.J. HARVEY and P.D.G. DEAN (1974) Eur. J. Biochem. *41*, 341.

LOWE, C.R. and K. MOSBACH (1975) Eur. J. Biochem. *52*, 99.

MARCH, S.C., I. PARIKH and P. CUATRECASAS (1974) Anal. Biochem. *60*, 149.

MAY, S.W. and O.R. ZABORSKY (1974) *In:* Separation and Purification Methods, Vol. I (Marcel Dekker Inc., New York) p. 1.

MESSING, R.A. (1969) J. Am. Chem. Soc. *91*, 2370.

MOSBACH, K., P.O. LARSSON and C.R. LOWE (1976) Methods Enzymol. *44*, 859.

MCCORMICK, D.B. (1965) Anal. Biochem. *13*, 194.

NICHOLAS, J.C., M. PONS, B. DESCOMPS and A. CRASTES DE PAULET (1972) FEBS Lett. *23*, 175.

NICHOL, L.W., A.G. OGSTON, D.J. WINZOR and W.H. SAWYER (1974) Biochem. J. *143*, 435.

NISHIKAWA, A.H. (1975) Chem. Technol., 565.

NISHIKAWA, A.H. and P. BAILON (1975) Anal. Biochem. *68*, 274.

NISHIKAWA, A.H., P. BAILON and A.H. RAMEL (1976) J. Macromol. Sci. Chem. *A10*, 149.

NISHIMURA, K., K. HIWADA, E. VEDA and T. KOKUBU (1976) Biochim. Biophys. Acta *445*, 158.

NORDBERG, M.E. (1944) J. Am. Ceram. Soc. *27*, 299.

O'CARRA, P. (1974) *In:* Industrial Aspects of Biochemistry, Spencer, B. (ed.) (North Holland, Amsterdam) p. 107.

O'CARRA, P. and S. BARRY (1972) FEBS Lett. *21*, 281.

O'CARRA, P., S. BARRY and T. GRIFFIN (1973) Biochem. Soc. Trans. *1*, 289.

O'CARRA, P., S. BARRY and T. GRIFFIN (1974) FEBS Lett. *43*, 169.

O'CARRA, P., S. BARRY and T. GRIFFIN (1974) Methods Enzymol. *34*, 108.

OHNO, Y. and M.A. STAHMANN (1972) Immunochemistry *9*, 1077.

OHLSSON, R., P. BRODELIUS and K. MOSBACH (1972) FEBS Lett. *25*, 234.

OKA, T. and Y.J. TOPPER (1974) Proc. Natl. Acad. Sci. U.S.A. *71*, 1630.

PAHUD, J.J. and K. SCHWARTZ (1976) Sci. Tools *23*, 40.

PAULSON, J.C., W.E. BERANEK and R.L. HILL (1977) J. Biol. Chem. *252*, 2356.

Subject index p. 519

PELLIZARI, E.D., J. LIU, M.E. TWINE and C.E. COOK (1973) Anal. Biochem. *56*, 178.

PETERSON, E.A. and H.A. SOBER (1956) J. Am. Chem. Soc. *78*, 751.

PETTERSSON, G. (1968) Arch. Biochem. Biophys. *123*, 307.

POLSON, A. (1961) Biochim. Biophys. Acta *50*, 565.

POONIAN, M.S., A.J. SCHLABACH and A. WEISSBACH (1971) Biochemistry *10*, 424.

PORATH, J. (1968) Nature (London) *218*, 834.

PORATH, J. (1972) FEBS Lett. *24*, 229.

PORATH, J. (1973) Biochimie *55*, 943.

PORATH, J. (1976) Symp. Proc., Advances in Chromatographic Fractionation Macromolecules (Birmingham) In Press.

PORATH, J., J. CARLSSON, I. OHLSSON and G. BELFRAGE (1975) Nature (London) *258*, 598.

PORATH, J. and N. FORNSTEDT (1970) J. Chromatogr. *51*, 479.

PORATH, J. and T. KRISTIANSEN (1975) *In:* The Proteins, Vol. I (Academic Press, New York) p. 95.

PORATH, J. and L. SUNDBERG (1970) Protides Biol. Fluids *18*, 401.

PORATH, J. and L. SUNDBERG (1972) Nat. New Biol. *238*, 261.

REGNIER, F. and W.H. SCOUTEN (1974) Int. Lab., Nov/Dec., 13.

RICKWOOD, D. (1972) Biochim. Biophys. Acta *269*, 47.

ROBBINS, J.B. and R. SCHNEERSON (1974) Methods Enzymol. *34*, 703.

ROBINSON, P.J., P. DUNNILL and M.D. LILLY (1972) Biochim. Biophys. Acta *285*, 28.

ROBINSON, N.C., R.W. TYE, H. NEURATH and K. WALSH (1971) Biochemistry *10*, 2744.

ROSENGREN, J., S. PÅHLMANN, M. GLAD and S. HJERTÉN (1975) Biochim. Biophys. Acta *412*, 51.

ROSNER, W. and H.L. BRADLOW (1971) J. Clin. Endocrinol. Metab. *33*, 193.

ROSSMANN, M.G., M.J. ADAMS, M. BEUHNER, G.C. FORD, M.L. HACKERT, P.J. LENTZ, A.MCPHERSON, R.W. SCHEVITZ and I.E. SMILEY (1972) Cold Spring Harbor Symp. Quant. Biol. *36*, 179.

RUBINSTEIN, M., Y. SHECHTER and A. PATCHORNIK (1976) Biochem. Biophys. Res. Commun. *70*, 1257.

RUIZ-CARRILLO, A. and V.G. ALLFREY (1973) Arch. Biochem. Biophys. *154*, 185.

RUOSLAHTI, E. (ed.) (1976) *In:* Immunoadsorbents in Protein Purification (University Park Press, Baltimore, U.S.A.).

RUSSEL, B., T.H. MEAD and A. POLSON (1964) Biochim. Biophys. Acta *86*, 169.

SANDERSON, C.J. and D.V. WILSON (1971) Immunology *20*, 1061.

SATO, K., E. HIEI, S. SHIMIZU and R.H. ABELES (1978) FEBS Lett. *85*, 73.

SCHABORT, J.C. (1972) J. Chromatogr. *73*, 253.

SCHALLER, H., C. NUSSLEIN, F.J. BONHOEFFER, C. KURZ and I. NIETZSCHMANN (1972) Eur. J. Biochem. *26*, 474.

SCHARSCHMIDT, B.F., P.H. PLOTZ, P.D. BERK, J.G. WAGGONER and J. VERGALLA (1974) J. Clin. Invest. *53*, 786.

SCHLOSSMAN, C.F. and L. HUDSON (1973) J. Immunol. *110*, 313.
SCHMIDT, J. and M.A. RAFTERY (1972) Biochemistry *12*, 852.
SCHNAAR, R.L. and Y.C. LEE (1975) Biochemistry *14*, 1535.
SCOUTEN, W.H. (1974) Int. Lab. Nov/Dec, 13.
SHALTIEL, S. and Z. ER-EL (1973) Proc. Natl. Acad. Sci. U.S.A. *70*, 778.
SHANBHAG, V.P. and G. JOHANSSON (1974) Biochem. Biophys. Res. Commun. *61*, 1141.
SHIU, R.P.C. and H.G. FRIESEN (1974) J. Biol. Chem. *249*, 7902.
SICA, V., E. NOLA, I. PARIKH, G.A. PUCA and P. CUATRECASAS (1973) Nat. New Biol. *244*, 36.
SIEGEL, L.M. and K.J. MONTY (1966) Biochim. Biophys. Acta *112*, 346.
SILMAN, I. and E. KATCHALSKI (1966) Annu. Rev. Biochem. *35*, 873.
SLAVÍK, K., W. RODE and V. SLAVÍKOVÁ (1976) Biochemistry *15*, 4222.
SLUYTERMAN, L.A.Æ. and J. WIJDENES (1974) Methods Enzymol. *34*, 544.
SO, L.L. and I.J. GOLDSTEIN (1968) Biochim. Biophys. Acta *165*, 398.
STAGE, D.E. and M. MANNIK (1974) Biochim. Biophys. Acta *343*, 382.
STASSEN, F.L.H. (1976) Biochim. Biophys. Acta *438*, 49.
STEERS, E.J.W., P. CUATRECASAS and H.B. POLLARD (1971) J. Biol. Chem. *246*, 196.
STELLWAGEN, E., R. CASS, S.T. THOMPSON and M. WOODY (1975) Nature (London) *257*, 716.
STEVENSON, K. and A. LANDMAN (1971) Can. J. Biochem. *49*, 119.
SUNDARAM, P.V. (1976) J. Solid Phase Biochem. *1*, 101.
SUNDBERG, L. and J. PORATH (1974) J. Chromatogr. *90*, 87.
SVENSSON, B. (1973) FEBS Lett. *29*, 167.
TAKEO, K. and S. NAKAMURA (1972) Arch. Biochem. Biophys. *153*, 1.
TESSER, G.I., H.U. FISCH and R. SCHNYZER (1974) Helv. Chim. Acta *57*, 1718.
THOMAS, D.B. and B. PHILLIPS (1973) Eur. J. Immunol. *3*, 740.
THOMPSON, S.T., K.H. CASS and E. STELLWAGEN (1975) Proc. Natl. Acad. Sci. U.S.A. *72*, 669.
THOMPSON, S.T. and E. STELLWAGEN (1976) Proc. Natl. Acad. Sci. U.S.A. *73*, 361.
TOPPER, Y.J., T. OKA, B.K. VON DER HAAR and M. WILCHEK (1975) Biochem. Biophys. Res. Commun. *66*, 793.
TOSA, T., T. SATO, R. SANO, K. YAMAMOTO, Y. MATUO and I. CHIBATA (1974) Biochim. Biophys. Acta *334*, 1.
TRAVIS, J. and R. PANNELL (1973) Clin. Chim. Acta *49*, 49.
TRAYER, I.P. and H.R. TRAYER (1974) Biochem. J. *141*, 775.
TRAYER, I.P., H.R. TRAYER, D.A.P. SMALL and R.C. BOTTOMLEY (1974) Biochem. J. *139*, 609.
TRUFFA-BACHI, P. and L. WOFSY (1970) Proc. Natl. Acad. Sci. U.S.A. *66*, 685.
TURKOVÁ, J. (1974) J. Chromatogr. *91*, 267.
TURKOVÁ, J., O. HUBÁLKOVÁ, M. KŘIVÁKOVA and J. ČOUPEK (1973) Biochim. Biophys. Acta *322*, 1.

Subject index p. 519

Wagner, A.F., R.L. Bugianesi and T.Y. Shen (1971) Biochem. Biophys. Res. Commun. *45*, 184.

Wallach, D.F.H., B. Kranz, E. Ferber and H. Fischer (1972) FEBS Lett. *21*, 29.

Weetall, H.H. (1976) Methods Enzymol. *44*, 134.

Weissbach, A. and M. Poonian (1974) Methods Enzymol. *34*, 463.

Werber, M.M. (1976) Anal. Biochem. *76*, 177.

Weston, P.D. and S. Avrameas (1971) Biochem. Biophys. Res. Commun. *45*, 1574.

Wide, L. and J. Porath (1967) Biochim. Biophys. Acta *130*, 257.

Wigzell, H. and O. Makela (1970) J. Exp. Med. *131*, 110.

Wilchek, M. (1973) FEBS Lett. *33*, 70.

Wilchek, M. (1974) Methods Enzymol. *34*, 182.

Wilchek, M., T. Oka and Y.J. Topper (1975) Proc. Natl. Acad. Sci. U.S.A. *72*, 108.

Wilchek, M., Y. Salomon, M. Lowe and Z. Selinger (1971) Biochem. Biophys. Res. Commun. *45*, 1177.

Wold, F. (1967) Methods Enzymol. *11*, 617.

Wolf, F.J. (1969) *In:* Separation Methods in Organic Chemistry and Biochemistry (Academic Press, New York) p. 177.

Wolpert, J.S. and M.L. Ernst-Fonberg (1973) Anal. Biochem. *52*, 111.

Woychik, J.H. and M.V. Wondolowski (1972) Biochim. Biophys. Acta *289*, 347.

Wykes, J.R., P. Dunnill and M.D. Lilly (1972) Biochim. Biophys. Acta *286*, 260.

Yamazaki, Y., H. Maeda and H. Susuki (1977) Eur. J. Biochem. *77*, 511.

Yon, R.J. (1972) Biochem. J. *126*, 765.

Subject index